Ch. 1-3

Guide to the Geology & Natural History
of the Blue Ridge Mountains

The Blue Ridge Mountains

Highlights from Maryland to Tennessee

The Northern Blue Ridge

1. Catoctin Mtn. Natl. Park & Cunningham Falls State Park - p. 93
2. Gambrill State Park - p. 99
3. Washington Monument State Park - p. 102
4. Harpers Ferry & The Potomac Water Gap - p. 104
5. Shenandoah National Park - p. 113
6. Blue Ridge Parkway – Waynesboro to the James River - p. 142
7. James River Traverse Across the Blue Ridge - p. 154
8. Blue Ridge Parkway – James River to Roanoke - p. 174

The Southern Blue Ridge

9. The Upland Plateau - p. 202
10. Blue Ridge Parkway – Roanoke to Doughton Park - p. 208
11. Philpott Reservoir & Fairy Stone State Park - p. 219
12. Stone Mtn. State Park - p. 230
13. Mt. Jefferson State Natural Area - p. 236
14. Mt. Rogers National Recreation Area - p. 238
15. Grandfather Mtn. Area & Linville Falls - p. 258
16. The Mars Hill Terrane at Roan Mtn. State Park - p. 268
17. Spruce Pine Mining District - p. 272
18. Mt. Mitchell State Park & Craggy Gardens - p. 273
19. Precambrian Basement Rocks on I-26 - p. 278
20. Blue Ridge Parkway – Asheville to Oconaluftee - p. 279
21. Great Smoky Mtns. Natl. Park - p. 290

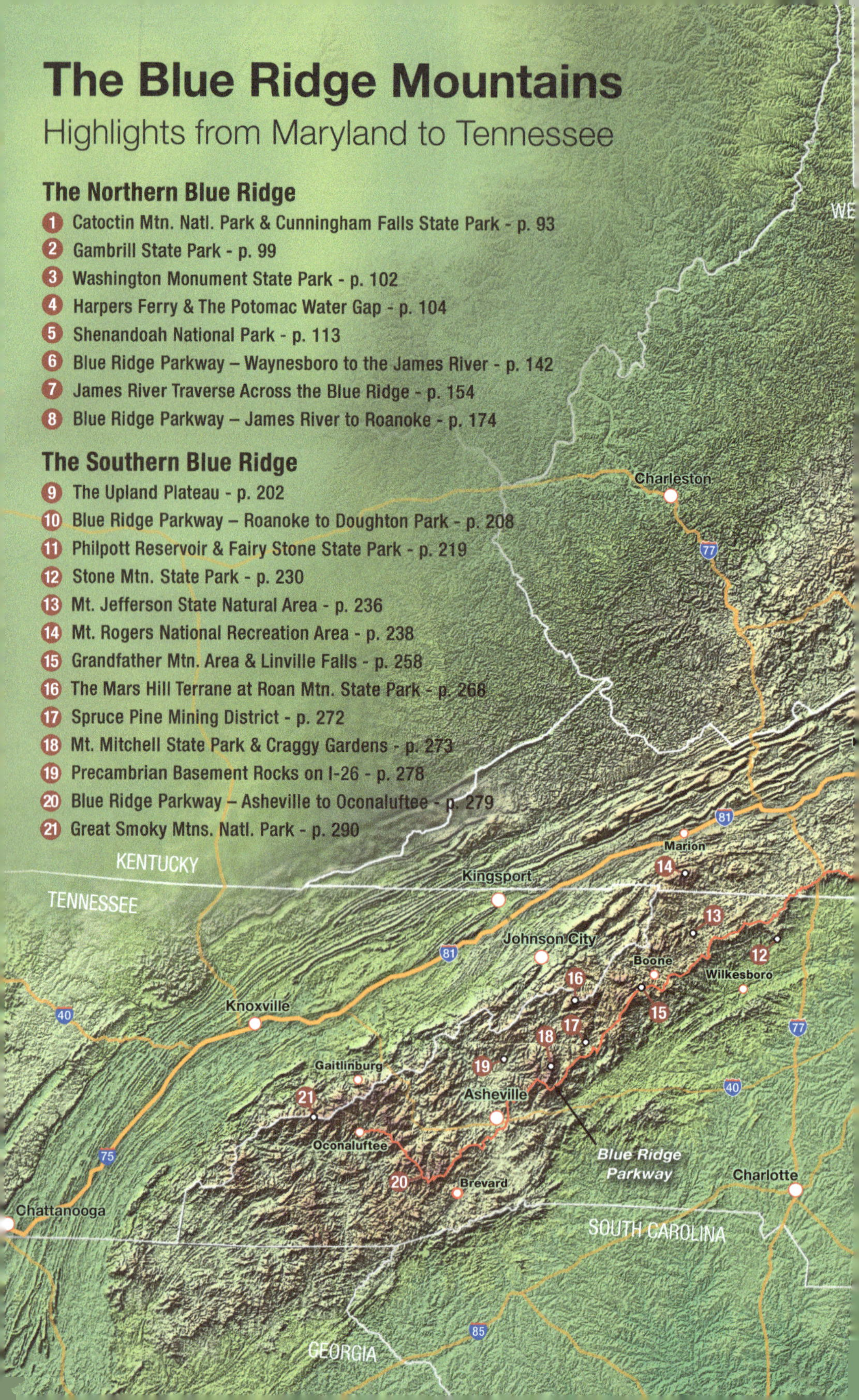

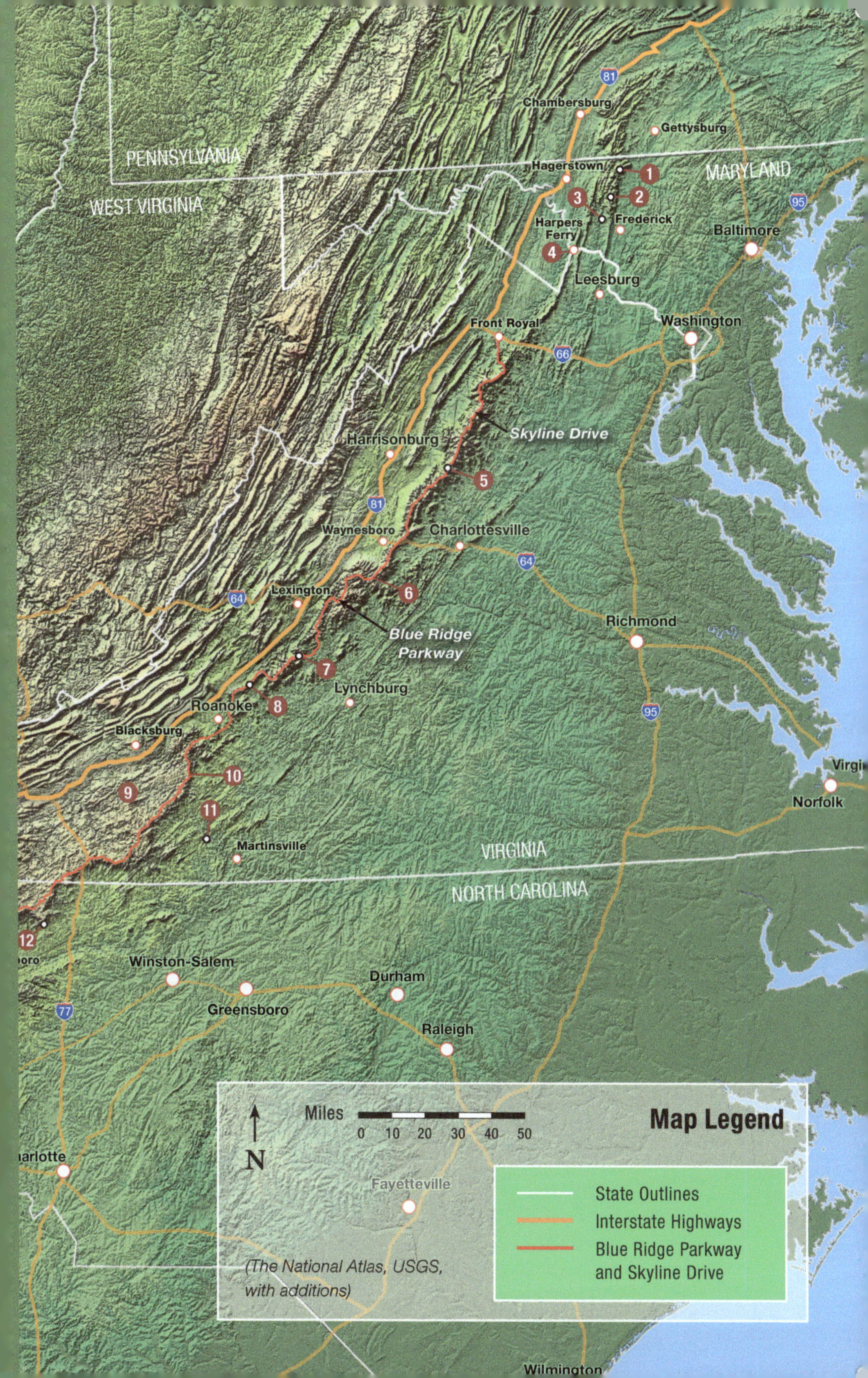

PENNSYLVANIA
WEST VIRGINIA
MARYLAND
VIRGINIA
NORTH CAROLINA
81
Chambersburg
Gettysburg
Hagerstown
1
2
3
Frederick
Harpers Ferry
4
Baltimore
95
Leesburg
Washington
Front Royal
66
Skyline Drive
Harrisonburg
5
81
Waynesboro
Charlottesville
64
64
Lexington
6
Blue Ridge Parkway
Richmond
7
Lynchburg
8
Roanoke
Blacksburg
95
10
9
11
Martinsville
Norfolk
12
Winston-Salem
Greensboro
Durham
Raleigh
77
Map Legend
Miles
0 10 20 30 40 50
N
Fayetteville
State Outlines
Interstate Highways
Blue Ridge Parkway and Skyline Drive
(The National Atlas, USGS, with additions)
Wilmington

Also by Edgar W. Spencer

Guide to Natural Bridge State Park and the Caverns at Natural Bridge
Coauthors Shannon and Shawn Spencer, Poorhouse Mountain Press

Geologic Maps: A Practical Guide to the Preparation and Interpretation of Geologic Maps, 3rd. edition, Waveland Press Inc.

Field Guide to the Chessie Nature Trail, "Trail Log,"
Edited by Larry Bland and Lisa Tracy, Mariner Publishing

Valley and Ridge and Blue Ridge Traverse, Central Virginia,
28th International Geological Congress: American Geophysical Union
Coauthors J.D. Bell and S.J. Kozak.

Earth Science - Understanding Environmental Systems, McGraw-Hill Book Co.

Introduction to the Structure of the Earth, McGraw-Hill Book Co.

Physical Geology, Addison-Wesley Publishing Co.

The Dynamics of the Earth, T.Y. Crowell Co.

Geology - A Survey of Earth Science, T.Y. Crowell Co.

Basic Concepts of Historical Geology, T.Y. Crowell Co.

All photographs and illustrations were created by the author, unless otherwise noted. Photographs requiring lengthy attributions (indicated by the symbol: †) are detailed in the additional photo attributions subsection of the reference section.

ISBN: 978-0-9837471-6-1

Printed by King Printing, Lowell, Mass.
First Edition/Fourth Printing

If you have suggestions to help keep this guide up to date
please contact us at: guidetotheblueridge@gmail.com

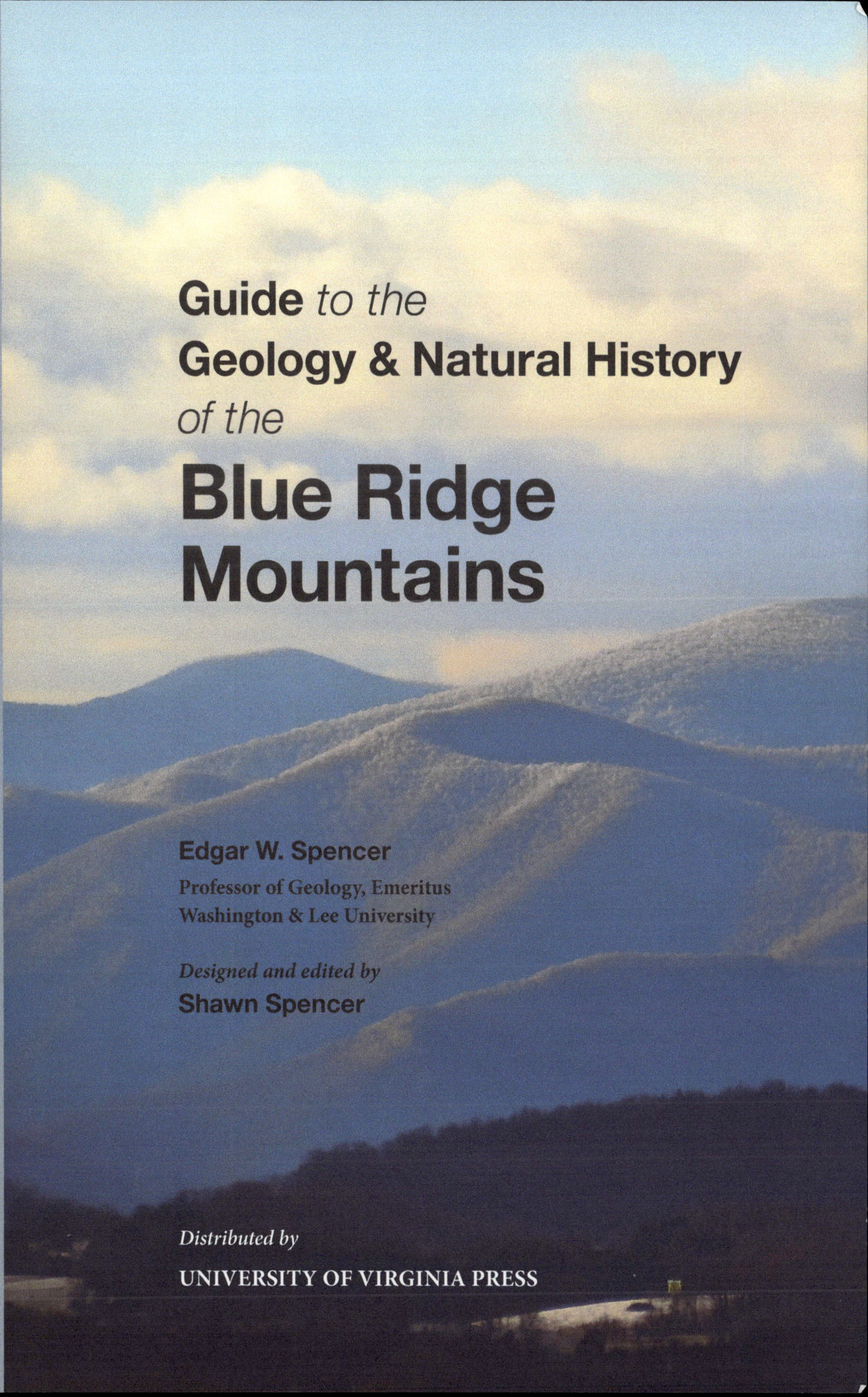

Guide *to the* Geology & Natural History *of the* Blue Ridge Mountains

Edgar W. Spencer
Professor of Geology, Emeritus
Washington & Lee University

Designed and edited by
Shawn Spencer

Distributed by
UNIVERSITY OF VIRGINIA PRESS

Table of Contents

Preface 1

PART 1. INTRODUCTION

1. **The Appalachian Mountains** 6
2. **Geological Evolution of the Blue Ridge** 16
3. **The Blue Ridge Landscape** 34
4. **Natural Environments in the Blue Ridge** 54

PART 2. FIELD GUIDES

5. **The Northern Blue Ridge** 72
6. **Places of Special Interest - Northern Blue Ridge**
   from Catoctin Mountain to Roanoke 90

   6A - Catoctin Mtn. Natl. Park & Cunningham Falls State Park - MD 93
   6B - Gambrill State Park - MD 99
   6C - Washington Monument State Park - MD 102
   6D - Harpers Ferry & Potomac Water Gap - WV/MD/VA 104
   6E - Shenandoah National Park - VA 113
   6F - Blue Ridge Parkway – Waynesboro to the James River - VA 142
   6G - James River Traverse Across the Blue Ridge - VA 154
   6H - Blue Ridge Parkway – James River to Roanoke - VA 174

7. The Southern Blue Ridge 184

8. Places of Special Interest - Southern Blue Ridge
from the New River Basin to the Great Smoky Mountains 200

8A - **Upland Plateau: New River Basin & B.R. Escarpment** - VA/NC 202
8B - **Blue Ridge Parkway – Roanoke to Doughton Park**
Roanoke Valley Overlook - VA 208
Buffalo Mountain Natural Area Preserve - VA 210
Rocky Knob Recreation Area - VA 212
Mabry Mill - VA 213
Stewarts Creek Wildlife Mgmt. Area - VA 214
Galax Area - VA 215
Doughton Park (Doughton Recreation Area) - NC 217
8C - **Philpott Reservoir & Fairy Stone State Park** - VA 219
8D - **Stone Mountain State Park** - NC 230
8E - **Mt. Jefferson State Natural Area** - NC 236
8F - **Mt. Rogers National Recreation Area** - VA 238
8G - **Grandfather Mountain Area & Linville Falls** - NC 258
8H - **The Mars Hill Terrane at Roan Mountain State Park** - TN 268
8I - **Spruce Pine Mining District** - NC 272
8J - **Mt. Mitchell State Park & Craggy Gardens** - NC 273
8K - **Precambrian Basement Rocks on I-26** - NC/TN 278
8L - **Blue Ridge Parkway – Asheville to Oconaluftee** - NC 279
8M - **Great Smoky Mountains National Park** - NC/TN 290

PART 3: IDENTIFICATION GUIDES

9. Rocks & Minerals *Commonly Found in the Blue Ridge* 312

10. Plants *Commonly Found in the Blue Ridge* 323

11. Birds *Commonly Found in the Blue Ridge* 354

References 373
Index 386

Acknowledgments

My interest in guidebooks about the Blue Ridge began in the early 1980s when Fred Schwab and I started taking students on extended field trips through the Appalachians. We depended heavily on the guides prepared by Byron Cooper, Ernst Cloos, Douglass Rankin, Gill Espenshade, Robert Newman, and Philip King among others but little information was available for many of the places we wanted to go. Fred became enthusiastic about doing research on some of these areas and became a great source of information about the region. Over the years many new guidebooks and professional papers, notably those by Robert Hatcher, Scott Southworth, Bill Henika, David Brezinski, and Arthur Merschat have been important sources of guidance. I have had the good fortune of spending time in the Blue Ridge with many of those mentioned above, as well as many others including David Bell, Thomas Gathright II, Gerry Wilkes, Marc Carter, Irving Brown, Christopher Bailey, Jeff Rahl, Howard Capito, Bob Root, Noell Potter, Michael Schafale, and Robert Whitemore, as well as my colleagues Ken Bick, Sam Kozak, David Harbor, Jeff Raul, Odell McGuire, Paul Low, and Chris Connors. I have also learned a lot from former students who worked with me in the Blue Ridge, notably Chris Bowring, Jon Kelaphant, Patrick Reynolds, Mike Follo, Chris Haley, Missy Eppes, Whit Morris, and Hal Newell III.

When coverage was expanded to include birds and plants I became much more heavily dependent on the help of others. John Knox, Peggy Dyson-Cobb, Lisa Tracy, Wendy Richards, Jim Warren, Jan Jarrard, Katie Letcher Lyle, Michele Fletcher, Dick Rowe, and Richard Wallace gave generously of their time and expertise to help me. A special note of thanks goes to those who supplied photographs for use throughout the guide. They are acknowledged in the captions.

Many people connected with state and national parks, the U.S. Geological Survey, and conservation agencies and organizations, especially those in Pennsylvania, Maryland, Virginia, and North Carolina have provided help, illustrations, and guided me to information about the region and images of birds and plants. I am grateful to them all, and give special thanks to Wendy Cass, Gary Fleming, Rickie White, and Thomas Jordan.

Many friends and colleagues took time to review and help me edit this book. I greatly appreciate the help of Bob Root, John Knox, Lisa Tracy, James Dick, Jeff Rahl, Scott Southworth, Bill Henika, Howard Capito, Greg Bank, Richard Wallace, my wife Liz, and my daughters Shawn and Shannon. Laura Parsons and Sarah Clayton were very helpful in advising me about publishing matters. I greatly appreciate the help and support the faculty and staff of the Washington & Lee University Geology Dept. have given over the years.

Preface

How to Use this Guidebook

Part 1 (Introduction) provides background for readers who are unfamiliar with or need a review of geological and/or environmental and ecological topics. It includes: a general introduction to the Appalachian Mountains; an account of the geological evolution of the Appalachians as interpreted in terms of plate tectonics; an overview of the factors and processes involved in the development of landscapes over vast periods of time; and an account of the environmental forces at work in the Blue Ridge as well as an overview of how ecologists define its diverse array of ecological habitats.

Part 2 (Field Guides) identifies and gives more detailed information about places one is likely to visit when traveling in the Blue Ridge. Most travelers will drive along the Blue Ridge Parkway and visit state and national parks. Special attention is given to these places. The character of the landscape changes south of where the Roanoke River cuts across the Blue Ridge. This provides a useful place to divide the Blue Ridge into southern and northern sections. Each has an overview chapter summarizing its major features: Ch. 5 for the northern section and Ch. 7 for the southern section. The reader should be sure to go over these chapters before using the more detailed accounts of localities described in Chs. 6 and 8. These accounts provide basic information about how to reach each locality, the accommodations available, and where to go to see many of the features of special interest to those who desire to learn about the geology and natural history of these areas. References for additional information are provided.

Part 3 (Identification Guides) contains information about how to identify some of the most common rocks and minerals; trees and plants; and birds found in the Blue Ridge. Readers with a serious interest in these subjects will want to refer to more specialized guides or websites, some of which are identified in the chapter openings and the reference section.

Abbreviations

To save space in the text, the following abbreviations have been used:

AT =	Appalachian Trail
Fm. =	Formation (rock formation)
Mtn. =	Mountain
Mt. =	Mount
NHP =	National Historic Park
NPS =	National Park Service
NRA =	National Recreation Area
PGC =	Penn. Game Commission
USF&WS =	U.S. Fish & Wildlife Service
USFS =	U.S. Forest Service
USGS =	U.S. Geological Survey

Road & Highway Abbreviations:

I-64 =	Interstate 64
US 50 =	United States Highway 50
SR 39 =	State Road 39
FS 701 =	Forest Service Road 701

Icons

= Park visitor centers and headquarters, museums, demonstration sites

= Naturalist interest

= Unreferenced photographs

Safety in the Field

Individuals who walk on trails, along creeks, or explore waterfalls need to be aware of hazards found in the Blue Ridge. Many of these dangers are present in the national parks and along the Blue Ridge Parkway as well as off the roads.

Fig. P-1. Walking along the sides of curvy mountain roads can be extremely dangerous.

1. Traffic is the greatest danger to those who are examining rock outcrops in roadcuts (Fig. P-1). Be extremely careful! Do not stop on interstate highways or other major routes. Very few rock exposures along dangerous highways are identified in this guide, but even rural roads can be dangerous. Be careful to park in pull-outs along the Parkway or off the road elsewhere. Do not park where your tires are on the pavement along any road. Wear bright-colored clothes in the field, especially along roads.

2. Falling rocks are a serious hazard along roadcuts, quarries, and on steep slopes. Stand clear of vertical rock faces and do not dislodge rocks.

Fig. P-2. Poison ivy has a cluster of three shiny leaves at the end of an often reddish-tinged stem. Note the green berries.

3. Use a geologic hammer when breaking rocks. Never use an ordinary hammer to break rocks. The steel in geologic hammers is a high-strength alloy, but even it too could break when hitting hard rocks, sending fragments of the rock or the hammer into your body.

4. Poison ivy is a vine found near the ground or twining up into trees and fences. It has a cluster of three leaves at the end of a long, often reddish-tinged stem (Fig. P-2). Touching the vines, even when the leaves and berries are not out may still infect people who are allergic. If you touch it, wash thoroughly with a strong soap or even just water as soon as possible. If a rash breaks out make a lather of soap and let it dry on the affected area. For more serious infections, especially those on your face, see a physician.

5. Watch for rattlesnakes and copperheads during summer and early fall. (Figs. P-3, P-4) Rattlesnakes are more common than copperheads in the Blue Ridge. Avoid tall grasses and rocky crevices. Boots and a walking stick are helpful. If bitten, it is important to get away from the snake and stay calm, keeping your pulse down. Dial 911

Fig. P-3. Rattlesnake (NPS)

Fig. P-4. Copperhead (Photo: Dane Conley)

and sit still or walk slowly back to your car and get to a hospital as soon as possible. The area will begin to swell in 1-2 min. Remove shoes/jewelry in the affected area. Keep it immobilized if possible and below the level of your heart. DO NOT cut the fang marks, apply ice, or use a tourniquet - this will cut off blood flow and the limb may be lost. Rattlesnake and copperhead bites are rarely fatal.

6. Yellow jackets and hornets are present throughout the Blue Ridge during the summer and before frost in the fall. Most yellow jackets have nests in the ground with access through a small hole. If you are stung and develop hives or have difficulty breathing, go immediately to a hospital for appropriate medications.

7. Black bears live in many parts of the Blue Ridge. Normally they avoid people. Do not get between cubs and their mother or try to feed them. If a bear approaches, remain calm, then together your party should make yourselves as large/imposing as possible, wave your arms around and shout loudly, making as much noise as possible. Persist until the bear retreats. Do not run away. If necessary, continue to face it while slowly backing away.

8. Do not dive into pools along streams. Although the water may appear inviting, rocks may be present at shallow depths even though they are not visible.

9. Be careful when walking or climbing in or near streams and waterfalls. Rocks in streams may be smooth and rounded. Add a coating of algae and they become very slippery. A number of people have died from accidents near waterfalls in the Blue Ridge.

10. Deer ticks are the primary source of lyme disease, a malady that may go undetected for months or even years. Deer ticks only grow up to .137 in. (3.5 mm) in diameter and have a distinctive coloration (Fig. P-5). Unable to jump or fly, they only pass onto humans in direct contact with tall grasses or shrubs. Check for ticks at the end of the day. Remove with tweezers by grasping the head close to your skin and slowly pulling. Remove any leftover parts. If a rash forms around a deer tick

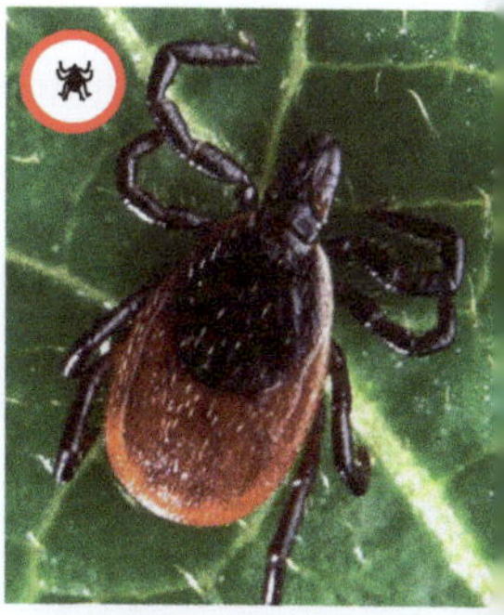

Fig. P-5. Deer tick. Inset shows actual adult size. (Photo: Scott Bauer)

bite, or if you experience flu-like symptoms after a bite, see a doctor.

11. Be careful about where you go during hunting season. Stay out of the woods when hunters are using rifles. Untrained hunters may shoot at any sign of motion in the underbrush. Check state websites for hunting season dates where you plan to hike off road. Usually seasons span from early fall through winter.

12. For more details about field safety refer to "Field Safety in Uncontrolled Environments" by S.R. Oliveri and Kevin Bohacs, 2005. This is available at **http://www.aap.org/**.

Maps

If you are traveling into a national or state park or along the Skyline Drive or Blue Ridge Parkway, be sure to obtain a copy of the maps published by the NPS and state parks. These are available at Parkway offices and visitor centers. These publications show routes of access to the parks or the Parkway. Some of the maps identify overlooks, gaps, visitor facilities, campgrounds, places of interest, and other valuable information. Many of these maps are available on the web. If you plan to take trails along the Parkway obtain a copy of the *Blue Ridge Parkway Outdoor Guide* published by the Park Service. This publication contains a list of trails and has maps showing the roads and trails at parks and other trail sites along the Parkway.

If you are traveling off the Parkway, you will need detailed maps showing highways and topography. The 1:100,000 scale (area overview) or 1:24,000 scale (land detail) topographic maps published by the USGS are good choices

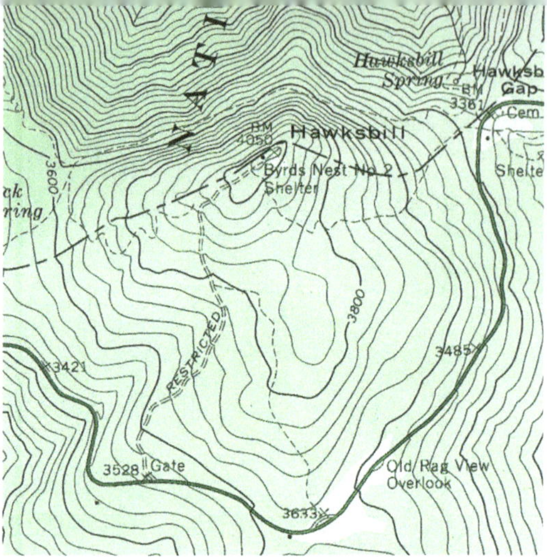

Fig. P-6. Part of a 1:24,000 topographic map showing land detail for part of the Shenandoah National Park. (USGS)

(Fig. P-6). These can be found at outdoor supply stores or may be bought or downloaded for free from the USGS website, **http://www.usgs.gov/pubprod/** (or call 1-888-275-8747). The U.S. Forest Service also publishes updated topographic 1:24,000 scale maps that show trails and forest service roads (see **http://www.fs.fed.us/maps/**). For some areas, the National Geographic Society has published a series of *Trails Illustrated* maps that are available on waterproof paper from their online store at **http://www.nationalgeographic.com/**. Geological maps are also available for many areas. Contact the USGS or state geological survey websites for information. For information about the interpretation of topographic and geologic maps, refer to *Geologic Maps: A Practical Guide to the Preparation and Interpretation of Geologic Maps* by Edgar Spencer, Waveland Press.

Google Earth is an application you can download to your computer or mobile that makes it possible to obtain images of areas over a great variety of scales, and allows you to obtain oblique as well as vertical aerial views. ❖

Part 1:
Introduction

Fig. 1-1. Shaded relief map of the Southern and Central Appalachians with a white dashed outline of the Blue Ridge. Most of the Appalachian Plateau is a high level (3,000 to 4,000 ft.) plateau that is deeply dissected by streams that flow west into the Gulf of Mexico. The Valley and Ridge Province is characterized by northeast-southwest trending ridges separated by deep valleys. North of Roanoke streams flow into the Atlantic. South of Roanoke the streams flow into the Gulf of Mexico. North of Roanoke the Blue Ridge has a prominent ridge along its western margin. The southern part of the Blue Ridge starts as a high level plateau south of Roanoke that changes into a region of high mountain peaks near the Great Smoky Mtns. At its southern end the Blue Ridge has low relief. Most of the Piedmont is characterized by low relief and rolling hill country that drops in elevation to the east until it disappears and is replaced by a nearly flat landscape in the Coastal Plain. Waterfalls and rapids are present where streams flowing eastward across the Piedmont pass into the Coastal Plain. (The National Atlas, USGS, with additions)

Chapter 1
Introduction to the Appalachian Mountains

The Blue Ridge Mountains in Context

While this guidebook is devoted to the Blue Ridge Mtns. it is important to view them part of a much larger mountain system—the Appalachian Mtns. The word "Appalachian" comes from the American Indian Creek language in which Apalachee means highland farmers. It seems probable that it was first applied to the mountain range in South Carolina by a French explorer in 1567. Today it is applied to the mountainous region that extends all the way from Newfoundland to Alabama and Georgia (Fig. 1-1). From a global perspective, geologists use the name "Appalachian Mtns." (which they refer to as a **mountain belt, mountain chain** or an **orogen**) to designate those parts of the Earth's crust that were involved in several periods of mountain building that took place along the eastern margin of North America between about one billion and 200 million yrs. ago.

During each of these mountain building periods portions of the Earth's crust were uplifted and compressed. We now think this process resulted from the collision of large sections of the crust, including whole continents or chains of volcanic islands, that moved relative to one another. These movements are explained by the theory of plate tectonics, which is described in Ch. 2. When these collisions occurred, compression between two plates caused continental margins to rise and become strongly deformed, leading to the formation of long mountain belts like the Alpine-Himalayan mountain chain that extends from Spain to Southeast Asia. This belt was formed as a result of collisions between the continental crust of what is now Africa, Arabia, and India with that of Eurasia. Plate movements also caused continents to slowly move apart allowing oceans to form. One such ocean is now forming in the Red Sea where Africa and Saudi Arabia are moving apart. The Atlantic Ocean is in a much more advanced stage that started about 165 million yrs. ago.

What we see today in the Appalachians are the remains of mountains that may have been as high as the Alps or the Himalayas (20-40,000 ft.) when they were forming. Over the last 200 million years erosion has reduced the height of the Appalachians to a region that is now much lower and only partially mountainous. Indeed some parts of it have been reduced to hilly terrain only a few hundred to a thousand feet above sea level. Mountain building along the continental margin was not a single event, nor was the entire belt affected at any one time. Periods of inactivity were long enough for erosion by streams and downslope movements of materials on steep slopes to remove the topographic features of what had once been high and rugged mountains. The products of this erosion

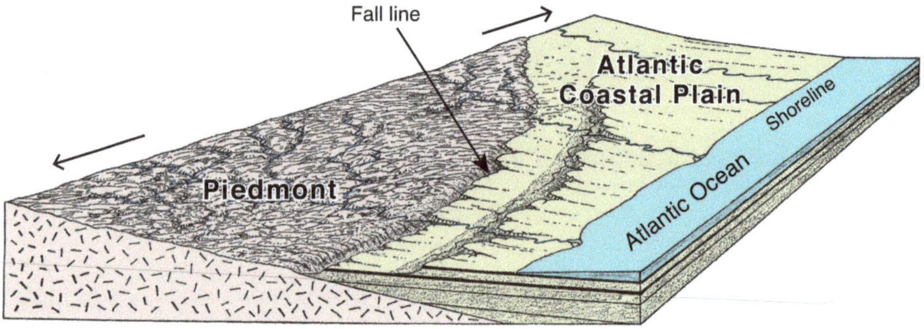

Fig. 1-2. Coastal Plain and Piedmont.

were deposited in the ocean as sediment along the North American continental margin. For long periods of time much of the land that had been high mountains was worn down to a surface close to sea level. When sea levels rose relative to the land the ocean advanced across the margin of the continent, leaving behind accumulations of marine sediments as it retreated. These sediments now form the **Atlantic Coastal Plain**, a region with low relief (differences in elevation) that extends from Cape Cod to Florida and continues from there to the west as the Gulf Coastal Plain. During the last 100 million years when the ocean covered these coastal plains (Figs. 1-1, 1-2) they were part of the continental shelf, a place where marine sediments settled on them as they do now on modern continental shelves. These marine sediments are very thin near the modern Appalachian Mtns. but become much thicker as they pass under the continental shelf and eventually under the much deeper waters farther off the coast (Fig. 1-3).

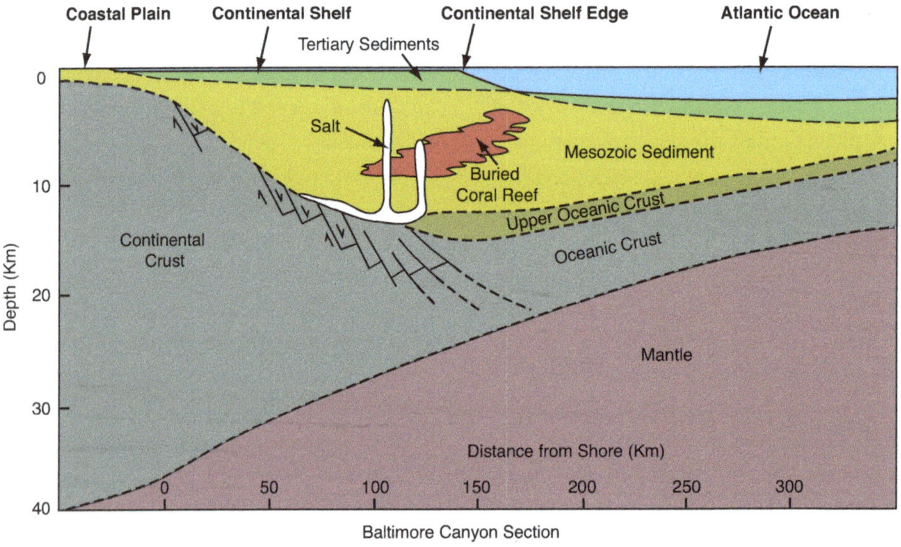

Fig. 1-3. Schematic cross-section of the eastern continental margin showing marine sediments passing under the continental shelf. (Modified after Grow, 1980)

Far Extensions of the Appalachian Mountain System

Although rocks deformed during mountain building in the Appalachians appear abruptly along the dramatic, rugged coastline of Newfoundland, similar rocks with closely related histories are present in Ireland. That mountain chain continues across Scotland and along the coast of Norway. This European chain, known as the Caledonide Mtns. is a northern continuation of what we know as the Appalachians. These mountains, so far removed from one another today, were all part of a single continuous system that split apart and separated as North America and Eurasia moved apart about 165 million years ago. The waters of what we know as the Atlantic Ocean filled the gap between the drifting continents.

To the south, the Appalachians end in central Alabama where the much younger and undeformed, flat-lying sedimentary rocks of the Gulf Coastal Plain lie on top of the older rocks that were deformed during the last episode of Appalachian mountain building. Nevertheless, similar deformed rocks continue below the surface of the coastal plain and emerge again far to the west in the Ouachita Mtns. of Arkansas. They disappear again beneath the Gulf Coastal Plain in Oklahoma only to reappear in western Texas.

The boundary between the unconsolidated sediments underlying the coastal plain and the older, hard, consolidated, and strongly deformed rocks that compose the mountains is commonly marked by a prominent change in landscape and the presence of rapids and waterfalls along streams that flow across the boundary (Fig. 1-2). During colonial settlement of this region, this boundary, often called the **fall line,** was used as the navigational **head** for ships traveling along the rivers. Consequently, many cities including Columbia, S.C., Richmond, Va., Washington D.C., and Philadelphia, Pa. were established at or just downstream from rapids.

Internal and External Parts of Mountain Belts

Geologists recognized long ago that most mountain systems could be divided into **internal and external parts**. The internal portions are usually located near the center of the system and contain older rocks which have been uplifted from deep in the Earth's crust where they were altered by the heat and pressure found at great depths. These alterations were caused by what is called **metamorphism** (p. 15). Because this middle portion of the Appalachian mountain belt was uplifted so long ago, erosion has effectively removed much of what was once the top of it. This process of erosion is now happening in the relatively young Alpine-Himalayan belt whereas it is in the advanced stages in the Appalachians.

The external portions of mountain belts are made up of sedimentary rocks, most of which were deposited on top of the internal parts of the system. Both the internal and external parts of the Appalachians are preserved on the western side of the belt, but the eastern part of the belt lies buried beneath sediments in the coastal plain and the waters that cover the continental margin.

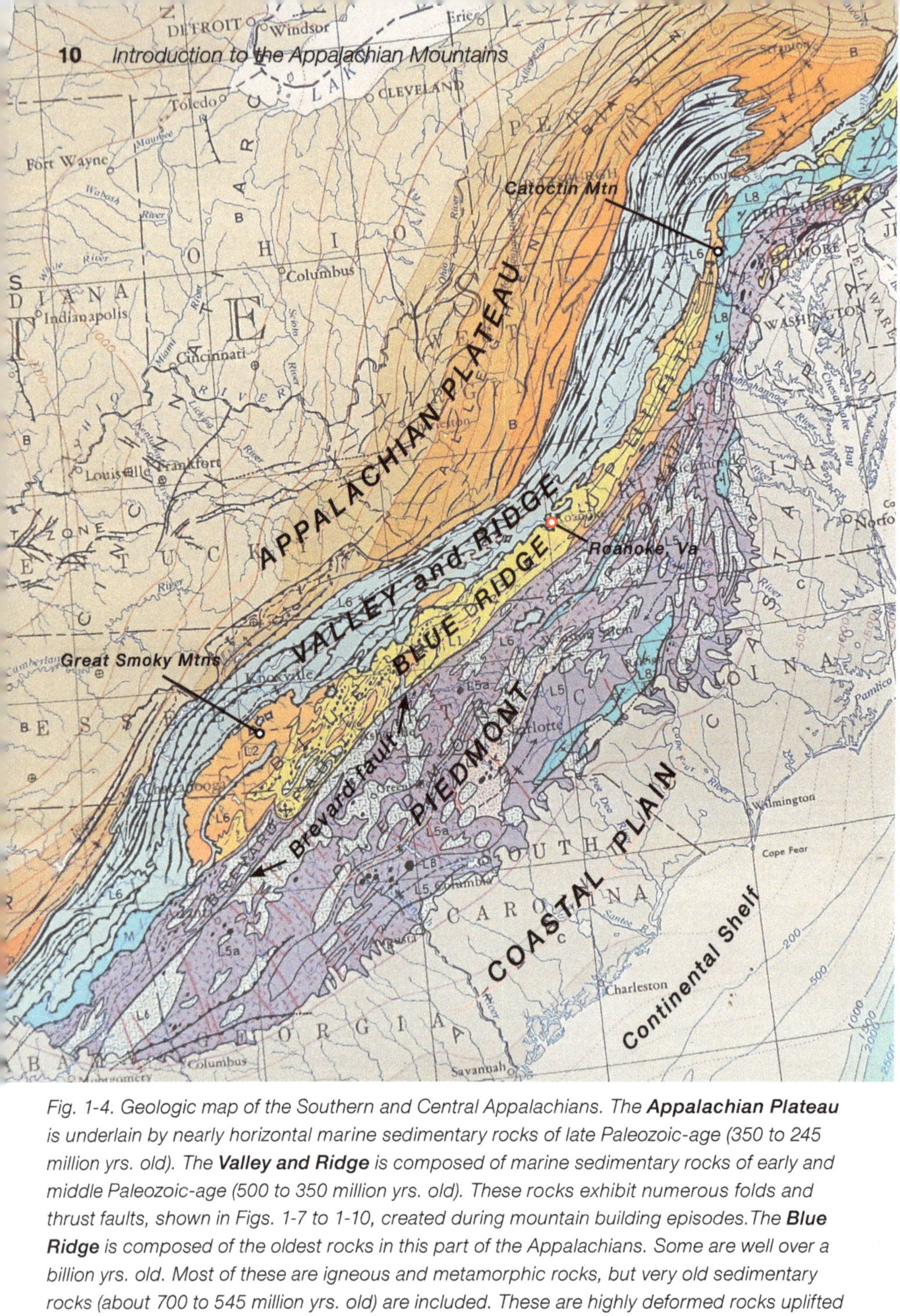

Fig. 1-4. Geologic map of the Southern and Central Appalachians. The **Appalachian Plateau** is underlain by nearly horizontal marine sedimentary rocks of late Paleozoic-age (350 to 245 million yrs. old). The **Valley and Ridge** is composed of marine sedimentary rocks of early and middle Paleozoic-age (500 to 350 million yrs. old). These rocks exhibit numerous folds and thrust faults, shown in Figs. 1-7 to 1-10, created during mountain building episodes.The **Blue Ridge** is composed of the oldest rocks in this part of the Appalachians. Some are well over a billion yrs. old. Most of these are igneous and metamorphic rocks, but very old sedimentary rocks (about 700 to 545 million yrs. old) are included. These are highly deformed rocks uplifted and moved to the northwest during mountain building episodes. The **Piedmont** includes igneous and metamorphic rocks of Paleozoic-age (545 to 350 million yrs. old), and Precambrian rocks (older than 545 million yrs). A few sedimentary basins of Mesozoic age (245 to 200 million yrs. of age) are present in the Piedmont. Much younger, undeformed marine sedimentary rocks underlie the **Coastal Plain**. Most of these are less than 150 million yrs. old and some are modern sediments. This map is taken from an early 1969 version of the Geologic Map of the United States. More recent and complex maps are currently available. (USGS, with additions)

Fig. 1-6. View across the Appalachian Plateau. Note that the rock layers are subhorizontal (nearly horizontal). (Photo: Jim Petretich)

Our Focus: The Central and Southern Appalachians

The western portions of both the external and internal portions of the Appalachian Mtn. system are exposed at land surface all the way from Newfoundland to Alabama. Not surprisingly, the landscape along this great distance changes significantly, reflecting variations in the structure and geologic history of the underlying rocks. Three major changes occur along this belt making it possible to divide it into three main parts: the northern, central, and southern sections. The northern section starts in Newfoundland and continues southward, close to New York City. The central and southern portions, which we will examine in this guidebook, are similar to one another in that they consist of four major physiographic provinces characterized by distinctive landscapes and internal structure. All four provinces are elongate and run from northeast to southwest. From west to east they are the **Appalachian Plateau**, the **Valley and Ridge**, the **Blue Ridge**, and the **Piedmont** (Figs. 1-1, 1-4, 1-5). The westerly provinces of the Appalachian Plateau and Valley and Ridge, which are both external parts of the Appalachians, are underlain by sedimentary rocks such as those formed by the deposition of sand, mud, and limestone in marine waters that geologists think covered the internal portions of the belt before mountain building began. The Blue Ridge and Piedmont, on the other hand, are internal parts of the Appalachians. The rocks now exposed in these areas were originally several miles deeper in the crust before mountain building episodes uplifted them and moved them to the northwest.

The Appalachian Plateau

This province that includes the Pocono, Alleghany, and Cumberland plateaus, has a high, nearly flat surface that has been dissected by streams flowing in valleys cut deeply into the upper surface (Fig. 1-6). The rocks exposed in these valleys are nearly horizontal layers of sedimentary rocks 200-400 million yrs. old. Their structure shows up in cross-sections across parts of the Plateau (Fig. 1-7).

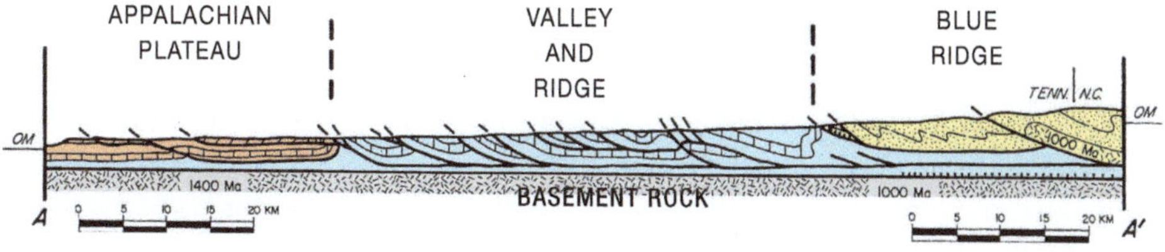

Fig. 1-5. Representative cross-section of the central Appalachians showing thrust faults carrying the Blue Ridge and parts of the Valley and Ridge to the northwest over little-deformed basement rock. (John Rodgers, 1970)

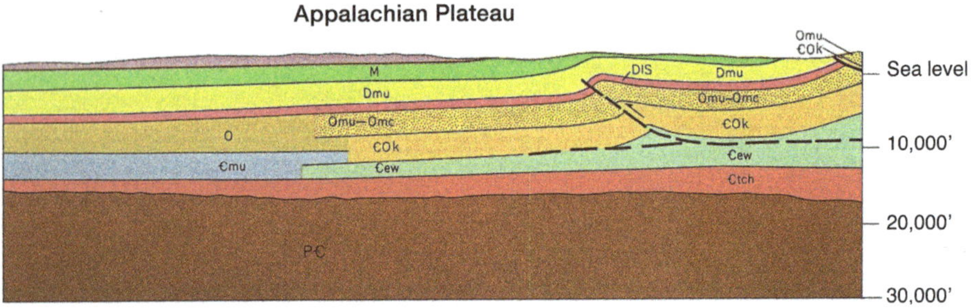

Fig. 1-7. Cross-section of part of the Appalachian Plateau. (From the Geologic Map of W. Va., 1968)

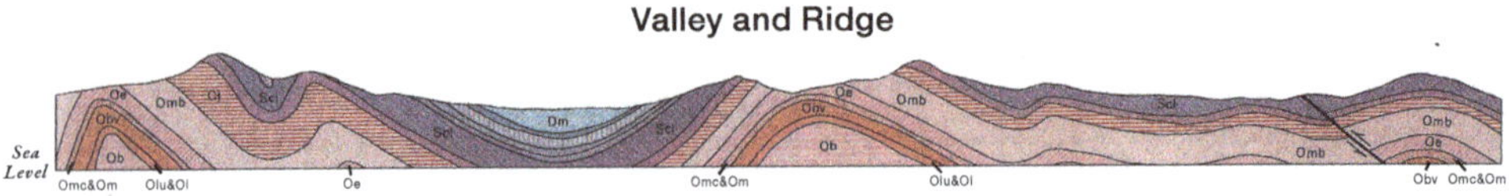

Fig. 1-8. Cross-section of part of the Valley and Ridge Province. (From the Geologic Map of W. Va., 1968)

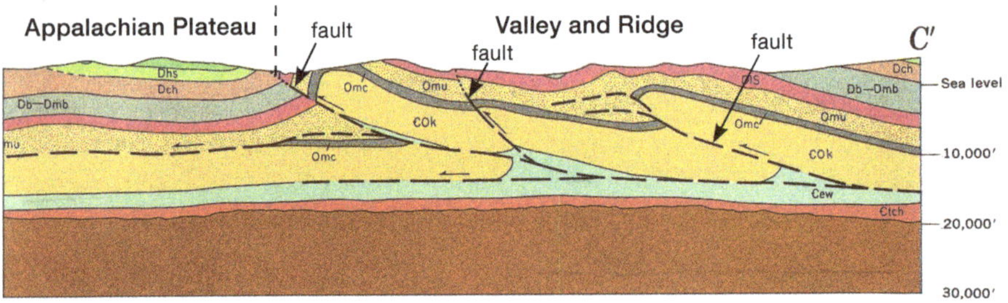

Fig. 1-9. Cross-section drawn across the western part of the Valley and Ridge Province and a small part of the Appalachian Plateau (left side). (From the Geologic Map of W. Va., 1968)

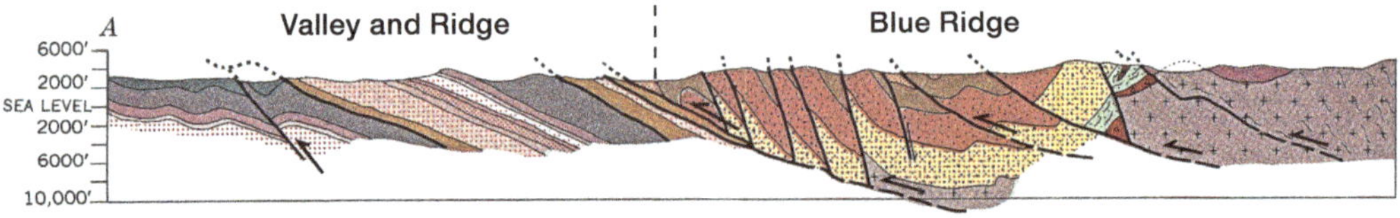

Fig. 1-10. Cross-section drawn across the eastern part of the Valley and Ridge and the western margin of the Blue Ridge. The purple rocks in the Blue Ridge are ancient (over a billion years old) igneous and metamorphic rocks. (Rankin, Espenshade, and Neuman, 1972)

Along the eastern edge of the Plateau, the structure of the sedimentary layers begins to change. In the north they begin to exhibit broad, wave-like folds. Farther south the structure of the layers changes from horizontal and gently folded to strongly folded (Fig. 1-8) or the layers may be broken by faults (Fig. 1-9). This change marks the transition from the Plateau to the Valley and Ridge (Figs. 1-10, 1-11). In the eastern part of the Valley and Ridge Province faults that have broken through and displaced the sedimentary rocks are exposed at the ground surface, but some faults lie hidden below the ground surface and extend under the Valley and Ridge and into the Plateau where they can be detected by samples from deep wells.

The Valley and Ridge Province

In the Valley and Ridge layers composed of sandstone that are resistant to erosion stand out in the landscape as ridges (Figs. 1-1, 1-8, 1-12). Intervening layers of shale and limestone, much less resistant to erosion, have been reduced by erosion to lower elevations and thus lie, as sediment, in the floors of valleys. The resulting topography consists of long ridges separated by valleys (Fig. 1-1, 1-8).

Fig. 1-11. Boundary between the Appalachian Plateau and the Valley and Ridge near Seneca Rocks.

Most of the rocks in the Blue Ridge and Piedmont Provinces are igneous and metamorphic rocks that were formed at very high temperatures and pressures deep in the Earth. They were uplifted thousands of feet and shoved many miles to the northwest over faults that are subhorizontal (nearly horizontal) and then later eroded to their present configuration. It is here that the most intense deformation took place during mountain building. Rocks in the Blue Ridge are the oldest rocks in the Appalachians and have been uplifted from greater depths than those in other provinces.

In most places, the landscape changes with an abrupt increase in elevation as you cross the boundary between the Valley and Ridge and Blue Ridge Provinces. This happens because the rocks present in the Valley and Ridge near this boundary are not very resistant to erosion while the rocks that occur along the western edge of the Blue Ridge are highly resistant to erosion. They are sedimentary in origin, but in a short distance much older igneous and metamorphic rocks, elevated from great depth are present at the surface. They have been uplifted by major faults that are exposed close to the boundary between the Valley and Ridge and the Blue Ridge, especially in the southern part of the Blue Ridge (Figs. 1-5, 1-10).

The Blue Ridge Province

The bluish color of the mountains from a distance (for which they are named) is produced as a result of the selective scattering of light in the atmosphere by tiny particles produced by a chemical reaction between hydrocarbon molecules released by vegetation and ozone. Initially the name Blue Ridge was applied to the continuous, prominent ridge that starts south of Carlisle, Pa. and continues to Roanoke, Va. with only two gaps formed where the Potomac and James rivers cut across the ridge (Fig. 1-13). Later, as mapping of the geology of the region progressed, geologists discovered that rocks on a western ridge at the northern end of the Blue Ridge could be traced around to the east into another ridge named Catoctin Mtn., that continues many miles to the southwest. These two ridges are part of a much larger **anticlinal structure** that appears between Harpers Ferry and Catoctin Mtn. in Fig. 1-1. Part of this anticlinal structure is similar to that shown in Fig. 5-2.

South of Roanoke, Va., this anticlinal structure is replaced by an elevated

Fig. 1-12. Photo of an anticlinal fold known as Rainbow Ridge located in the Valley and Ridge at Clifton Forge, Va.

plateau surface, called the **Blue Ridge Upland Plateau** (Fig. 1-1). Farther south this plateau gradually changes into a large oval shaped area of high mountains, referred to as the **Blue Ridge Highlands** (Fig. 1-1). Mt. Rogers and the Great Smoky Mtns. are part of this high, mountainous region. Still farther south the Highlands give way to a region of lower topography that ends where the sedimentary rocks of the Gulf Coastal Plain cover it.

The name Blue Ridge is applied to all three of these sections combined. The Skyline Drive and Blue Ridge Parkway pass continuously through them, connecting the Shenandoah and Great Smoky Mtns. national parks.

The Piedmont Province

The boundary between the Blue Ridge and the Piedmont is more difficult to define. To the south a major fault zone, known as the Brevard fault (Fig. 1-1), is often used as a boundary. Farther north, other faults that appear to be continuations of the Brevard fault zone are used. They coincide with the edge of the eastern ridge (Catoctin Mtn.) described above.

Looking Ahead

Before turning to the interesting places one can visit to get a first hand look at the geology of the Blue Ridge, it will be useful to review the geological history of this central feature of the Appalachian Mtns. When we talk about the age of mountains it is important to recognize that several very different answers may emerge. Age may refer to the time that the rocks found in the mountains were formed, or it may indicate the time during which mountain building took place, or the time at which the mountainous landscape took on its present form. In the case of the Blue Ridge, the oldest rocks have an age of about 1.3 billion years. However, the present landscape is "young," which to a geologist means a few million years old, and it is continually changing. The age of the mountain building here turns out to be far more complex than one might expect. The Blue Ridge contains a fascinating record of multiple mountain building episodes separated by periods during which mountains formed earlier were eroded away and seas advanced across the region before renewed mountain building took place. This history is now interpreted in terms of plate tectonic theory, the focus of the next chapter.

Travelers through the Blue Ridge will be impressed by the immense diversity of flora and fauna enlivening the landscapes of the region. We will take a look at these environments in Ch. 4. ❖

Fig. 1-13. James River Water Gap near Glasgow, Va.

Geological Terms

Anticline: a fold that is convex on the upside (Fig. 3-2a). Due to erosion, rock beds exposed at the core of the fold are older than those exposed on the edges.

Escarpment: a steep slope or long cliff that occurs from erosion or faulting that separates two relatively level areas of differing elevations.

Fall line: the break between an upland region of relatively hard basement rock and a coastal plain consisting of softer sedimentary rock.

Fault: a crack or shear zone in the Earth's crust along which one side has been displaced in relation to the other side.

Igneous: a term meaning "born from fire" that is applied to both volcanic rocks and rocks that crystallize slowly within the Earth's crust.

Metamorphism: the processes acting beneath the ground by which rock is so extensively altered that it obtains a new texture, mineral composition, or both.

Orogen: a mobile mountain belt in the Earth's crust that has been subjected to folding and other deformations due to movement in the Earth's tectonic plates.

Orogeny: the process of mountain building caused by the movement of tectonic plates.

Relief: relative variations in elevation between areas on the Earth's surface.

Shearing: deformation that occurs when the internal parts of rock material slide relative to one another under stress. Brittle rock tends to fracture while more ductile rock compresses or stretches.

Subhorizontal plane: a plane that is nearly horizontal in orientation.

Talus: (also called ***scree***) Loose rock that forms at the base of cliffs or slopes where fallen boulders and/or smaller rock fragments accumulate.

Type locality: the place where rock units such as formations were originally found and described by geologists.

This reconstruction shows the initial stages in the breakup of the supercontinent, Pangaea which was formed by continental collision late in the Paleozoic Era (about 250 million years ago). The Appalachian Mtns. and their continuation in the Caledonide Mtns. through Scotland and Scandinavia were by this collision between Laurentia (North America) and Gondwana (encompassing Africa and South America). All areas shown in light blue are underlain by continental crust. The bodies of water along the center of the Appalachian Mtns. occupied rift basins such as the Connecticut Valley, the New Jersey Lowlands, and the Gettysburg and Culpeper basins. (Reconstruction graphic: Ron Blakey, Colorado Plateau Geosystems)

Chapter 2

Geological Evolution of the Blue Ridge

The Interior of the Earth

For those unfamiliar with the theory of **plate tectonics** and the long scope of geologic time, it is hard to imagine that the mountains of eastern North America resulted from the collision of North America with Eurasia (Fig. 2-1). Earth's crust is made up of continents and ocean basins. The term "crust" was first used when geologists thought that a shell of solid rock covered a hot, liquid interior. As seismologists learned how to decipher the records produced by shock waves generated by earthquakes, they learned that Earth's interior consists of several concentric shells of material. A solid ball of metallic material forms the **inner core**. It is overlain by a thick shell of liquid metal, rich in iron, called the **outer core** that extends halfway to the surface. The next shell, known as the **mantle**, is composed of rock hot enough to flow slowly. The mantle is rich in magnesium and iron and rigid enough to transmit seismic waves that travel through solids.

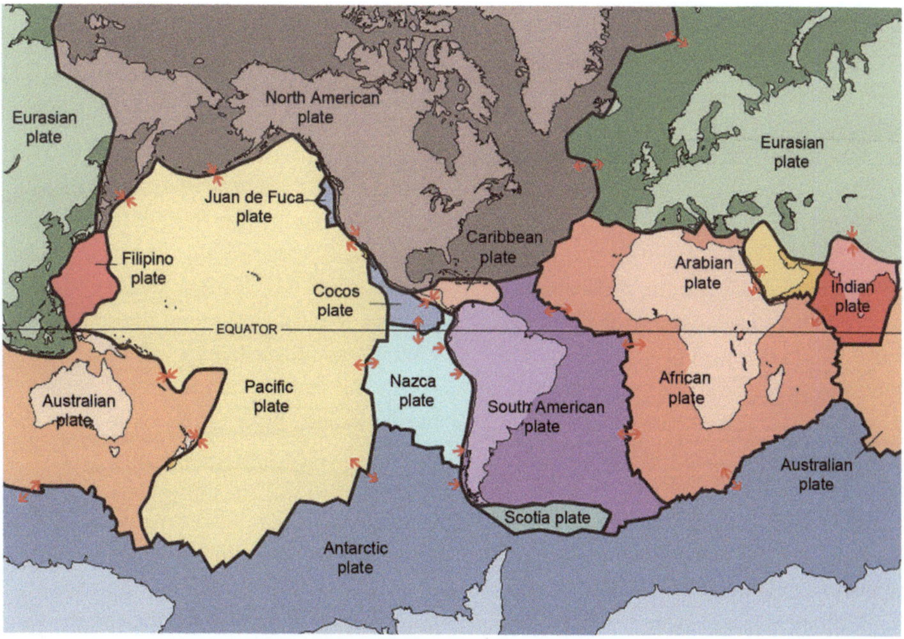

Fig. 2-1. Modern tectonic plates. Plate boundaries are shown in black. Red arrows indicate the direction of plate movement. Spreading ridges where new crust is formed are shown as arrows pointing away from one another. Subduction zones, where the oceanic crust sinks into the Earth, are represented by arrows pointing inward into plate boundaries. Only one major strike slip fault, the San Andreas, located along the coast of California, is shown. (USGS)

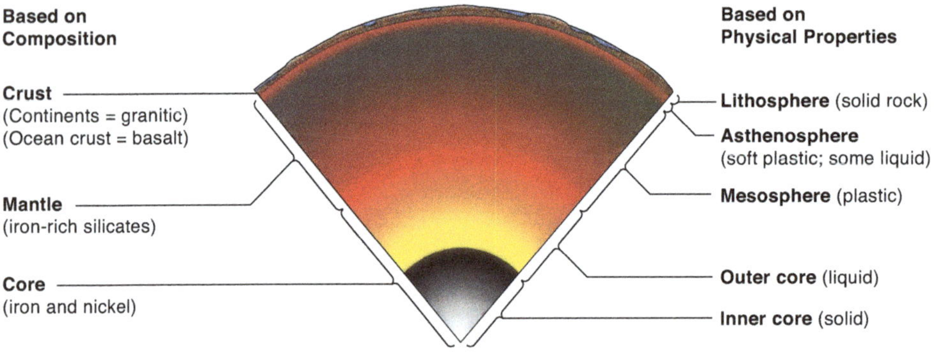

Fig. 2-2. Divisions of Earth's interior based on seismic analysis. (E.W. Spencer, 2003)

Near the top of the mantle the temperature-pressure conditions are very close to those needed to melt rock. This nearly liquid part of the mantle is called the **asthenosphere** (Fig. 2-2). Above it the outer most shell of Earth, the **lithosphere**, is composed of solid rock. Only the outermost part of the lithosphere, the **crust**, is composed of the types of rocks we see at the surface. Some of the rocks in the Blue Ridge came from deep within the crust, which extends 12-25 miles under the continent. The asthenosphere is so soft that the much cooler and more rigid overlying lithosphere can slip over it or sink into it.

Continental crust is quite different in composition and density from **oceanic crust**, which is found beneath the seafloor. The higher-density oceanic crust is heavier than the continental crust and thus sinks deeper into the asthenosphere. This difference in density is reflected in differences in elevation. The average elevation of continents is about half a mile above sea level, while the surface of the seafloor averages nearly three miles below sea level. The rocks exposed on continents contain much more silicon and are of lower density than those found in oceanic crust. On average, the composition of continental rocks is similar to granite. In contrast, the rocks that make up most of the oceanic crust are basalts (Ch. 9) containing more magnesium and iron than granite. The crust and rigid part of the upper regions of the mantle are essentially floating on top of a deeper layer of ductile (pliable) material that makes up the asthenosphere.

Plate Tectonic Theory

Since the 1960s, geologists have interpreted the formation of mountain belts in terms of plate tectonic theory. According to this theory the lithosphere (Fig. 2-2) is a mosaic of relatively thin sheets of rock, called plates, which move relative to one another and interact along their margins. These movements are driven by the rising and sinking of hot masses of the highly viscous material that makes up the deeper parts of Earth's interior.

In the early 1900s, Alfred Wegener, a German meteorologist, proposed that continents could move. He suggested that South America and Africa were once part of a single continent that broke and drifted apart. Most North American geologists ignored the idea of continental drift because Wegener did not explain how continents could move through the solid rock that makes up the oceanic crust. This changed dramatically in the early 1960s with the suggestion that new oceanic crust was forming along ridges such as the Mid-Atlantic Ridge where volcanic centers line a ridge crest that is slowly rising and spreading apart as new oceanic crust forms along the ridge (Fig. 2-3). The consensus view now is that continents do not move *through* the oceanic crust. Instead, they move *with* the oceanic crust until the heavier, more dense, oceanic crust sinks back into the mantle.

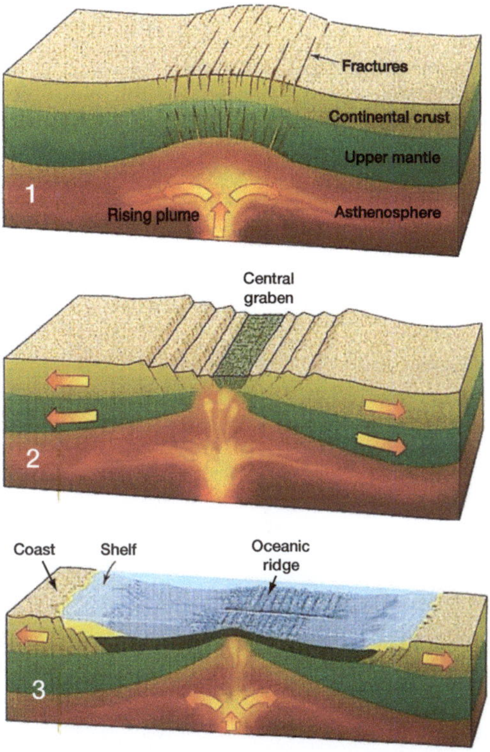

Fig. 2-3. Seafloor spreading takes place as new oceanic crust forms along oceanic ridges. (E.W. Spencer, 2003)

A variety of geophysical studies subsequently established that new oceanic crust forms along **oceanic ridges** and then sinks back into Earth's interior along certain continental margins. Many of these areas are located around the rim of the Pacific Ocean basin. Deep-sea trenches, active volcanoes, and deep focus earthquakes mark places where the oceanic crust is sinking back into Earth's interior, a process known as **subduction** (Fig. 2-4). Currently, these zones often occur in chains of volcanic islands like those found along the northern and western margins of the Pacific Ocean (e.g., the Aleutian Islands, the Philippines, and the islands north of New Zealand). Subduction zones are also present along the west coast of South America, in the Caribbean, and along the coast of Java and Sumatra.

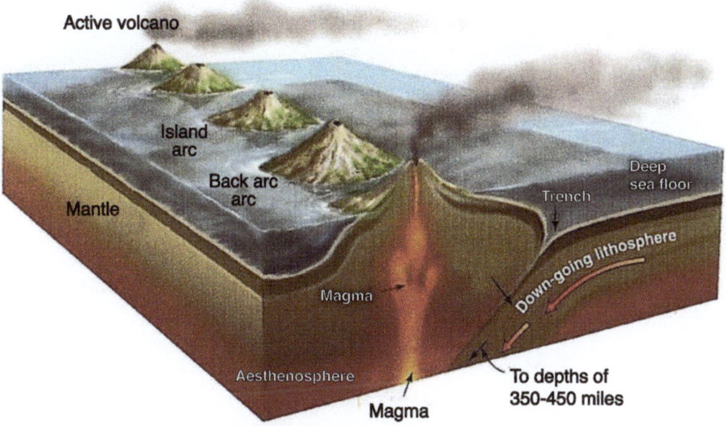

Fig. 2-4. Island arcs develop where volcanic islands form above subduction zones in which the seafloor sinks into the mantle. (E.W. Spencer, 2003)

The continental margin off Oregon and Washington is similar, but differs in that sediments carried into the ocean by the Columbia river fill what would otherwise be a trench. Subduction is *not* taking place along many other margins such as those around the Atlantic and Indian Oceans, nor is it along plate margins defined by major faults such as the San Andreas.

Many of the rocks now found along the eastern side of the Blue Ridge and in the Piedmont originated from fragments of continental crust or from the types of volcanic rocks found in modern island arc systems. These fragments of continental crust that separated from larger continents such as Laurentia contain granitic rocks as well as marine sediments that were deposited on or along the margins of the fragments. Volcanic island arcs contain lava flows, ash deposits, seafloor sediments mixed with volcanic materials that moved into the trenches found at subduction zones, and marine sediments deposited in the shallow bodies of water behind the volcanic arcs.

Oceanic ridges and subduction zones define the margins of most tectonic plates. Ridges are commonly found where plates have split and spread apart—these are called **divergent boundaries**. One of these, the Mid-Atlantic Ridge, is the eastern boundary of the modern North American plate. Subduction zones are plate boundaries where one plate sinks under another plate—these are called **convergent boundaries**. Major faults that extend vertically through the lithosphere define a third type of plate boundary. Plates slip by one another along these fault zones that are called **transform boundaries**. The San Andreas fault zone in California is an example of this type of plate boundary. Note that continental margins, the broad zones along the edges of continents, are not always plate boundaries. Many continental margins such as the eastern margin of North and South America as well as the margins of Africa are not currently plate boundaries. They are passive margins left behind where continents broke apart as new oceanic crust formed.

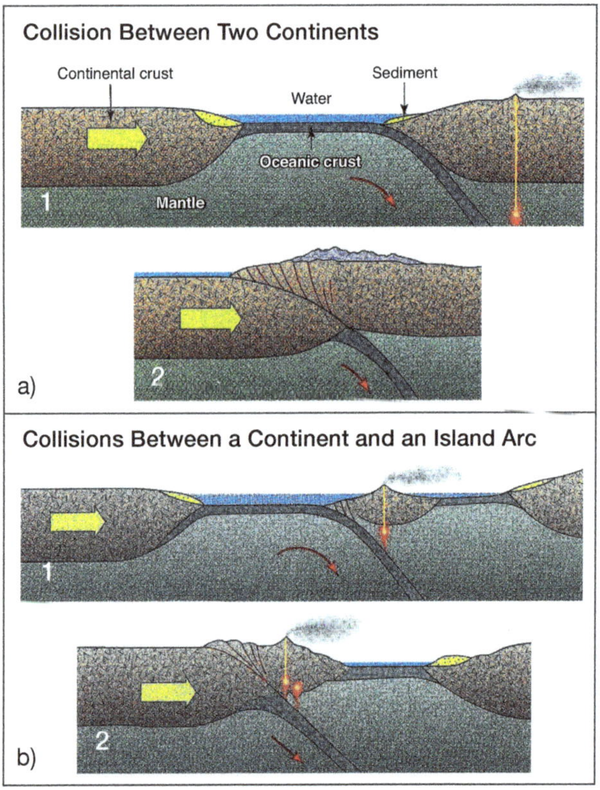

Fig. 2-5. Schematic representations of a) collisions between continents and b) between a continent and an island arc. Both types of collisions took place in the Appalachians during the Paleozoic Era. (E.W. Spencer, 2003)

Fig. 2-6. Fold formed during continental collision. This one is in the Pyrenees Mtns. of Spain.

Mountain Building

Mountain belts like the Appalachians form where plates converge (Fig. 2-5). The mountain building event is called an **orogeny**. We refer to the large areas and long zones where the crust has been deformed during these events as **orogenic belts**. They may contain volcanoes and blocks of crust that have been uplifted, but they are much larger and more complex than individual mountains. Orogenies take place where plates collide. These collisions may involve the smashing together of a **volcanic island arc** with a continent or the jamming together of two continents. During these collisions the heavy, high-density part of the lithosphere—the oceanic crust—sinks under the less dense continental crust. When two plates carrying continental crust collide the lithosphere becomes greatly thickened as one plate overrides the other. The pressures that develop along these zones of convergence are so great that the sediment and rocks along the margins of the plates are deformed as one might expect them to be if they were caught in a gigantic vise. As the margins become compressed, rock masses rise as mountains. At depth the rock is so hot and under such great pressure that metamorphism takes place; it yields like a plastic and becomes folded. Here, the rock gets squeezed, forming huge folds like those seen on the sides of the deep valleys in the high Alps and Himalayas (Fig. 2-6). The layered sedimentary rocks that occur at shallower depths in the continental crust are forced into folds resembling those seen in a tablecloth when pushed across a table. Faults inclined at low angles form as the rock mass is compressed (Fig. 1-5). We see these folds and faults in the Valley and Ridge Province in the Appalachian Mtns. Slabs of crustal rocks are shoved away from the zone between the converging plates. These slabs may move as much as a hundred miles laterally. All of these processes were involved in the creation of the Appalachian Mtns. Today parts of this system that were at shallow depths during mountain building are present in the Appalachian Plateau. From these external parts of the mountain belt, we cross the Valley and Ridge Province where the sedimentary rocks formed high in the crust are folded and faulted. Farther east, in the Blue Ridge and Piedmont, erosion that has taken place over more than 200 million years has exposed rocks that were deeply buried during mountain building. The history of mountain building in the Appalachians involved repeated episodes of plate collisions and took place over periods of time involving many millions of years, time spans known as **geologic time**.

Geologic Time

Efforts to decipher Earth's history began long before there were ways of determining the age of events in terms of years, referred to as **absolute age** (Fig. 2-7). Instead, geologists determined the **relative age** of events (Fig. 2-8). Many events in Earth's history were put into sequence by scholars in the 1700s. The next major step in developing a time scale took place during the mid-1800s, when William Smith, a prominent English engineer, discovered that many sedimentary layers

Geologic Time Scale

Time in millions of years (Ma)	Eras	Periods
0 Ma		
	Cenozoic	Quaternary Neogene Paleogene
66 Ma		
	Mesozoic	Cretaceous Jurassic Triassic
252 Ma		
	Paleozoic	Permian Carboniferous Devonian Silurian Ordovician Cambrian
541 Ma		
	Precambrian	

Fig. 2-7. Eras are subdivided into Periods that are further subdivided into Epochs (not shown).

could be identified by their embedded fossils. This led to the recognition that the fossils found in younger layers of rock are different from those found in older rocks. Using the relative ages of sequences of events in combination with the key to relative ages provided by fossils, it became possible to develop a geologic time scale.

The major divisions, called **Eras**, are named for the types of animals and plants found as fossils. **Cenozoic** refers to modern life forms, **Mesozoic** to middle life forms, **Paleozoic** to ancient life forms, **Proterozoic** to primitive life forms, and **Archean** to rocks that contain only the

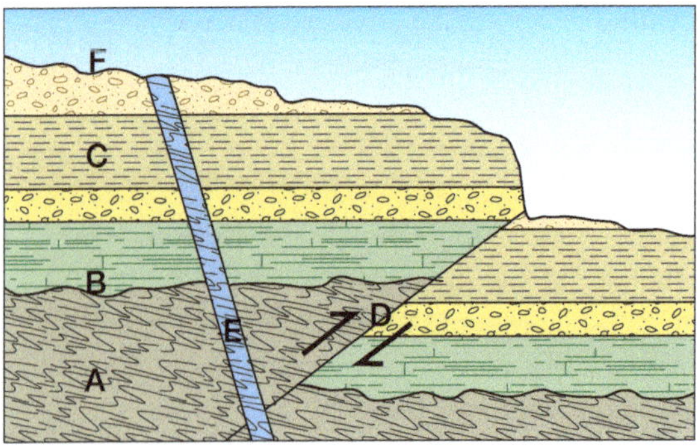

A - 1st: Formation of the metamorphic rocks (shown at the base).

B - 2nd: Development of the erosion surface on top of the metamorphic rocks.

C - 3rd: Deposition of layers of sediment on top of the erosion surface.

D - 4th: Displacement along the fault.

E - 5th: Intrusion of the dike.

F - 6th: Uplift of the area and erosion of the present land surface.

Fig. 2-8. Cross-section with relative ages of events.

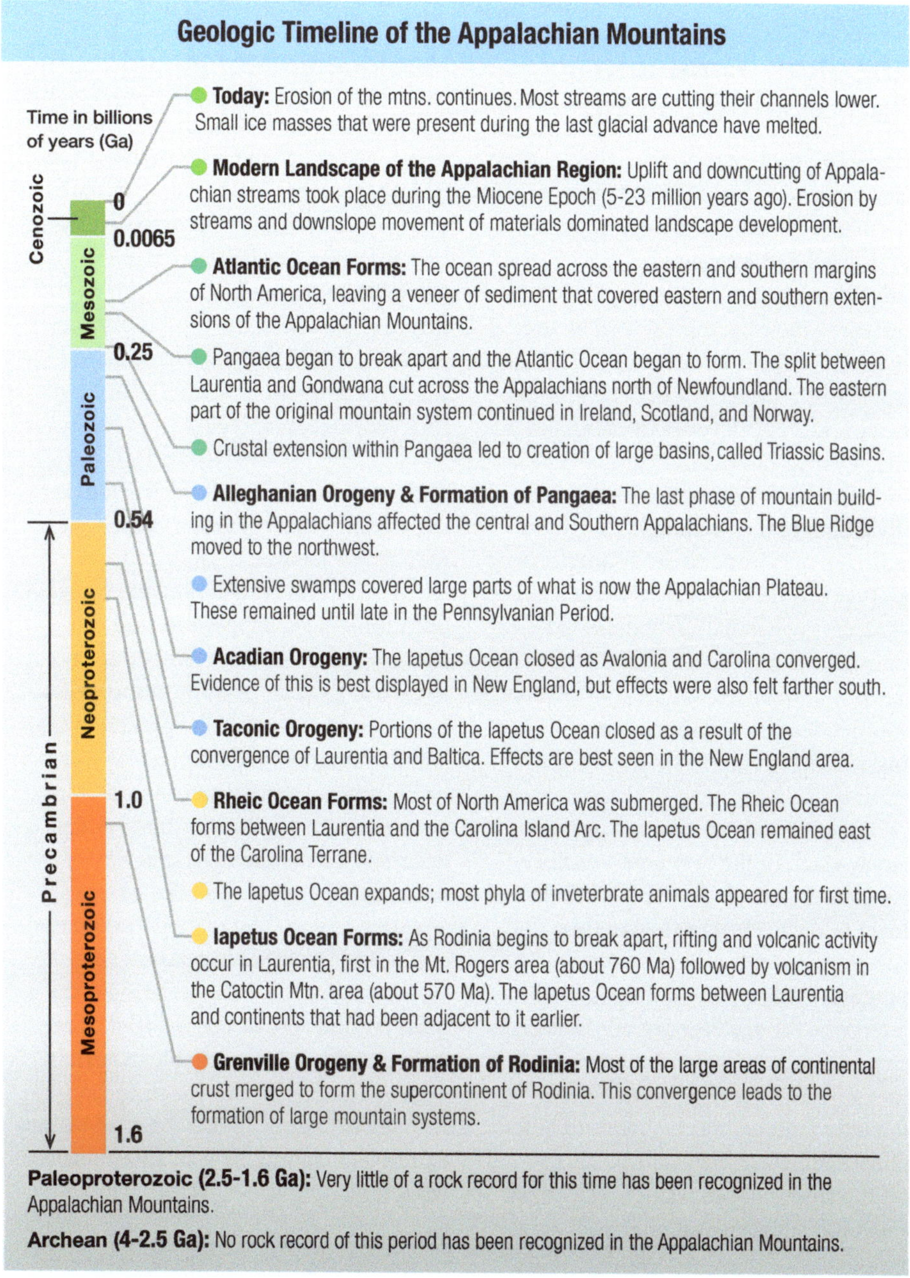
Geologic Timeline of the Appalachian Mountains
Time in billions of years (Ga)
Cenozoic
Mesozoic
Paleozoic
Neoproterozoic
Mesoproterozoic
Precambrian
0
0.0065
0.25
0.54
1.0
1.6
Today: Erosion of the mtns. continues. Most streams are cutting their channels lower. Small ice masses that were present during the last glacial advance have melted.
Modern Landscape of the Appalachian Region: Uplift and downcutting of Appalachian streams took place during the Miocene Epoch (5-23 million years ago). Erosion by streams and downslope movement of materials dominated landscape development.
Atlantic Ocean Forms: The ocean spread across the eastern and southern margins of North America, leaving a veneer of sediment that covered eastern and southern extensions of the Appalachian Mountains.
Pangaea began to break apart and the Atlantic Ocean began to form. The split between Laurentia and Gondwana cut across the Appalachians north of Newfoundland. The eastern part of the original mountain system continued in Ireland, Scotland, and Norway.
Crustal extension within Pangaea led to creation of large basins, called Triassic Basins.
Alleghanian Orogeny & Formation of Pangaea: The last phase of mountain building in the Appalachians affected the central and Southern Appalachians. The Blue Ridge moved to the northwest.
Extensive swamps covered large parts of what is now the Appalachian Plateau. These remained until late in the Pennsylvanian Period.
Acadian Orogeny: The Iapetus Ocean closed as Avalonia and Carolina converged. Evidence of this is best displayed in New England, but effects were also felt farther south.
Taconic Orogeny: Portions of the Iapetus Ocean closed as a result of the convergence of Laurentia and Baltica. Effects are best seen in the New England area.
Rheic Ocean Forms: Most of North America was submerged. The Rheic Ocean forms between Laurentia and the Carolina Island Arc. The Iapetus Ocean remained east of the Carolina Terrane.
The Iapetus Ocean expands; most phyla of inveterbrate animals appeared for first time.
Iapetus Ocean Forms: As Rodinia begins to break apart, rifting and volcanic activity occur in Laurentia, first in the Mt. Rogers area (about 760 Ma) followed by volcanism in the Catoctin Mtn. area (about 570 Ma). The Iapetus Ocean forms between Laurentia and continents that had been adjacent to it earlier.
Grenville Orogeny & Formation of Rodinia: Most of the large areas of continental crust merged to form the supercontinent of Rodinia. This convergence leads to the formation of large mountain systems.
Paleoproterozoic (2.5-1.6 Ga): Very little of a rock record for this time has been recognized in the Appalachian Mountains.
Archean (4-2.5 Ga): No rock record of this period has been recognized in the Appalachian Mountains.

Fig. 2-9. Geologic timeline with events in the formation of the Appalachian Mtns.

most primitive plants (Fig. 2-9). Each of the eras is subdivided into time intervals known as **Periods**, many of which are named for geographic areas where rocks of that age are well exposed or where they were initially recognized and described. For example, the Pennsylvanian Period is named for rock layers exposed in the state of Pennsylvania; the Jurassic Period is named for the Jura Mtns. of France and Switzerland; and the Permian Period is named for the Russian province of Perm.

Absolute Time and Radiometric Dating

The geologic time scale based on relative ages was well established long before the theory of evolution was advanced by Darwin and long before we had a reliable method to assess absolute time, measured in years, represented by the divisions of the time scale. In the 1950s geochemists succeeded in determining the absolute age of rocks by using the rate at which radioactive elements decay into other non-radioactive elements. The length of time required for half of a **radioactive isotope** to decay and form another isotope or a different but stable element is constant. Carbon-14, uranium-235 and 238, and potassium-40 are commonly used to date rocks. These isotopes decay spontaneously by emitting radiant energy.

Two types of radiometric dating are widely used. In the first type a radioactive isotope, called the **parent**, breaks down to form another radioactive isotope or a stable isotope, called the **daughter**. This type of dating is used to determine the age of inorganic materials such as minerals. Several changes from one radioactive isotope to another radioactive isotope may be involved before a final stable daughter isotope appears, but the amount of the daughter isotope present at any time is directly related to the amount of the parent isotope that is present and to the length of time of the decay process. If radioactive elements are present in the minerals of an igneous rock when it solidifies, the length of the decay process measures the time that has elapsed since the cooling and crystallization of the magma took place. Rocks composed of minerals that contain uranium, potassium, and rubidium have been dated using this technique. Because very small quantities of the parent and daughter are needed it is possible to determine the age of many igneous rocks. The same technique is used to determine how long ago a radioactive mineral crystallized during metamorphism. In this case, however, the crystallization is related to the temperature needed to cause the mineral to form, the original rock may be much older.

The second type of radiometric dating assumes that the rate at which the radioactive isotope is produced and enters a natural system remains constant until something happens to stop this process. For example, a living tree takes in carbon dioxide from the atmosphere, but when the tree dies the process stops and the quantity of carbon in the tree is fixed. When a system stops in this way, any radioactive isotope in that system decays at a fixed rate. The quantity of the isotope remaining in the system at any time after it "closes" depends on the amount of time that has passed since the system closed. Carbon-14 dating is the most familiar use of this method. It can be used to date most organic materials. Unfortunately, the half-life of carbon-14 is so short that it can only be used to date materials that are less than about 50,000 years old.

By measuring absolute ages it is possible to determine how long-ago events, dated using the relative time scale, took place. Based on these methods we are now determining the age and sequence of events that have transpired during the development of the Appalachian Mtns. and the oldest rocks found in its once deeply buried core, the Blue Ridge.

Geologic Evolution of the Southern Appalachian Mtns.

Deciphering the sequence of events that have transpired during the history of the Appalachians is a fascinating subject. The complexity of these events is a challenge to which many geologists have devoted their careers. Though many questions have been answered, many remain. Most of the events in this long and complex history, especially those revealed in the sedimentary rocks that occur in the Valley and Ridge and along the western edge of the Blue Ridge, have been known for decades. However, attempts to devise a unified model that includes all events recorded in the core of the Blue Ridge and in the Piedmont have been difficult. Modern interpretations of the history of mountain building in the Appalachians are based on plate tectonic models.

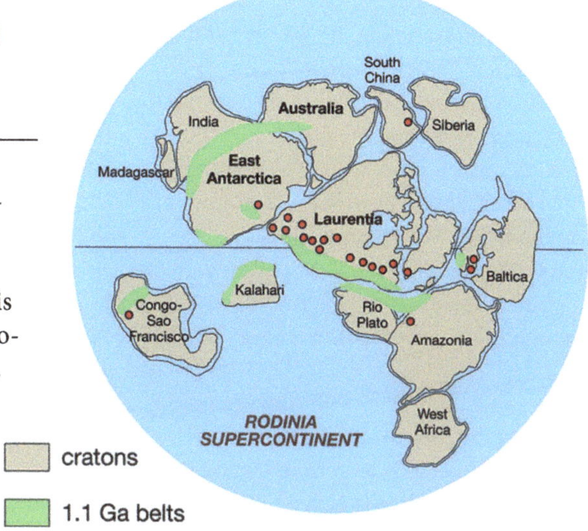

Fig. 2-10. One of the modern schematic restorations of the primitive supercontinent Rodinia as it may have existed in late Precambrian time when it began to break up. (After John Goodge, 2011)

It is easier to reconstruct recent geologic events than those that took place much earlier. Since most of this reconstruction involves the interpretation of rock exposures, gaps occur where rocks of particular ages are missing. Two conditions cause most of these gaps. The first is when younger rocks cover older rocks, making the record left in the older rocks hard to see, even though it may be exposed in scattered rock outcrops and preserved at depth where it can be revealed by drilling. The second condition occurs when no sediments are deposited because the area is above sea level and unaffected by volcanic activity.

Rodinia and the Grenville Orogeny: Rocks that were metamorphosed as early as 1.3 Ga (1.3 billions of years ago), have been found in the core of the Appalachian Mtns. Although older rocks have been identified in the Blue Ridge, their history has yet to be well defined. A more widespread metamorphism took place about 1.1 Ga during a mountain building episode known as the Grenville Orogeny that involved much of the Blue Ridge as well as the region farther to the northeast and into eastern Canada. All of these rocks were subjected to high temperatures, causing most of the minerals containing radioactive elements to recrystallize. (The mineral zircon is an exception, requiring much higher temperatures to recrystallize.) The recrystallization of these metamorphic rocks "reset" the radiometric clocks in most of them, making it difficult to use absolute dates to establish the age of earlier events. During the metamorphism that took place about one billion years ago the older rocks were highly deformed under pressures many miles below the surface. This deformation was accompanied by the intrusion of large amounts of molten rock. Most geologists think that the Grenville Orogeny took place during continental collisions that produced a supercontinent called **Rodinia** (Fig. 2-10).

A significant portion of Rodinia remains intact in the interior portions of North America, especially in central Canada. Most of the rocks of that age in the United States lie buried beneath younger sedimentary rocks in the Great Plains. After millions of years of erosion, the mountains that rose during the Grenville Orogeny disappeared. Only those rocks that had been deeply buried during mountain building remain exposed at the surface along the eastern margin of North America. Much later, after several additional orogenies, some of those rocks were brought to the surface during the last major mountain building in the Appalachians, and are now exposed along the Blue Ridge Parkway and other roads.

Breakup of Rodinia: Late in the Precambrian Era, Rodinia began to break apart. Several continents resulted. Two of these were destined to play a major role in the history of the Appalachians. One of these, called **Laurentia**, included most of modern North America. The other is known as **Gondwana**. Geologists still debate the history of the other parts of Rodinia. As Rodinia split apart, large rift basins (Fig. 2-11) formed along its continental margins, and lavas poured out onto the surface of the eastern margin of Laurentia (now the northern part of the Blue Ridge) as a new ocean called the **Iapetus** began to form. Some of these lavas are present on Catoctin Mtn. in Maryland and form Mt. Rogers in southern Virginia. Sediments, notably sands, eroded from Laurentia and carried into the ocean by streams, were deposited in these rift basins. These sediments are also present in many places along the western edge of the Blue Ridge. Most of the rocks found in the Great Smoky Mtns. were originally sediments deposited in one of these rift basins.

Several large continental fragments split away from Laurentia (North America) during the breakup of Rodinia (Fig 2-10). The Iapetus Ocean grew as seafloor spreading continued for many millions of years. Finally, expansion of the ocean stopped and the Iapetus began to close as seafloor subduction began along the North American (Laurentian) margin. Volcanic island arcs formed along the coast. About the same time a large fragment of continental crust split away from Gondwana, and a new ocean, the Rheic Ocean (also Theic), formed on the eastern side of this fragment (between

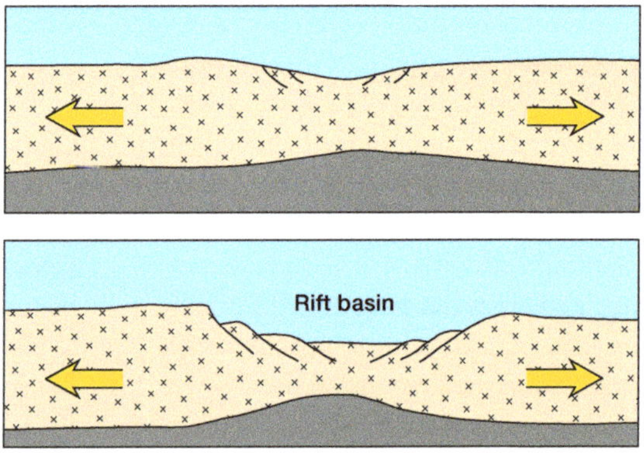

Fig. 2-11. A rift basin forms as a result of the pulling apart of continental crust.

Laurentia and Gondwana). Geologists continue to debate details about the location of island arcs and fragments of continents, the timing of collisions, and the direction of movements that took place in the Iapetus and Rheic oceans during the Paleozoic Era. Nevertheless, the record of some of the major orogenies that took place along the eastern margin of Laurentia (North America) is clear, as is the role they played in the evolution of the Blue Ridge and Piedmont.

Collisions Along the Laurentian Margin: During the Ordovician Period, limestones and dolomites were deposited in the shallow waters that covered much of the eastern part of Laurentia. These sediments are now exposed in the Great Valley of Virginia and in Pennsylvania. Farther offshore, island arcs, one known as **Carolinia**, formed within the Iapetus Ocean. Subduction along this margin pulled the volcanic arcs and some of the continental fragments formed during the splitting of Rodinia into Laurentia. The rocks that were part of these island arcs and continental fragments are now recognized as **terranes** in the Appalachians (Fig. 2-12). The compression squeezed the rocks, sediments, and volcanic materials caught in this collision. Mountains rose as the continental fragments and island arcs pressed into the continent. The Taconic Mtns. of New York emerged during this orogeny, which bears the name **Taconic Orogeny**. Evidence of this mountain building also shows up to the north in Newfoundland and in the Southern Appalachians, especially in the Great Smoky Mtns. region where thrust faulting and metamorphism took place at this time. As the mountains rose, streams carried sediments eroded from them to the west where they were deposited in what would later become the Valley and Ridge Province. By the middle of the Silurian Period the mountains had been largely removed by erosion and the ocean began to flood across the eastern edge of Laurentia (North America). Quartz sandstone beaches and sediments bearing iron minerals formed along the Silurian shoreline. These iron-bearing sandstones are prominent in the Valley and Ridge, and became the ore for the iron produced in this region. Because the quartz-rich rocks are resistant to erosion they now make up many of the ridges in this province. As the depth of the sea increased, the sandy beaches were covered by deeper water and limestones began to form. Still deeper water and muds that came to the sea from lands that were rising to the north and east, buried these limestones during the Devonian Period. The resulting shales, exposed in many of the valleys in the Valley and Ridge, hint at the mountain building that was then taking place in the Appalachians north and east of the Blue Ridge. These mountains may have been located in what is now the Atlantic Coastal Plain and continental shelf.

Acadian Mountain Building: During the Devonian and early Carboniferous Periods subduction began again along the eastern margin of Laurentia drawing more island arcs and continental fragments into the continent. Huge volumes of volcanic ash, lava flows and igneous intrusions were added to the eastern side of the Appalachians and are now well exposed in New England. These collisions gave rise to mountain building known as the **Acadian Orogeny**. The new mountains rose in eastern New England and off the modern coast of the Atlantic Ocean. Streams carried sediments from the uplifted mountains westward, building

Terrane* Map of the Blue Ridge

Appalachian Plateau Province

This high plateau (3,000 ft. +/-) makes up the western side of the Appalachian Mtns. It is underlain by mid to late Paleozoic rocks derived from the mountains to the east and deposited on the Laurentian (North American) continent. These rocks include many of the coal deposits of the Appalachians.

Valley and Ridge Province

This long belt of valleys and ridges lies between the continental interior and the Blue Ridge. Sedimentary rocks of lower Paleozoic-age were deposited along the margin of Laurentia. Thick sections of limestone, shale, and sandstone accumulated on this continental margin as it subsided during the Paleozoic Era. These rocks were folded and faulted, especially during the late stages of Appalachian mountain building, the Alleghanian Orogeny.

Atlantic Coastal Plain

The flat landscape of the coastal plain reflects the shape of the underlying sedimentary rocks. These sediments were deposited when the Atlantic Ocean advanced across the continental margin and across the eastern portions of the Appalachian Mountains which had been eroded after mountain building ended over 250 million years ago. The sediments range in age from about 140 Ma (Cretaceous) to recent along the modern coast. The layers are slightly wedge-shaped, growing thicker as they extend out under the continental shelf, Fig.1-3.

Triassic Rift Basins

These basins formed when Gondwana began to break apart about 200 million years ago. The Atlantic Ocean formed as Laurentia (North America) separated from Africa and Eurasia. Mountain building in the Appalachians had ended, but the mountains were still high as the fault bounded basins developed.

Laurentian Basement Cover

Laurentia was one of the large continental fragments that formed when Rodinia broke apart late in the Precambrian (about 570 Ma). Laurentia included the central part of North America. Its cover in the Blue Ridge is composed of volcanic and sedimentary rocks.

Laurentian Continental Basement

The Laurentian basement rocks crop out in the Blue Ridge. They are the igneous and metamorphic rocks formed during the Grenville Orogeny about 1.1 Ga when Rodinia was formed. The largest exposures of Laurentian basement occur in central Canada. They are buried under younger sedimentary rocks in the interior of North America.

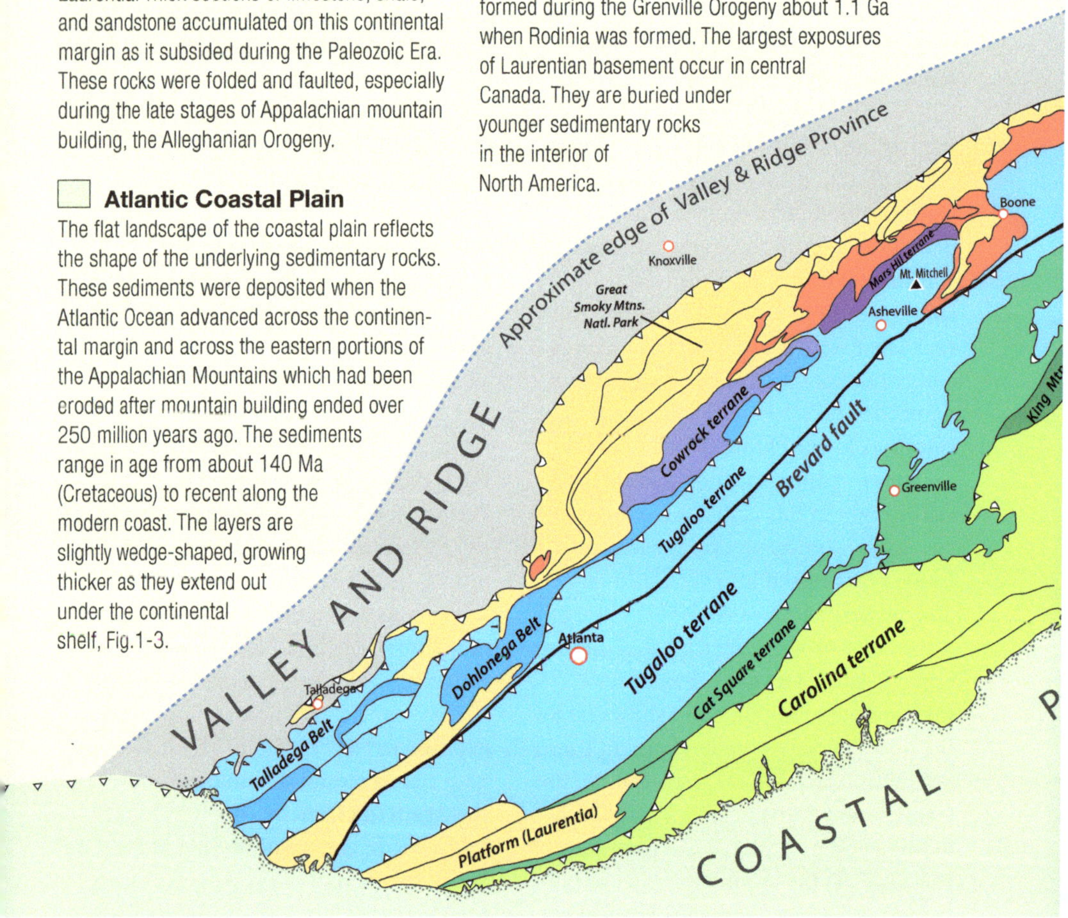

Fig. 2-12. This terrane map shows where rocks found in the Blue Ridge formed, and when they became part of the Blue Ridge. (Modified after Hatcher, 2007; and Merchant and Hatcher, 2007)

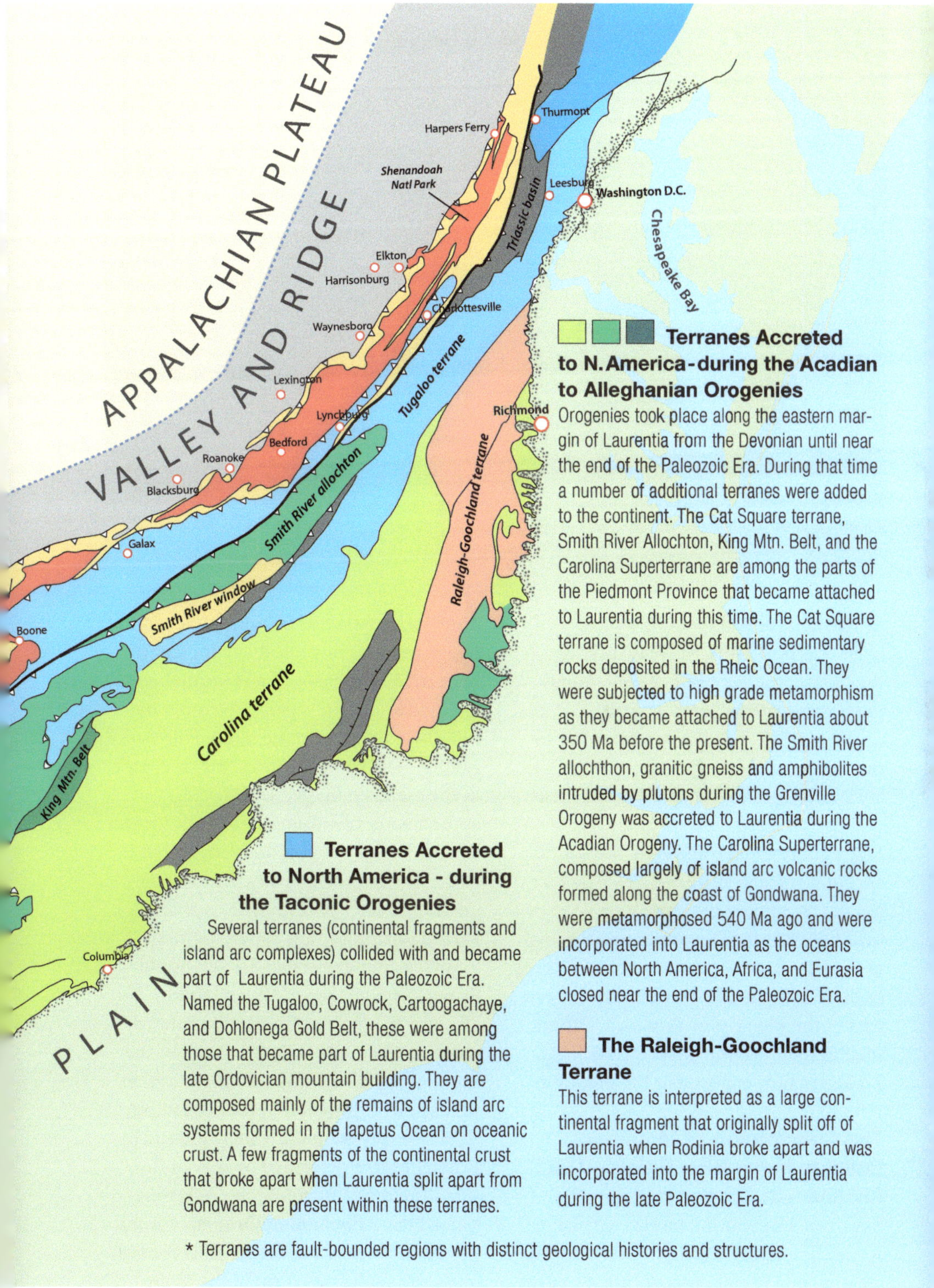

Terranes Accreted to N. America - during the Acadian to Alleghanian Orogenies

Orogenies took place along the eastern margin of Laurentia from the Devonian until near the end of the Paleozoic Era. During that time a number of additional terranes were added to the continent. The Cat Square terrane, Smith River Allochton, King Mtn. Belt, and the Carolina Superterrane are among the parts of the Piedmont Province that became attached to Laurentia during this time. The Cat Square terrane is composed of marine sedimentary rocks deposited in the Rheic Ocean. They were subjected to high grade metamorphism as they became attached to Laurentia about 350 Ma before the present. The Smith River allochthon, granitic gneiss and amphibolites intruded by plutons during the Grenville Orogeny was accreted to Laurentia during the Acadian Orogeny. The Carolina Superterrane, composed largely of island arc volcanic rocks formed along the coast of Gondwana. They were metamorphosed 540 Ma ago and were incorporated into Laurentia as the oceans between North America, Africa, and Eurasia closed near the end of the Paleozoic Era.

Terranes Accreted to North America - during the Taconic Orogenies

Several terranes (continental fragments and island arc complexes) collided with and became part of Laurentia during the Paleozoic Era. Named the Tugaloo, Cowrock, Cartoogachaye, and Dohlonega Gold Belt, these were among those that became part of Laurentia during the late Ordovician mountain building. They are composed mainly of the remains of island arc systems formed in the Iapetus Ocean on oceanic crust. A few fragments of the continental crust that broke apart when Laurentia split apart from Gondwana are present within these terranes.

The Raleigh-Goochland Terrane

This terrane is interpreted as a large continental fragment that originally split off of Laurentia when Rodinia broke apart and was incorporated into the margin of Laurentia during the late Paleozoic Era.

* Terranes are fault-bounded regions with distinct geological histories and structures.

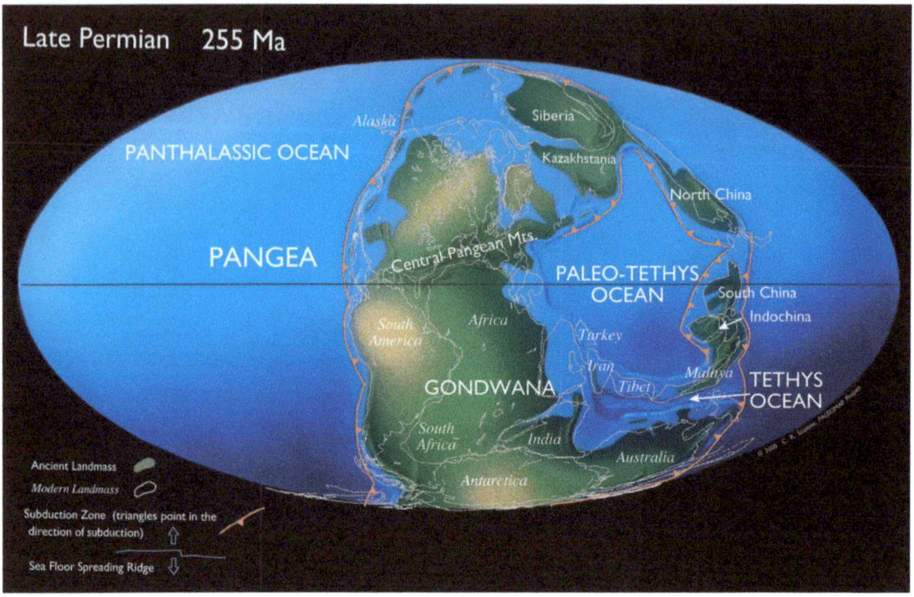

Fig. 2-13. A reconstruction of Pangaea which formed as the oceans between Laurentia and Gondwana closed. (C.R. Scotese, Paleomap Project)

a huge delta and alluvial fans preserved in the Catskill Mtns. of New York. Farther south, mainly black shales containing organic matter were deposited on the continent. These are preserved today in the valleys in the western part of the Valley and Ridge Province in Va. and Penn.

Many continental fragments and island arc complexes were added onto the margin of Laurentia during the long span of time between the Devonian and the end of the Paleozoic. Rocks once part of some of these complexes lie within the eastern part of the Blue Ridge Province. Others, such as the Carolina Superterrane (Fig. 2-12), are now exposed in the Piedmont.

The Culmination of Mountain Building in the Blue Ridge – The Alleghanian Orogeny: During the Paleozoic Era South America, Africa, Antarctica, Australia, and India merged, creating one huge supercontinent known as Gondwana. By that time much of modern Eurasia had become attached to Laurentia, forming another supercontinent called **Laurasia.** It was located north of Gondwana and separated from it by the Rheic Ocean. By late in the Paleozoic Era the west coast of Africa (by then part of the greatly enlarged Gondwana) was approaching Laurasia as the Rheic Ocean closed. The two supercontinents came closer together and finally began to collide in the early part of the Mississippian Period forming a new supercontinent known as **Pangaea** (Fig. 2-13). The collision started in the north and gradually moved south as Gondwana rotated counter-clockwise (Fig. 2-14). All of the rocks of the Blue Ridge and Piedmont as well as the sedimentary cover of Laurentia now exposed in the Valley and Ridge Province were deformed by this collision. The resulting mountain building, named the **Alleghanian Orogeny** for the Alleghany Mtns., was responsible for the folds and faults now seen in the Valley and Ridge and along the western edge of

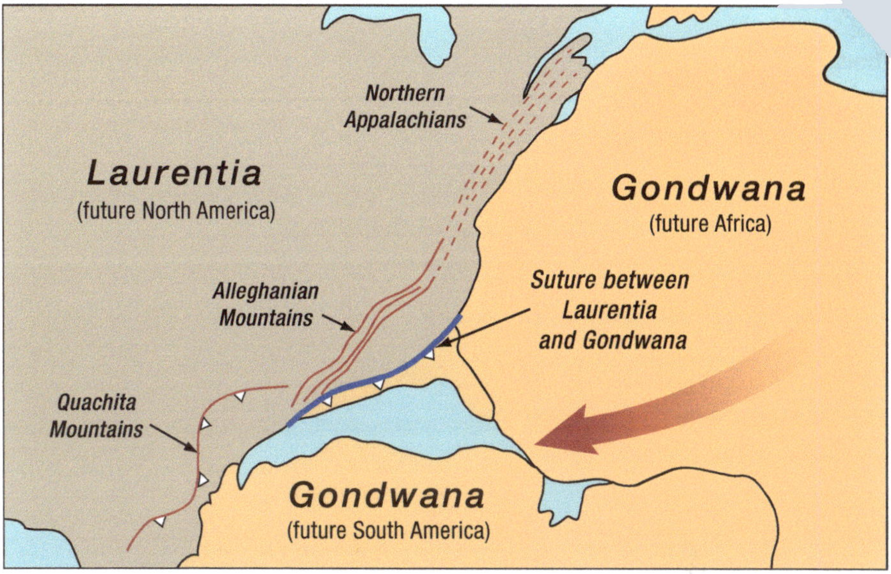

Fig. 2-14. Closing of the ocean by the fitting together of the the continental margins which took place near the end of the Paleozoic Era as Gondwana and Laurentia collided. In this interpretation Gondwana is rotating counter-clockwise. This explains evidence that the region southeast of the Brevard fault moved to the southwest relative to that on the north side of the Brevard fault. The Brevard fault zone is used as the eastern boundary of the Blue Ridge south of Va. (After Hatcher, 2002)

the Blue Ridge. At this time the collision with Gondwana (Africa) pushed the rocks now exposed in the Blue Ridge upward and moved them westward. Some geologists estimate that the rocks of the Blue Ridge moved as much as 186 miles to the northwest. The terranes in the Piedmont that had been added to the edge of Laurentia during the Paleozoic also shifted to the southwest as a result of the rotation of Gondwana relative to Laurentia.

It is likely that the mountains formed near the end of the Paleozoic were once spectacular—much higher than any part of the modern Appalachian chain. They could have been as high as today's Alps or Himalayas. We lack the tools needed to be certain how high they were, but we do know that the huge volumes of sediment that lie buried off the east coast under the continental shelf and slope (Fig. 1-3) originally came from these mountains.

Post-Paleozoic Events in the Appalachian Mtns.

Early in the Mesozoic Era, Pangaea began to rift apart. As in the breakup of Rodinia, the break between North America and Africa took place along the Appalachian Mtn. belt. Today, Triassic-age, fault-bound basins (Fig. 2-12) filled with sediment from the nearby high mountains, are prominent features along the east coast from Connecticut to Florida. The Connecticut Valley, New Jersey Lowlands, and the Gettysburg and Culpeper basins are among these. Other basins lie farther south in the Piedmont and are buried under younger sediments in the coastal plain. By the middle of the Jurassic Period a new ocean, the **Atlantic** was forming. That was about 165 million years ago. It continues to expand today as North America and Africa drift farther apart.

ic deformation in the sed at the end of the ozoic, uplift has continued in much of the region, including the Blue Ridge. The higher parts of the oldest mountains have been reduced in elevation by the work of streams, the downslope movement of rocks and soil on steep slopes, and the solution of limestone in the Valley and Ridge. Through these processes the modern landscape gradually evolved to its present shape. Chapter 3 will focus on this issue. ❖

Features in the Appalachian Region

Oceans

Iapetus Ocean: The ocean that began to form about 550 Ma when the supercontinent Rodinia split apart. The ocean opened between North America (Laurentia) and Africa-Eurasia (Gondwana). A number of continental fragments separated from Rodinia and island arcs formed along the margins of the Iapetus Ocean.

Rheic (also Theic) Ocean: The ocean that formed in the late Cambrian/early Ordovician between Gondwana (Africa-Eurasia) and an island arc named Carolinia, that had formed in the Iapetus Ocean.

Atlantic Ocean: The modern ocean that started to form at about 165-200 Ma as a result of the splitting apart of the supercontinent Pangaea. This was accompanied by the definition of modern North America, Africa, Eurasia, and South America.

Continents, Continental Fragments, and Supercontinents

Rodinia: The largest, most primitive, and least known of supercontinents. It is thought to have included the oldest parts of all modern continents. Geologists apply the name Grenville Orogeny to the mountain building that took place when Rodinia formed about one billion years ago.

Laurentia: A continent that had been part of Rodinia. The central part of North America composed most of Laurentia. A number of other continents as well as several smaller fragments of continents also resulted from the breakup of Rodinia (Fig. 2-10).

Gondwana: A supercontinent that included what we now know as Africa, South America, Australia, Antarctica, and most of India that existed during the Paleozoic Era after the breakup of Rodinia. Gondwana collided with Laurasia late in the Paleozoic to form Pangaea. In the early Mesozoic Era, Pangaea broke apart leading to the development of modern continents.

Taconica: The volcanic island arc that was located east of Laurentia in the early Paleozoic. The collision between Taconica and Laurentia (North America) caused the Taconic Orogeny at about 450 Ma in New England. The volcanic arcs of Carolinia collided with North America farther south at about the same time.

Piedmontia: One of the continental fragments that separated from Laurentia (North America) during the breakup of Rodinia. This fragment was located along what is now the Southern Appalachians and collided with North America in the Ordovician, about the same time Taconica collided farther north, in the New England area. The orogeny resulting from these collisions is known as the Taconic Orogeny.

Avalonia: Microcontinent formed during the breakup of Rodinia. Avalonia collided with North America (Laurentia) about 380 Ma, causing the Acadian Orogeny (which is best recorded in the New England area.)

Carolinia: An volcanic island arc that collided with the southern part of Laurentia (North America) at about the same time that Avalonia collided with the northern part of North America causing the Acadian Orogeny.

Laurasia: The supercontinent that was formed by the coalescence of Laurentia and Eurasia during the Paleozoic Era. Later, it collided with Gondwana toward the end of the Paleozoic, forming Pangaea.

Pangaea: The supercontinent that formed when Laurasia and Gondwana collided late in the Paleozoic Era. The Appalachian Mtns. formed between Gondwana and the Laurentian part of Laurasia.

Mountain Building Events

Grenville Orogeny: Mountain building associated with the assembly of Rodina about one billion years ago.

Taconic Orogeny: Mountain building best preserved in New England and Newfoundland from about 450 Ma, when a volcanic arc collided with Laurentia (North America). Evidence of this mountain building event is also found in the Southern Blue Ridge and Piedmont.

Acadian Orogeny: Mountain building that took place at about 380 Ma (late Devonian) when the continental fragment called Avalonia collided with Laurentia (North America) This event is best preserved in New England where many igneous intrusions of this age are present. Metamorphism of this age is present in the Blue Ridge and Piedmont.

Alleghanian (also Alleghany) Orogeny: The last major mountain building episode in the Appalachians took place as a result of the collision between Gondwanaland and Laurentia. This orogeny was responsible for the last major uplift and for most of the folds and faults seen in the Southern Appalachians.

Major Volcanic Events Found in the Blue Ridge

Catoctin Volcanism: Lava flows spread over large areas in the northern part of the Blue Ridge. These volcanics were generated during the breakup of Rodinia about 570 Ma. They occur beneath the Chilhowee Group.

Mt. Rogers Volcanism: These volcanic rocks, which formed about 760 Ma, are found in southwestern Virginia. They were generated during the early stages in the breakup of Rodinia.

Terrane

The term "terrane" is commonly applied to parts of a continental fragment or island arc that have been accreted to a continent. Terranes are separated from adjacent terranes by faults, most of which are thrust faults. Thus, in the Blue Ridge, the Carolina Terrane is composed of rocks formed in the island arc known as Carolinia.

Note: *Excellent animated maps showing the evolution of the continents over geologic time are available at websites for "The Paleogeographic Atlas Project of the University of Chicago" and "Paleogeography on Wikipedia." Many fine paleogeographic maps created by Ron Blakey are also available on the web.*

Chapter 3

Development of the Blue Ridge Landscape

This chapter provides background information that the reader may find useful in understanding more detailed discussions of particular parts of the Blue Ridge later in the guide. From the human perspective, the landscape of the Blue Ridge seems unchanging. Occasionally we witness the effects of severe storms, but for the most part the mountains seem to remain the same from generation to generation. However, over long periods of time, measured in millions and hundreds of millions of years, the shape of the land surface undergoes dramatic changes. Upward warping of the crust and mountain building affected the Appalachians repeatedly in the past. Today a number of processes are slowly reducing the height of the Blue Ridge. The decay and disintegration of rocks at the ground surface (weathering), the downslope movement of loose materials, and erosion by streams are the primary causes of this degradation.

What Controls Landscape Development

The striking contrast among the various landscapes of the high mountains, the nearly-flat land of the coastal plain, the rolling hills of the interior lowlands in the central United States, and many other localized variations result from the complex interaction of:

1) The structure of the rock beneath the ground surface.
2) Differences in the resistance of rocks to weathering and erosion.
3) Processes of erosion and deposition that have taken place.
4) Length of time that processes of erosion and deposition have operated.

Often one or two of these factors is primarily responsible for the development of isolated peaks, ridges, plateaus, or escarpments. Everywhere the materials at the surface are affected by their contact with the atmosphere, the climate, and the organisms that live there.

The Underlying Rock Structure

Structure refers to the shape of rock bodies and to their internal features. Most of the sedimentary rocks found in the Blue Ridge were originally laid down on the seafloor or along coasts. They occur as layers referred to as strata or beds. They may have great lateral extent relative to their thickness. Some layers are nearly uniform in thickness for great distances, others are wedge-shaped, and some layers interfinger with others of a different composition (Fig. 3-1). In mountain belts such as the Appalachians, most of the rocks have been deformed. Layers that were once flat have been tilted, bent, or **folded** into a great variety of forms depending on the type of stress that is applied to them and the temperature and pressure conditions under which

Fig. 3-1. Most sedimentary rocks are stratified (layered). Some layers of uniform composition occur with uniform thickness and extend laterally for many miles. Others overlap or interfinger in wedge shapes where one type of sediment meets another.

the folding takes place. Sometimes the amount of deformation is so great that originally horizontal layers are folded until parts of them are turned upside down. Under such conditions all types of rock may break and become displaced along **faults**. Some common structural forms seen in the sedimentary rocks exposed along the western edge of the Blue Ridge are illustrated in Fig. 3-2.

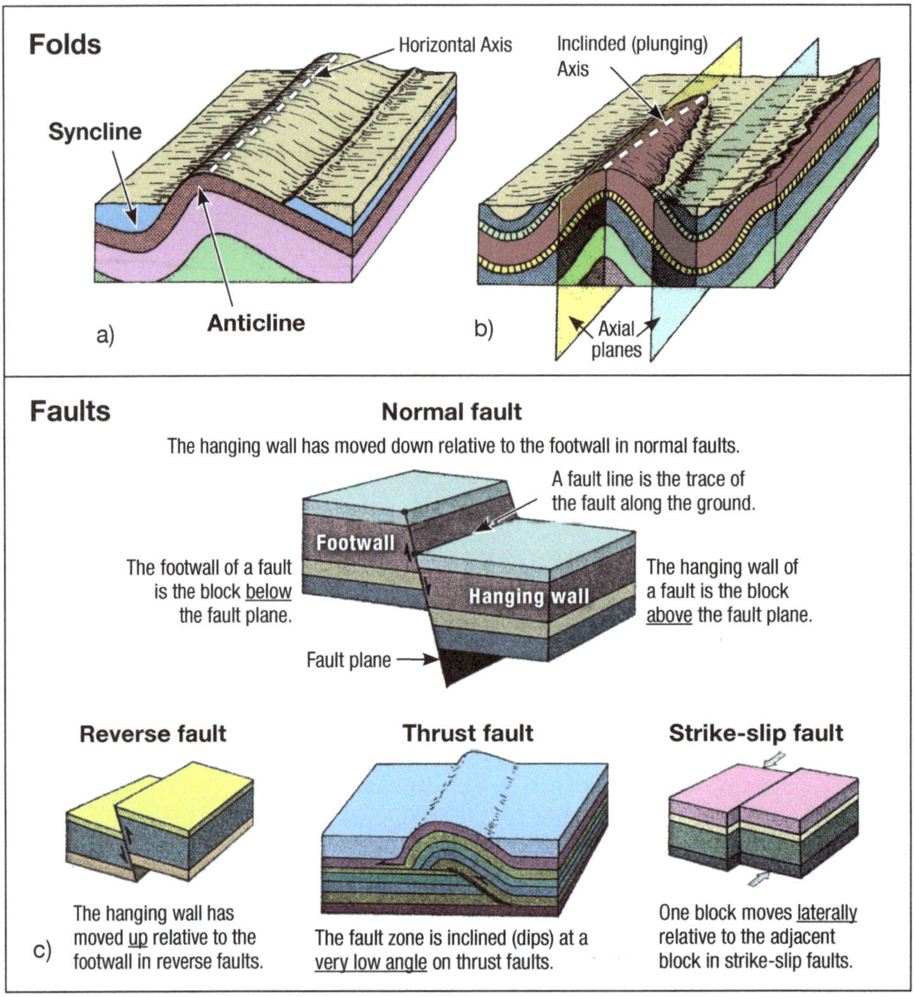
Folds
Horizontal Axis
Syncline
Anticline
a)
Inclinded (plunging) Axis
b)
Axial planes
Faults
Normal fault
The hanging wall has moved down relative to the footwall in normal faults.
A fault line is the trace of the fault along the ground.
Footwall
The footwall of a fault is the block below the fault plane.
Hanging wall
The hanging wall of a fault is the block above the fault plane.
Fault plane
Reverse fault
The hanging wall has moved up relative to the footwall in reverse faults.
c)
Thrust fault
The fault zone is inclined (dips) at a very low angle on thrust faults.
Strike-slip fault
One block moves laterally relative to the adjacent block in strike-slip faults.

Fig. 3-2. Folds and faults shown in block diagrams. a) An anticline (up-fold, as seen in Fig. 3-5) and a syncline (down-fold) with horizontal axes. b) Folds that have inclined (plunging) axes. c) Four common faults: normal, reverse, thrust, and strike-slip faults. (E.W. Spencer, 2003)

Fig. 3-3. A massive layer of quartzite, inclined (dipping) to the northwest, has been cut by streams, producing many v-shaped valleys (Fig. 3-4) along the northwestern flank of the Blue Ridge.

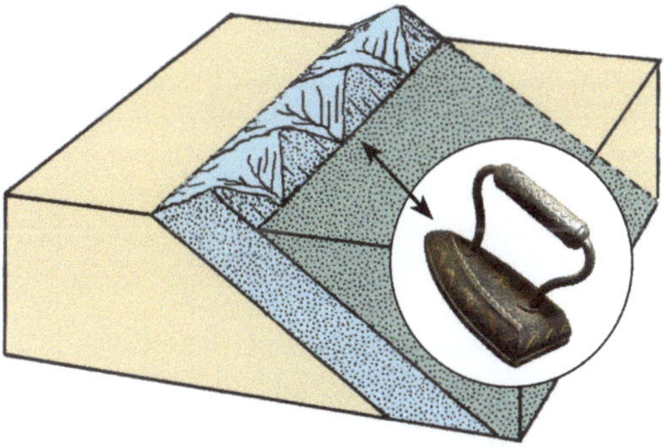

Fig. 3-4. Block diagram showing the configuration of the layer of quartzite that forms the flat irons in Fig. 3-3 (inset: antique flat iron).

Landforms resembling old-fashioned flat irons (once used to iron clothing), have formed where streams have cut their channels across the tilted layers of sandstone (Fig. 3-3).

Igneous and metamorphic rocks are widely exposed in the central and eastern portions of the Blue Ridge. Some of these were lavas so hot that they spread out great distances from the fissures where they originated. They filled the valleys and in some places built up great thicknesses. We can see them at Catoctin Mtn. State Park and in the Shenandoah National Park. The volcanic centers from which most emanated have long since been destroyed by stream erosion, but ash deposits, often those deposited in oceans, remain. We also find tabular-shaped intrusions called **dikes** (Fig. 3-6) that cut across the rock into which they were injected. It is more difficult to define the original shape of **plutonic bodies**, those masses of igneous rock formed and solidified deep below the ground surface. In the Blue Ridge most of these **plutons** have been deformed and altered (metamorphosed) during episodes of mountain building. In the Appalachians, soil cover and forests make it difficult to trace contacts between different bodies of rock.

Fig. 3-5. The massive layer of quartzite shown in Fig. 3-3 is folded in some places. Folds of this type are present in many of the sedimentary rocks of the Blue Ridge and Valley and Ridge.

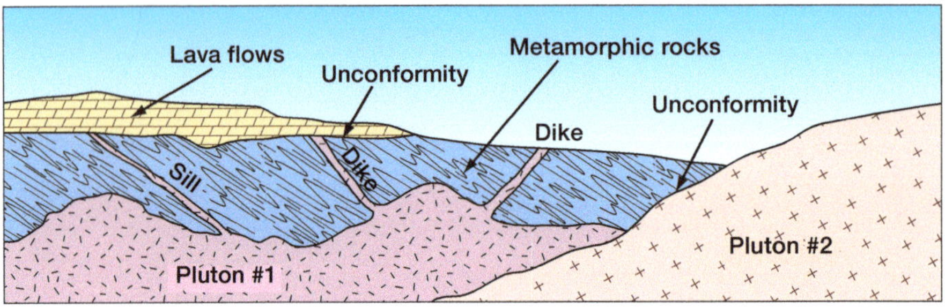

Fig. 3-6. Common types of igneous intrusions found in the Blue Ridge. The oldest rocks (blue with wavy lines) are metamorphic rocks. The pattern indicates that they have been strongly deformed. Both metasediments and metaigneous rocks are present in the Blue Ridge. In this schematic drawing two large plutons are shown intruding the metamorphic rocks. Depending on their size they may be called ***batholiths*** *(40 sq. miles or more) or* ***stocks*** *(less than 40 sq. miles). Two* ***dikes*** *are shown rising from Pluton #1. Dikes cut across the layering or borders of other rock bodies. One* ***sill*** *is shown. Sills are intrusions that invade the rocks parallel to pre-existing layers. These intrusions may break through to the ground surface and produce lava flows or volcanoes. Although volcanoes may have existed when some of the lavas were extruded in the Blue Ridge many millions of years ago, their forms are no longer clearly evident.*

Weathering

Rocks are altered when they come into contact with the atmosphere, plants, and animals close to the ground surface. These processes, called weathering, include chemical reactions that lead to the **decomposition** of the minerals and physical **disintegration** of the rock and soil at the surface. The effects of weathering make rocks more susceptible to **erosion**, which involves the movement of materials downslope by the effects of gravity, stream action, and the solution of carbonate rocks by groundwater.

Chemical weathering includes all of the processes that cause rocks to break down or decompose as a result of chemical reactions that take place when rocks come into contact with the liquids or gases at the Earth's surface. For example, water in the atmosphere and rainwater and is slightly acidic and will react with many of the common rock-forming minerals. Water in the atmosphere may also combine with pollutants such as sulfuric acid or volcanic gases to form much stronger acids. All of these acids dissolve rocks that are composed of calcium carbonate, such as limestone ($CaCO_3$) and dolomite ($CaMgCO_3$). The acidic water carries the constituents dissolved from the rock into the ground, where groundwater transports them into streams that eventually flow into the ocean.

The **minerals** composing rocks vary greatly in their **resistance** to chemical weathering. The common mineral, quartz, is highly resistant to chemical weathering. For this reason, in humid regions like the Appalachians, rocks like quartz sandstone resist chemical weathering and thus tend to form ridges. In con trast, valleys form where soluble or easily eroded rocks are present. This is very common in the Valley and Ridge Province, where the valleys are underlain by soluble limestone or easily eroded shale.

Chemical weathering is also very effective in decomposing rocks that contain feldspar, the most common mineral found in

Fig. 3-7. Granitic rock that has been weathered. The feldspar minerals have broken down into clay which still preserves the original texture of the rock.

Fig. 3-8. Water moving along fractures in otherwise solid rock freezes in winter and expands, breaking the rock apart.

granitic rocks. When this reaction takes place, the feldspar slowly changes into clay minerals, while the quartz remains unaltered. In the Blue Ridge you will see exposures of weathered granite in which the original texture of the granite and the quartz minerals are preserved, but the feldspars have changed to clay (Fig. 3-7). This accounts for much of the weathering of rocks containing feldspar minerals in the core of the Blue Ridge (Ch. 9).

Some minerals, especially those containing iron, react with oxygen in a process called **oxidation.** The oxygen reacts with iron to form an iron oxide, called limonite or "rust." Small amounts of oxidized iron are responsible for most of the red, orange, and yellow stains found in rocks. Often these stains are confined to the surface of the rock or cracks in the rock. If you break the rock open with a geologic hammer, you will see the true color of the rock, rather than the surface that has been altered by weathering.

Physical weathering includes all of the natural processes that cause rocks to disintegrate. Prominent among them in the Blue Ridge are the effects of **freezing water.** Once trapped in cracks in rocks, water expands as it freezes, creating sufficient pressure to expand the crack and eventually break the rock apart (Fig. 3-8). Once this happens, chemical weathering begins on the newly exposed fracture surfaces. Repeated freezing and thawing causes cracks to grow and dislodges blocks of rock that may then begin a journey downslope. When water freezes in the soil, layers of ice may form within the soil or between it and the solid rock underneath. Expansion of the ice heaves the soil upward. This type of upward pressure also occurs when water freezes below loose rocks that are in the soil. Gradually the rock is pushed up through the soil to the surface. The effects of freezing and thawing were much more active during the last episode of glaciation in North America, which ended about 10,000 years ago, and probably account for the development of many of the accumulations of loose blocks of rock commonly seen on steep slopes in the Appalachians (Fig. 3-9).

Fig. 3-9. Fragments of rock, dislodged by freezing and thawing, fall down steep slopes and accumulate in a sheet of talus.

Fig. 3-10. Plant roots growing in cracks in otherwise solid rock can exert enough pressure on the rock to further open the crack.

Organic activity is also responsible for the physical breakdown of rocks. This may take many forms ranging from the pressure exerted by roots that grow in cracks in rocks (Fig. 3-10), to the burrowing of worms and other organisms, and the effects of paws, feet, and hooves on soil. However, on a global scale, human activities—notably construction and farming—are responsible for the most rapid and extensive physical alteration of Earth's surface in recent time.

While weathering causes rocks to disintegrate and decompose, **erosion** refers to the movement of the products of weathering and other surficial materials. Gravity, surface runoff, streams, glaciers, wind, and water circulating underground are the agents that cause erosion.

Differential Weathering

Rocks differ greatly in their resistance to weathering as well as to erosion. Some of the isolated peaks in the Blue Ridge exist because the underlying rock is more resistant to weathering and erosion than those surrounding the high features. Mt. Rogers, Va. stands high because it is composed of resistant lavas. The sandstones and quartzites at Grandfather Mtn., N.C. are also very resistant to alteration. The Great Smoky Mtns. owe their elevation to the presence of thick layers of sandstone. Old Rag Mtn., a peak composed of granite in Shenandoah NP, has also resisted the effects of weathering. Differential weathering and erosion are responsible for the elevation of the Blue Ridge relative to the valleys in the Valley and Ridge Province. Highly resistant quartzites, quartz sandstones, lavas, and metamorphic rocks hold up many of the ridges in the Blue Ridge. In contrast, carbonate rocks such as limestones and dolomites underlie most of the valleys of the Valley and Ridge (Fig. 3-11).

Although plutons make up large parts of the ancient basement exposed in the core of the Blue Ridge, these bodies do not always stand out in the topography. In many cases their resistance to erosion is not significantly different from that of the enclosing rocks, most of which are metamorphic. In general, igneous and

Fig. 3-11. This view across the Great Valley of Virginia, which is underlain by carbonate rocks (limestone and dolomite) and the Blue Ridge (foreground) which is largely composed of rocks containing quartz (sandstone and quartzite) illustrates the effects of differential weathering. Ridges composed of sandstone are seen to the northwest of the Great Valley.

metamorphic rocks that contain a lot of quartz are more resistant than rocks that contain magnesium and iron-bearing minerals (e.g. hornblende and pyroxene, see Ch. 9).

Erosion Processes that Influence Landscape Development

Erosion on the Earth's surface is largely determined by the action of streams, the downslope movement of materials on the surface (especially where the slopes are steep), the dissolution of rocks by water as it percolates into the ground, the forces of wind, and the action of glaciers. All of the processes associated with these erosion agents cause the movement of materials. All of them remove materials from one place on the surface and deposit it in another. Among these agents, the action of streams and downslope movement (commonly referred to as mass wasting) are the most important in the Blue Ridge. Groundwater, wind, and ice play a lesser role. These will be examined first.

Groundwater, Wind, and Glaciers

Water that percolates into the ground moves through the spaces between grains of sand or through cracks in bedrock. This **groundwater** contributes to the weathering of the rock, but because most of the rocks in the Blue Ridge are composed of insoluble minerals, groundwater is not responsible for nearly as much erosion here as it is in the limestone-floored valleys west of the Blue Ridge.

Wind is an effective agent of erosion and deposition where it is strong and where loose, fine-grained materials such as dust and fine sand are present on the ground surface. Such surfaces are rare in the Blue Ridge where vegetation covers and protects most fine-grained materials.

Glaciers were present in the Northern Appalachians during the most recent Ice Ages, which reached their peak about 18,000 years ago. At that time, the last of many ice sheets that advanced into the United States during the Pleistocene, came as far south as central Pennsylvania and the Ohio River, but stopped short of the Blue Ridge. Small masses of ice almost certainly occupied areas high in the Blue Ridge, and long cold seasons increased the amount of physical weathering caused by frost action. This contributed to the formation of **talus** (**scree**) slopes found wherever steep slopes occur in the Blue Ridge.

Geologists currently debate the importance of ice in the development of piles of large blocks of rock found in a number

Fig. 3-12. Geologists look at huge rocks displaced by accumulations of ice in Shenandoah National Park (Ch. 6E).

of places along steep slopes in the Blue Ridge (Fig. 3-12). The blocks in some of these piles are as big as cars. They appear to be relatively stable today, but they are loosely packed. During the cold phases of the Ice Ages, the spaces between the blocks in these piles were probably filled with ice, transforming them into **rock glaciers**. The ice would have lubricated the contacts between the blocks and facilitated their downslope movement. An example of these rock piles is described in the section about Blackrock Summit in Shenandoah National Park (p. 138). These rock glaciers and glacial features that originated beyond the edge of the continental ice sheet are referred to as periglacial features.

Downslope Movements

Known as **mass wasting**, the movement of materials downslope (especially loose materials), is the most universal of all processes of erosion and deposition. You will see evidence of this on most steep slopes, especially those capped by outcrops of bedrock. Most rock masses, especially those of hard, consolidated rock, are broken by fractures. These cracks define blocks of rock that are dislodged by water that freezes and then expands as it turns to ice. Once the blocks are separated they may slide or roll downslope and accumulate as cone-shaped masses or sheets of talus (Fig. 3-9).

Mixtures of soil and rock fragments cover most of the steep slopes in the Blue Ridge. These mixtures, termed **colluvium**, are often unstable. Following heavy rains, colluvium saturated with water may move as a landslide. Colluvium may also move downslope as slides or slumps where streams or roads undercut the bottom edge of the slopes. One such slide interfered with traffic on I-81 for several years after road construction cut the toe of an ancient landslide near Roanoke, Va. Landslides are not uncommon in the Highlands portion of the Blue Ridge. For this reason, it is a good idea to check the condition of the Blue Ridge Parkway when you plan to travel in the Highlands.

Small-scale Landforms Caused by Streams

The slope of a stream channel affects the velocity of the stream and its capacity to move materials in the channel. Most of these materials—rocks, soil, or the products of weathering freed from the bedrock—get into the stream by moving downslope along the sides of valleys. Once in a stream, the finest particles—clay and silt—are carried in suspension, but most of the material in mountain streams moves as part of the bedload—the sand, pebbles, and rocks that roll or bounce along at the bottom of the stream. Gradually the edges of angular fragments are chipped or worn away, producing rounded pieces, called **gravel**. The roundness of the gravel indicates how far it has traveled. Ultimately, fragments of gravel are reduced in size to such a degree that they become sand. Because many of the layers of rock, especially those found on the western margin of the Blue Ridge, are composed of consolidated and cemented sand, weathering alone may free them, producing the loose sand that moves in mountain streams.

A slight increase in stream velocity greatly increases the size of the rocks that can be moved by the stream. The weight of the rocks that can be moved increases to the sixth power of the velocity. Thus when stream velocity doubles, the largest rocks potentially moved by the stream increases 64-fold. For this reason, huge rocks can be moved during floods.

Many of the mountain streams in the Blue Ridge have bare bedrock exposed in their channels. This indicates that the capacity of the stream to move the material being carried in the stream, known as its **load**, exceeds the amount of load supplied to it, and that the stream is actively cutting its channel downward. Much of the load gets into the stream by moving down steep slopes on the sides of the stream channel, but bedrock exposed in the channel is also eroded by the impact and abrasion from rocks carried in the stream. Gradually the bedrock erodes and the elevation of the channel is lowered. Circular holes in the bedrock, called **potholes** (Fig. 3-13) are produced where sand, silt, and gravel moving in the stream channel get caught in an eddy. These features are found in many streams that flow across bedrock throughout the Blue Ridge.

Erosion in the bottom of the channel in one part of a stream causes the upstream portion of the stream channel to have a steeper slope. This increases the capacity of the stream to erode its channel upstream and eventually all the way to the head of the stream. Over thousands of years the stream becomes longer as a result of **headward erosion**, which leads to the process known as stream piracy, described below.

Fig. 3-13. Streams that flow on bedrock commonly deepen their channels into the solid rock by cutting potholes in which loose rocks further erode the bedrock through the effects of abrasion and impact.

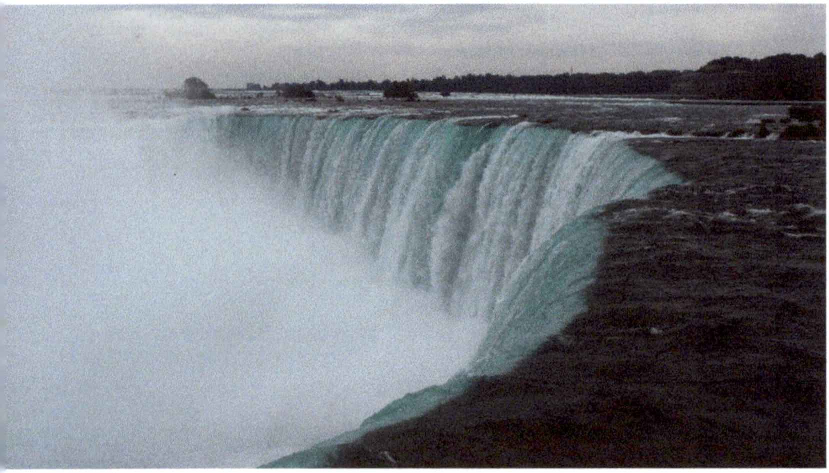

Fig. 3-14. Niagara Falls, N.Y.

Knickpoints and Waterfalls:

Most of the streams in the Blue Ridge are in a continual process of cutting their channels deeper. Long channels can be seen flowing over solid rock or shallow gravel beds that are being moved downstream. As streams cut their channels they may encounter rocks that differ in their resistance to erosion (Fig. 3-14). More resistant layers take longer to erode. While the erosion of these resistant layers is "delayed," the stream continues to deepen the less resistant layers directly downstream, eventually causing a drop in the slope (Fig. 3-15). Geologists call the places where sudden changes in the slope of the channel are formed in this way **knickpoints**. Rapids or waterfalls form at these places. Eventually the stream succeeds in cutting through the resistant rock. The waterfall or rapids will continue to gradually work their way upstream, toward the head of the stream.

Found throughout the Blue Ridge, a number of waterfalls: Crabtree, Stony Run, Blue Suck, Stone Mtn., Linville, and Raven Cliff are notable and well worth the effort required to reach them. If you are interested in hiking to falls, see *Waterfalls of the Southern Appalachians* by Brian A. Boyd.

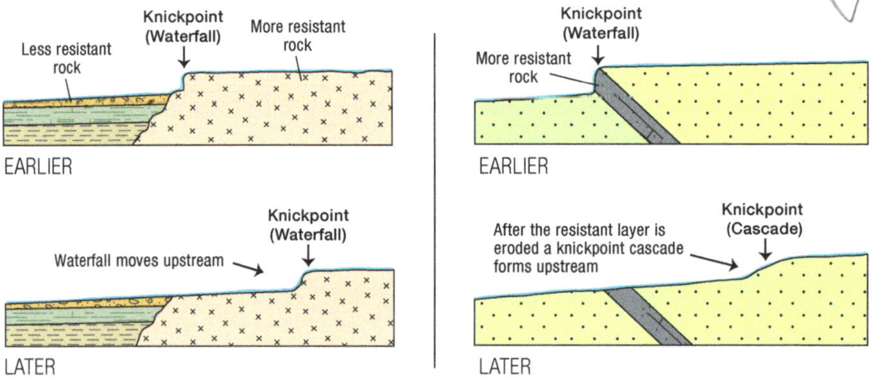

Fig. 3-15. A waterfall evolves where a stream gradually erodes its channel upstream at knickpoints.

Fig. 3-16. Floodplain along the James River in the Blue Ridge of central Virginia.

Floodplains: When a stream stays at the same level for a long period of time, it tends to undercut the bank on the outside of curves and deposit rock and sand bars on the inside of curves. Over long periods of time the channel shifts back and forth leaving a flat ground surface underlain by sand and gravel, called a **floodplain**, off to the side of the channel. During floods this flat surface is covered by water (Fig. 3-16).

Where the stream channel is being cut lower, as many are in the Blue Ridge, former floodplains are left above the level of floods. The resulting flat area composed of stream deposits is called a **stream terrace.** These are commonly found in the Blue Ridge and Valley and Ridge Provinces. In some places multiple terraces are present, and some are located high above modern streams.

Meanders: Most streams flow in courses that are curved. Because erosion is concentrated on the outside of curves where stream velocity is highest, while deposition takes place on the low-velocity inside of curves, the curves gradually expand and may become loop-shaped. These loops are features called **meanders** (Fig. 3-17). Most meanders form in places where the slope of the stream channel is nearly flat. Loose, unconsolidated sediment also facilitates the development of meanders. Meanders are not commonly found in mountains, but they are present in parts of the Blue Ridge. Their origin is discussed below.

Fig. 3-17. A meander located along the James River in central Virginia near Purgatory Mtn.

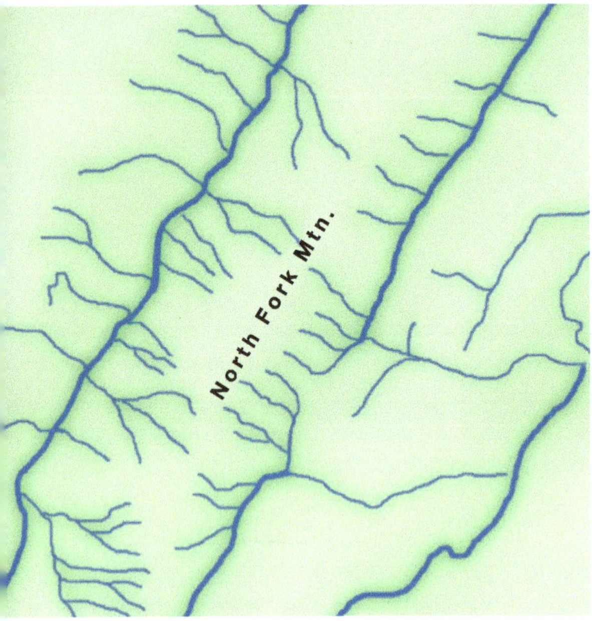

Fig. 3-18. Trellis drainage pattern in a part of the Valley and Ridge Province. The underlying bedrock structure responsible for this pattern is not found in the Blue Ridge.

Drainage Patterns

The Mississippi River is famous for its beautifully developed meanders found especially along the southern part of its course where it flows across a broad expanse of nearly flat-lying semi-consolidated sedimentary rock. The underlying bedrock is quite different in the Valley and Ridge Province where layers of rock that differ greatly in their resistance to weathering and erosion have been folded and faulted. Throughout the Valley and Ridge Province streams form a trellis-shaped drainage pattern (Fig. 3-18). Smaller streams that flow off of the ridges formed by sandstones join larger streams that flow in the northeast-southwest trending valleys underlain by shales and limestones.

This trellis-shaped stream pattern is dictated by the resistance of the underlying rock and by its structure. The larger streams in the trellis pattern flow in the valleys between ridges because the rocks that underlie the valleys are much less resistant to erosion than the insoluble sandstone and quartzite rocks in the ridges. Geologists call streams that follow paths of lesser resistance **subsequent streams**. Most of the subsequent streams in the Blue Ridge follow zones of weakness caused by faults, fracture zones, or layers of easily eroded rocks. Bedrock structures similar to those in the Valley and Ridge Province do not occur in the Blue Ridge.

Transverse Drainage in the Southern Appalachians

Although much of the drainage pattern in the Valley and Ridge is clearly subsequent, many major streams still flow across ridges in the Valley and Ridge and even directly across the highly resistant rocks of the Blue Ridge. Several ideas have been advanced to explain how the Potomac, James, and Roanoke rivers came to flow directly across the Blue Ridge—across rocks that are highly resistant to erosion. Two schools of thought developed to explain why this type of pattern evolved.

The first idea was put forth in the 1930s by William Morris Davis and Douglas Johnson. They proposed that the streams that flow through water gaps and across ridges had once flowed across a land surface characterized by low relief and located above the level of the present land surface. They envisioned that the major streams flowed from northwest to southeast across this surface. After the northwest to southeast drainage had been developed, the region was then warped upward and the streams began to cut their channels deeper into the more resistant layers of rock that they encountered. This type of drainage is known as **superimposed drainage**.

Since the earliest studies of the Appalachians, geologists have recognized that the present mountains have been eroded for millions of years, and that the drainage we see today resulted from the gradual lowering of mountains that were once much higher. The mountains may have been as high as the Alps or Himalayas. If this idea is correct, Appalachian drainage has changed dramatically as streams, while cutting downward, encountered rocks of varying composition and structure. Davis and Johnson were impressed by the observation that the elevation of the crests of the northern part of the Blue Ridge is similar to that of the ridges in the Valley and Ridge. They concluded that following the last major episode of mountain building near the end of the Paleozoic, the whole region was eroded until a nearly flat surface existed near sea level. They called this surface a **peneplain**. They further speculated that the ocean advanced across this plain, extending the flat-lying sediment now found in the Coastal Plain. Eventually the region now making up the Southern Appalachians emerged from the sea and a surface drainage pattern developed with major streams flowing northwest to southeast. Later this sedimentary cover was stripped away. As the drainage system evolved, the major streams became entrenched in the underlying rocks while tributaries became adjusted to the structure and composition of the rocks in the Valley and Ridge, resulting in the trellis pattern with a few water gaps, as seen in the Valley and Ridge Province. Davis and Johnson placed the highest-level peneplain at the top of the Blue Ridge and ridges to the west. They identified two lower level plains in the Great Valley of Virginia. These were interpreted as erosion surfaces that formed when uplift in the Appalachians paused for long periods of time.

Because marine sediments younger than the mountain building in the Appalachians have not been found on any of the peaks in the Appalachians, most geologists gave up the original idea of peneplains as having been formed near sea level. Some replaced this older concept with the idea that, following mountain building, streams in combination with mass wasting gradually reduced the elevation of the mountains to the subdued landscape we see today. It might have resembled the modern Blue Ridge Upland area south of Roanoke (Fig. 8A-5). As this degradation of the land surface took place, streams gradually changed their courses cutting their channels into the least resistant rocks and gradually etching out the topography we see today.

Transverse Drainage Caused by Stream Piracy

The second idea advanced to explain how major streams developed their courses across rocks of highly resistant rocks is that the streams gradually lengthened their channels by headward erosion, eventually cutting across the ridges. Over long periods of time, all streams gradually extend their length. The erosion that accomplishes this takes place at the head of streams where the slope of the stream channel is often steep. This causes the stream to flow rapidly and to erode its

Fig. 3-19. Topographic relief map section showing the head of a stream and the direction toward which it will migrate in the future. (USGS map, with additions)

channel quickly. Where a drainage divide between two valleys separates streams that flow on different slopes, the stream flowing down the steeper slopes will cut its channel across the divide and eventually intercept the stream located in the other valley (Fig. 3-19). When this happens the drainage from the higher valley will be diverted into the stream flowing in the lower valley. This process is called **stream piracy**. The higher stream is **"beheaded"** by the **pirate stream**. We see many places in the Appalachians where stream piracy is taking place. Excellent examples are seen along the eastern margin of the Blue Ridge south of Roanoke, Va. (Fig. 3-20). Some geologists think this process has also been responsible for the development of the water gaps formed by the Potomac, James, and Roanoke rivers. As a result of headward erosion, these streams flowing on the eastern side of the

Fig. 3-20. Stream capture is taking place along this escarpment at the eastern margin of the Blue Ridge in North Carolina.

Blue Ridge slowly cut westward across the Blue Ridge at Roanoke and to the north, intercepting streams that flowed at higher levels in the Valley and Ridge Province, and diverted their flow to the Atlantic. South of Roanoke, where the Blue Ridge is much wider than it is to the north, streams from the eastern side of the Blue Ridge have not yet succeeded in cutting through the high mountains to the streams in the Valley and Ridge, but they are active today and best illustrated along an escarpment located on the east side of the Blue Ridge south of Roanoke (Fig. 3-19).

The Effects of Time on the Evolution of Blue Ridge Landforms

Processes of weathering and erosion have gone on throughout Earth's history. Gradually they have etched out the features we now see in the landscape. In a way, we are looking at one frame of a *very* long movie showing the evolution of the land surface. Volcanic activity, mountain building, and the slow uplift of the Earth's crust have kept portions of the land high, but when these processes slow down or stop erosion gradually reduces the level of the land surface. Features of the present landscape are remnants of former landscapes waiting their turn to be removed by erosion. The processes that will accomplish this, notably erosion by streams and downslope movement of surficial materials, are active today.

We do not know how fast the mountains rose or how high they may have been when active mountain building took place in the Paleozoic Era. If the rise was as fast as it has been in the Himalayas, elevations in the Appalachians may have been similar to those in that highest mountain belt in the world. If the mountains rose slowly processes of erosion may have been fast enough to maintain much more subdued mountains. In any case, the quantity of sediment deposited along the continental margin since the Atlantic Ocean first began to form (Fig. 1-3) is sufficient to account for an amount of uplift comparable to that now seen in the Himalayas.

Based on the rate at which modern streams carry sediment into the ocean, sufficient time has elapsed, since the last major mountain building episode in the Appalachians, for erosion to reduce the high mountains to a surface near sea level. Some recent evidence suggests that this happened. Marine sedimentary rocks, more than 100 million years old are present in the Atlantic Coastal Plain (Figs. 1-1, 1-3). Some sediments of this age have been found in the modern Appalachian Mtns. at elevations of 1,000 ft., and similar sediments have also been found in sinkholes and fault-bounded depressions at elevations of up to 2,250 ft. This evidence suggests that the current elevations in at least some parts of the Appalachians are much younger than those formed during the mountain building near the end of the Paleozoic Era (Fig. 2-9). This type of upward movement of the crust is commonly seen in mountain belts. It is a response of the Earth's crust to the removal of weight, in this case by erosion of the mountains and the removal of the eroded sediment to the continental margin. A similar type of response is seen today in areas like central Canada that were covered by thousands of feet of ice 18,000 years ago. As the ice melted the crust that had been bent downward by the weight of the ice began to rebound. This regional uplift continues today.

Based on measurements of the rate at which erosion is currently reducing the elevation of the Appalachians (estimated to be about 9 ft. for every million years), upward warping of the Southern Appalachians is less than 33 million years old. This relatively young upward movement of the mountains is confirmed by observations that most of the streams in the region are cutting their channels downward at the present time, and that the sedimentary layers in the coastal plain are inclined toward the ocean and away from the Appalachians.

Shifting of the Appalachian Drainage Divide from East to West

The drainage divide between streams flowing into the Atlantic and those that flow west into the Ohio or Tennessee Rivers and thence into the Gulf of Mexico (Fig. 3-21) is gradually migrating from east to west.

North of Roanoke, Va. only the largest streams flow southeast across the Blue Ridge into the Chesapeake Bay. The Potomac River flows through the northernmost gap across the Blue Ridge at Harpers Ferry. In central Virginia, the James River cuts across the Blue Ridge between Glasgow and Lynchburg, Va. (Fig. 1-13). A short distance farther south the Roanoke River cuts across the Blue Ridge on its way to the Atlantic. A significant change in the direction of drainage from both the Blue Ridge and the Valley and Ridge Province takes place just south of Roanoke, Va. The Blue Ridge Upland Plateau and Highlands form barriers that block eastward drainage from the Valley and Ridge Province to the Atlantic Ocean. None of the major streams south of the Roanoke River flow completely across the Blue Ridge. The New River in Virginia, the French Broad, Tennessee, and Little Tennessee rivers all flow out of the Blue Ridge Highlands and continue to the west, their waters eventually reaching the Gulf of Mexico.

From the Roanoke River to the north the drainage divide between east- and west-flowing streams lies along the eastern edge of the Appalachian Plateau (Fig. 1-1). Elevations in the Blue Ridge north of Roanoke are generally lower than those to the south, and there is no sign of an upland plateau or a counterpart to the southern highlands in the Northern Blue Ridge. The northern section is more deeply eroded and the higher part lies farther west in the north than it does farther south. The Upland Plateau south of Roanoke may have once extended much farther north. If it did, then it has been removed as stream erosion progressed farther to the west in the north where streams in the Valley and Ridge have been captured and diverted to the Atlantic. If this idea is correct, we may envision the evolution of the landscape of the Appalachians as proceeding in a manner somewhat similar to that initially proposed by early students of the Appalachian landscape.

A modified model might envision the evolution of the landscape as follows:

1) Mountain building culminated toward the end of the Paleozoic Era, about 250-300 million years ago, as plate collisions caused Pangaea to form. Uplift and erosion started immediately and continued for millions of years.

2) The breakup of Pangaea and gradual development of the Atlantic Ocean that started about 150 million yrs. ago were

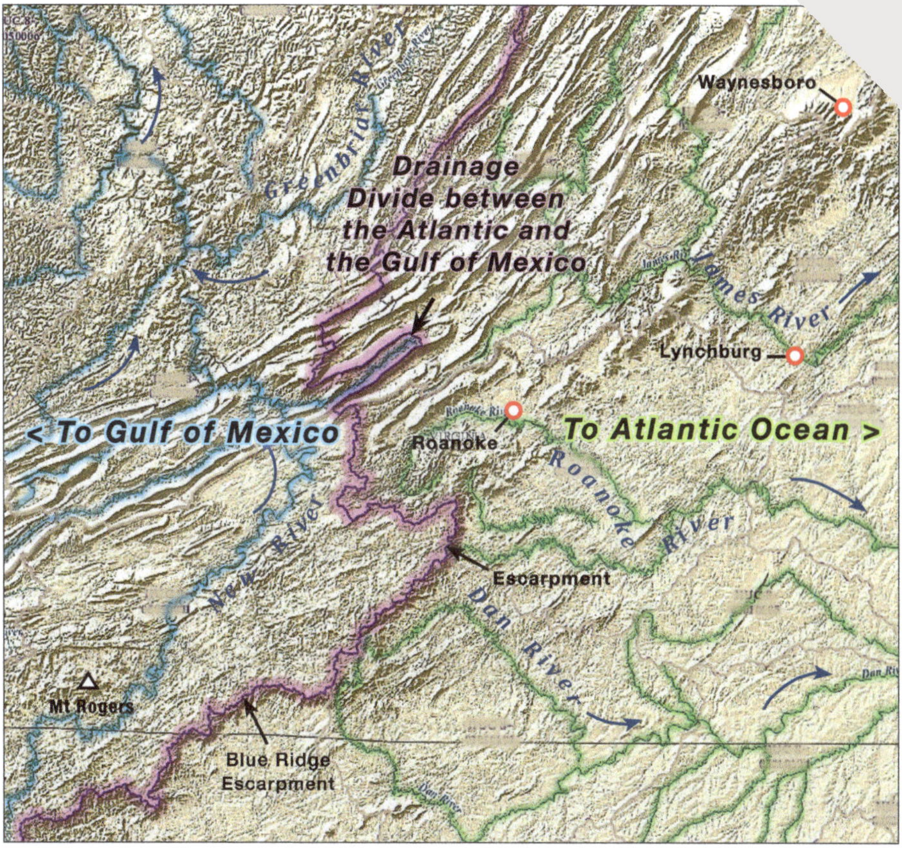

Fig. 3-21. The drainage divide (purple tint) between streams flowing into the Atlantic Ocean (green tint) and those that flow into the Ohio River (blue tint) will continue to shift from the eastern edge of the Ridge Uplands and Highlands to the west. (Modified after The National Atlas, USGS)

accompanied by major changes in the landscape. Huge rift basins formed along the east side of the Blue Ridge Province. The location of rocks then deposited in these Triassic basins is shown in Fig. 5-9.

3) During the late Mesozoic Era (65-150 million yrs. ago), an erosion surface, perhaps an elevated plateau with low relief similar to that of the Blue Ridge Upland formed. Part of this surface is covered by marine sediments in the Coastal Plain. How far inland this extended is unclear.

4) Much of the region now comprising the Blue Ridge was uplifted (warped upward) during the Miocene Epoch (part of the Cenozoic Era extending from about 23 to 5 million years ago).

5) As uplift took place and continued to the present, streams eroded their channels, cutting them deeper, exposing bedrock in stream channels, and cut across the Blue Ridge north of the Roanoke River, forming steep slopes, and causing meanders to become entrenched. This process led to an etching out of the topography as less resistant rock units eroded more rapidly than resistant units. Stream piracy accompanied these processes leading to the present landscape.

…rn part of the Blue …, a section east of the …ent (Fig. 3-20), and …he Blue Ridge north of Roanoke have been reduced in level (Fig. 3-22). The current stage in this landscape evolution is described and illustrated in the sections about the Shenandoah National Park (Ch. 6D) and the Upland Plateau (Ch. 8A). ❖

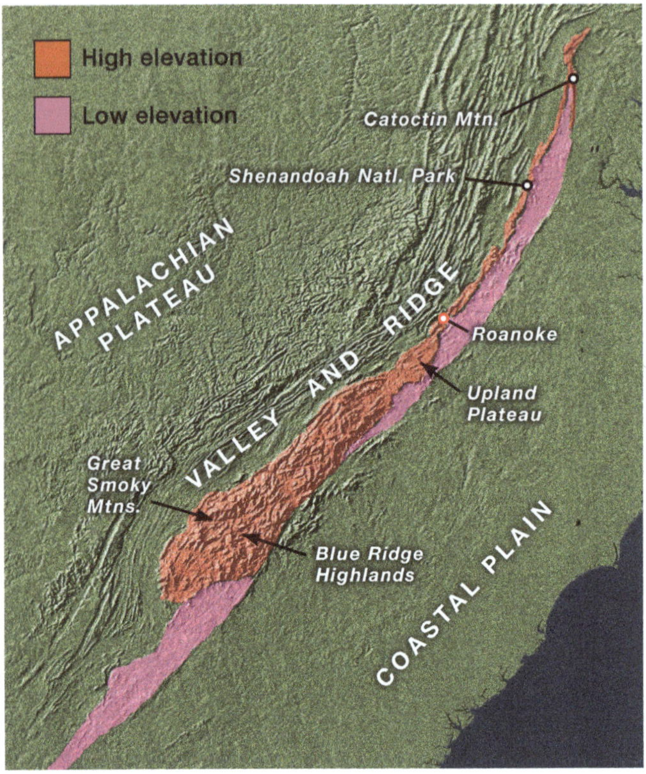

Fig. 3-22. Topography of the Southern Appalachians. The areas of high relief, mountains and ridges above 2,000 ft. in elevation, have been colored orange while sections of the Blue Ridge with elevations below 2,000 ft. are shown in pink. From this it is clear that the eastern and southern sections of the Blue Ridge Province have been eroded to much lower elevations than the western portions. (Modified after The National Atlas, USGS)

Additional Techniques Used to Determine the Age of Landscapes

Involving Helium:

In recent years, it has become possible to determine how long rocks on or very close to the ground surface have been exposed to radiation from space. Small quantities of helium are produced when cosmic radiation penetrates atoms such as silicon located at the Earth's surface. The concentration of noble gases in the atoms is a measure of how long they have been exposed to cosmic radiation. Higher concentrations indicate longer exposures. By knowing the rate at which helium reaches the Earth, it is possible to calculate how long the rock has been exposed.

Involving Uranium:

Another technique involves the use of the radioactive element uranium that occurs in the minerals apatite and zircon, both of which are commonly found in granitic rocks. The uranium breaks down to form thorium and helium. The ratio of the two is constant. The ratio of uranium to thorium can be used to obtain an age for the mineral being dated. Helium produced during this decay process escapes from the mineral structure when the temperature reaches about 70° C. Thus, it is possible to determine how long the mineral was exposed close enough to the surface to allow the helium to escape.

Geological Terms

Rock and Mineral Cleavage:

Rock cleavage is a property of rocks that causes sthem to break along smooth surfaces. Rock cleavage may result from the presence of closely-spaced fractures in the rock or by the strong alignment of platy minerals such as the micas in the rock. ***Mineral cleavage*** is caused by the internal arrangement of atoms that causes them to break along certain planes. Some minerals such as micas have a single cleavage; others such as calcite and salt (halite) have as many as three cleavages.

Klippes and Windows:

Klippes and windows are features associated with thrust faults, which are usually inclined at low angles, in areas where erosion has removed part of the sheet of rock on a hanging wall, called a thrust sheet (p. 36). ***Klippes*** are remnants of a thrust sheet that have been separated from the main part of the thrust sheet. ***Windows*** are places where erosion has cut through a thrust sheet to reveal the rocks on the footwall below it (p. 36).

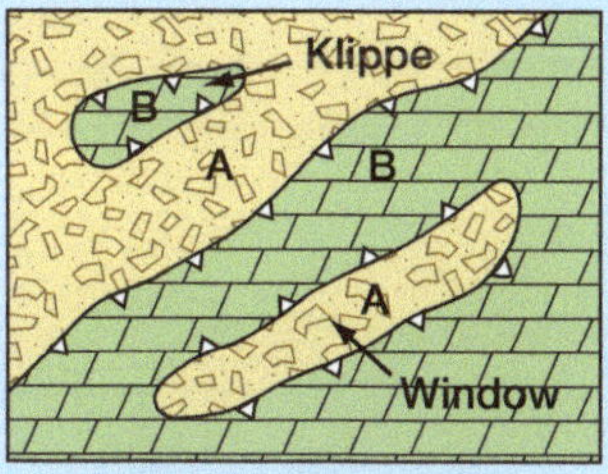

Sketch map showing a klippe and a window.

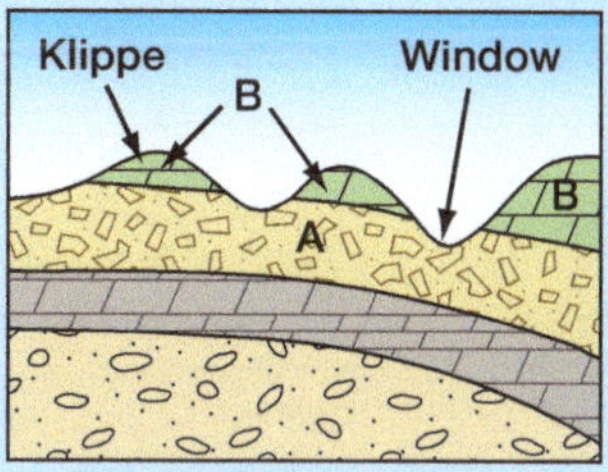

Cross-section showing a klippe and a window.

Unconformities:

Unconformities are places where erosion has removed part of the original rock sequence (the chronologically ordered rock layers). If the rocks below the unconformity are igneous or metamorphic in origin then the term ***nonconformity*** is applied. If stratified rocks below the erosion surface do not have the same inclination as those above the unconformity, then the break is called an ***angular unconformity***. If the angular inclination of the rocks below the unconformity is the same as those above, but part of the sequence is missing, then the erosion surface is called a ***disconformity***.

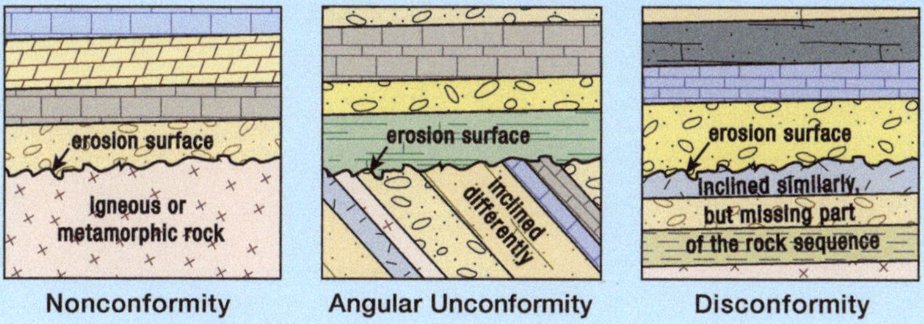

Cross-sections showing various types of unconformities.

Chapter 4

Natural Environments of the Blue Ridge

An extensive web of interactions and connections exist between the physical environment of the Blue Ridge and the organisms living there. Interactions between the underlying rocky foundation, the atmosphere and resident microorganisms creates soil, which in turn provides nutrients needed to sustain the growth of plants and animals. Animals and plants go through their own cycles of growth and decay, competing for survival and the opportunity to generate new life. The study of ecosystems, including the patterns of life through which organisms pass, the relationships among them, and their connection to the surrounding environment constitutes the field of ecology.

The physical environment of the Blue Ridge is a vast tapestry of subtle variations in **climate** (temperature, precipitation, amount of solar radiation), altitude, slope of the land, and various types of soil and underlying bedrock. Many ecosystems exist in the Blue Ridge, each containing a particular assemblage of organisms that interact with one another and with their physical environment.

Although animals are an important part of the natural environment, they are much too mobile and range across too many different habitats in their search for food and shelter to serve as a way of defining ecosystems. Hence, plants, especially types of forests, are used as the basis for the classification of ecosystems.

Ecologists currently mapping vegetation patterns in the national parks have concluded that "as a whole, bedrock parent material, soil fertility, elevation, and topographic position are the most important and interrelated environmental factors influencing major vegetation patterns in the Blue Ridge." Climate, including temperature and the amount of solar radiation received, are implicit in these factors.

Fig. 4-1. Waterfall with plants and mosses surrounding it at Crabtree Falls in Virginia.

Climate of the Blue Ridge

Long-term averages of precipitation and temperature define climate. Usually we express these as annual averages. A more precise description of climatic conditions includes information about averages and extremes for each month, as well as snowfall and wind conditions. Because the Southern Appalachians extend nearly 800 miles from Latitude 32° 30' N to 40° 20' N, significant differences in the amount of solar radiation occur at the northern and southern ends of the mountains. Winters are longer and colder in the higher latitudes and temperature extremes tend to be greater in the northern part of the mountains. Elevation also has a dramatic effect. The mountain peaks toward the southern end of the Blue Ridge, especially in the Smoky Mtn. region are several thousand feet higher than those to the north. The temperatures at the top of Mt. Mitchell in North Carolina (the highest mountain in the eastern U.S.), are more like those found in Southern Canada than those in the Piedmont or Valley and Ridge Provinces. While not quite as extreme as the effects on Mt. Mitchell, the elevation of the Southern Blue Ridge near this high point is still over 2,000 ft. higher than that encountered in the mountains farther north in Virginia, Maryland, and Pennsylvania.

The entire Southern Appalachian region lies in the belt of westerly winds that sweep across North America. These winds are accompanied by low-pressure storm centers that commonly move from the west or southwest across the mountains especially in winter months. Few tornadoes occur in the Appalachians, but hurricane winds and accompanying heavy rains have had dramatic effects on the mountains (p.148).

Table 4-1 illustrates the variation in climatic conditions found in the northern and southern parts of the Southern Appalachians. Extremes are seen in the measurements made on Mt. Mitchell in 2003.

Climate has important effects on the ecology of the mountains. In part these effects are regional in character, but microclimates that vary from one side of a mountain to another, or with elevation, are also critically important factors in the ecology of the Blue Ridge. Microclimates are determined by variations in the amount of sunlight, the strength and direction of prevailing winds, and the amount of rain and cloud cover in any given area. The growth pattern of tree limbs bent by local wind patterns and the preponderance of rhododendron and shrub thickets on northern and western slopes, as compared with the open oak forests found on many east facing slopes, demonstrate the effects of microclimatic conditions in many parts of the Blue Ridge.

Conditions	Pennsylvania (Harrisburg)	N.C. Piedmont (Mt. Airy)	Mt. Mitchell, N.C.
Avg. High temp in Jan. (F)	38	51	34
Avg. Low temp. in Jan. (F)	23	28	14
Avg. High temp in July (F)	85	89	68
Avg. Low temp in July (F)	66	67	53
Total annual Precipiation (in.)	41	42	75
Mean annual snowfall (in.)	35	9	91

Table 4-1. Comparison of climatic averages for the northern and southern parts of the Southern Appalachians. (Compiled from various sources)

Fig. 4-2. Radiation fogs often form in the Blue Ridge early in the morning after clear, cold nights.

Fogs

The classic postcard view of the Blue Ridge often features a long vista of blue-colored mountains enfolding great swaths of mist. Often these mists are actually **fogs,** dense ground-level clouds that reduce visibility down to under 0.62 miles. Many fogs in the Blue Ridge result from the lateral movement of clouds that form at low levels and simply run into the mountains as the air mass of which they are a part encounters higher ground. But fogs also form at ground level when air that is humid cools and begins to condense. This can happen in several ways.

Radiation fogs form when air at ground level cools due to the loss of heat caused by long wavelength solar radiation (Fig. 4-2). Favorable conditions for radiation fogs are a clear sky, light winds, long nights, and high relative humidity. Fog layers 30-100 ft. thick are common. These fogs form in valleys in the autumn and winter. Some continue throughout the day, but most break up by noon as a result of convection. **Rain fogs** form when warm moist air rides over cold ground. As rain falls from the warm air into the cold air near the ground it evaporates, becoming fog. **Advection fogs** also form when warm air moves over cold ground. Common in the east, where warm Gulf air moves into the cooler Central Atlantic States, it is also found in the far north during summer months when warm air flows over cold, slowly-thawing ground. Newfoundland has advection fogs that last for weeks. Finally, **steam fogs** form over lakes and river bottoms during summer months where cool air moves over warm water or ground.

Dew and Frost: Generally, water droplets, called dew, form when the temperature of the thin layer of air surrounding an object drops to the dew point, at which point condensation occurs. Dew forms on objects radiating heat (such as fences and grass), where atmospheric moisture condenses at a higher rate than it can evaporate (Fig. 4-3). During the colder months, when the air temperature drops below freezing, condensing air will freeze, forming frost.

Fig. 4-3. Dew collects on pine needles.

Fig. 4-4. A temperature inversion in the Valley and Ridge traps a thin layer of smoke from a fire.

Atmospheric Pollution in the Blue Ridge

Views from the Blue Ridge, especially toward the valleys to the west in the morning often reveal a low-level brownish-yellow cloud cover that stops abruptly at the flank of the ridge. While the valley appears to be filled with clouds, the air at the top of the ridge is crystal clear and long distance views may be perfect (Fig. 4-4). Temperature inversions formed in the valleys are usually responsible for these conditions as well as the resulting layer of pollution at low elevations.

A **temperature inversion** occurs when a layer of warm air lies over a layer of colder air. Normally, air temperature decreases with elevation. Under these conditions, hot air like that coming from a smoke stack would rise to higher altitudes where it might be carried out of the area by high-level winds in the troposphere, the lower six miles of the atmosphere. In a temperature inversion however, cold air produced during clear, cold nights from the cooling of mountain and hill slopes (often during fall and winter), slips down into the valleys, filling the valley floor with cold air. As this cold air reaches the dew point, condensation occurs and a dense fog may develop close to the ground. Often perfectly clear air overlies such a ground fog in the Blue Ridge. Once a temperature inversion has formed, hot air introduced at ground level can only rise to the level of the inversion where it spreads laterally, as if under an invisible ceiling (Fig. 4-4).

In this case, the hot air and any pollutants it may contain remain below the level of the inversion until the temperature of the ground-level air increases enough to break up and eliminate the inversion. This usually happens late in the morning after the sun has warmed the air. Temperature inversions have little effect on the large-size particulate matter that falls near the point of emission but the more dangerous, fine particles and gases such as those emitted by vehicles and industry are trapped and held near the ground until the temperature inversion dissipates.

There are other atmospheric conditions that also produce stagnant air over the region. During summer months a high-pressure cell known as the **Bermuda High** may cover parts of the Southern Appalachians, causing the air to be still while cool air at high altitudes acts like a cap holding pollutants close to the ground. Under these conditions, as well as when air pollution is high, one looks out from the Parkway into a haze that severely reduces visibility of landscapes in the distance.

However, not all of the atmospheric pollution occurs at low elevations in the Blue Ridge. Prevailing winds over much of the Southern Appalachians flow in from the west or southwest, bringing emissions from industrial centers in the Midwest and South Central States. From the Blue Ridge Parkway, this polluted air may appear as a high-level layer of brownish air.

Acid Rain

Carbon dioxide, respired by plants at night and exhaled continuously by animals, reacts with water vapor in the atmosphere to produce a weak acid known as carbonic acid (H_2CO_3). As a result, all forms of precipitation—rain, snow, ice, and fog—are at least slightly acidic. Natural, unpolluted rainwater has a pH of about 5.5 whereas distilled water is neutral with a pH of 7.

Other acids, notably those formed from sulfur dioxide (SO_2) and nitrogen oxides (NO_x), are formed in the atmosphere as a result of volcanic activity and the burning of sulfur-bearing coal in power plants, industrial boilers, and ore smelters. Vehicles are also a major source of nitrogen oxides. These compounds reduce the pH of water in the atmosphere and increase the acidity of the rain. As the concentration of these contaminants increases over time, rainwater can become devastatingly acidic. Rainwater near some industrial centers has reached pH levels between 1 and 2—about the same acidity as lemon juice. The highly industrialized areas of Europe, the United States, and China have the most severe problems with acid rain. Carried by weather patterns, the effects on lakes and streams may be experienced at great distances from the sources of the pollution. Streams in the Blue Ridge have been adversely affected, and some lakes in the Adirondack Mtns. have had pH measurements as low as 4.

In 1963, the U.S. Congress took an important step toward controlling acid rain and air pollution by passing the Clean Air Act. This act charges the federal Environmental Protection Agency with gathering information, conducting research, and planning ways to control air pollution. The act also establishes control mechanisms that state governments implement. It sets emission standards for automobiles and limits the amounts and types of emissions permitted from stationary sources like smoke stacks.

The effects of acid rain at ground level depend largely on where the rain falls. If buffering agents such as limestone are present, they neutralize the acid. However, where the soil and bedrock do not contain buffering agents, the pH of surface streams and lakes drops, and the acid disrupts the biological processes of many plants and organisms. Some cannot survive after the pH reaches certain critical levels. For many species of fish, these levels have already been reached in some lakes in the Adirondacks and in Scandinavia where the bedrock is granite and

Fig. 4-5. Woodsmen pose in a stand of American chestnuts (optically enlarged in the foreground) in Graham County, N.C, circa 1910. (Photo courtesy the Forest History Society, Durham, N.C.)

no buffering takes place. The lakes in these areas can no longer support fish. Acid rain is also a problem in the Blue Ridge where large areas of bedrock are granitic in composition and limestone buffers are absent. Acid precipitation also adversely affects forests. An estimated 75% of Europe's commercial forests suffer from damaging levels of sulfur and nitrogen deposition. The World Resources Institute estimates that the cost of pollution damage to European forests is about $30 billion per year. In addition to the loss of timber, these damaged forests cannot absorb carbon dioxide as they should, and this contributes to problems with greenhouse gases.

Natural Environments Change Over Time

Most ecological communities in the Blue Ridge are no longer in their native state. When the first European settlers arrived massive Am. Chestnut trees (Fig. 4-5) covered the mountains. Forests were felled to make charcoal and potash. Initially, hunting had the biggest impact on the Blue Ridge, but later with the establishment of farms came the introduction of new plants. Additionally, certain tree species were vulnerable to imported disease and blight. Some ecologists estimate that less than 20% of the land surface of the Blue Ridge remains in its original state as seen by the European settlers.

Of course, environments also change over time as a result of natural processes. The geologic record documents dramatic natural climatic changes that have occurred repeatedly throughout Earth's history. One of the most notable of these took place over the past 18,000 years. Early during this period, ice sheets covered the northern parts of the North America, including the mountains of New England, extending southward nearly to the northern tip of the Blue Ridge. Although the ice sheets did not reach the Blue Ridge, patches of ice probably remained throughout the year at higher elevations here. Annual rainfall was much higher and temperatures were much lower throughout the Central and Southern Appalachians. Since the end of the last advance of ice, about 10,000 years ago, the climate and local environments in the mountains have continued to undergo changes, leading to the present ecological conditions. Only remnants of the ecological communities that were dominant in the Highlands remain today.

Many climatologists have concluded that human activities, particularly the introduction of greenhouse gases into the atmosphere since the start of the industrial revolution have contributed to recent changes in climate, notably the increase in global temperature. Changes in the amount of solar radiation and influx of cosmic particles may have also played a role. Whether natural or man-made, the climate has changed and this has serious implications for future environments in the Blue Ridge.

Changes on the Ground

To the casual observer, however, the most obvious changes in natural systems in the Blue Ridge are being brought about by the construction of roads and dams, fires (Fig. 4-6), the expansion of cities, building of second homes, cultivation of crops, harvesting of timber (Fig. 4-7) and the introduction of non-native plants and animals, such as the **woolly adelgid**. This bug imported from East Asia sucks sap out of hemlock and spruce trees, killing them (Fig. 4-8). Little remains of the natural environments that existed prior to the arrival of European settlers. The best remnants of them are present in wilderness areas, national forests, and parks, and especially on the steepest slopes of the higher mountains where logging is difficult, leaving old growth forests largely untouched.

Fig. 4-6. Smoke from a forest fire.

Fig. 4-7. Clearcut

Fig. 4-8. Woolly adelgid infestation on a hemlock branch.

Fig. 4-9. Saprolite is chemically weathered rock that is common in regions with humid climates. The deep soils shown in this photo are found in the western Blue Ridge along SR 501.

Soil – The Plant-Rock Connection (Geobotany)

Although climate is a primary controller of plant life, nutrients also play an important role in determining which plants prosper. Generally, nutrients come from soil produced by the weathering of the underlying rock, which may be residual or transported in by erosion. Over large parts of the Blue Ridge, the soil was developed from the decomposition of the underlying bedrock. These residual soils may be tens of feet thick. Geologists often refer to such soils as saprolite (Fig. 4-9). On steep slopes where downslope movement of surficial materials is common and in stream valleys where alluvium is deposited in stream channels or on floodplains, soil may develop from transported rock. On floodplains, the soil itself may be transported. In a few places, plants grow directly out of cracks in solid rock (Fig. 4-10). In these instances, nutrients reach the plant roots through water that moves along the fractures or bedding planes. Some nutrients may come directly from the rock where it is beginning to break down. A few plants derive nutrients directly from the atmosphere. Since most plant nutrients ultimately come from the underlying rock, we might expect to find a close correlation between the types of plants that grow on the surface and the type of rock from which the nutrients come. Some such connections can be traced.

Certain distinctive plants prosper on rocks composed of minerals containing certain elements. For instance, soils with slightly higher concentrations of manganese, which is derived from the weathering of amphibolite, basalt, gabbro, and ultramafic rocks, support maples in the understory and blackberry as a shrub.

Fig. 4-10. Plants growing out of cracks where water seeps out to the surface.

Fig. 4-11. Pine growing out of rock exposed near the Skyline Drive.

A wide variety of plants grow on limestone (calcium carbonate), but there are few areas in the Blue Ridge that are underlain by such calcareous rocks.

In some places **endemic plant species** (those of limited geographic distribution) are restricted to outcrops of certain rock types. However, rocks containing a variety of minerals, especially clays, which are sedimentary rocks formed in marine environments and hence are composed of many elements, may support a great variety of plants. For this reason, maps of forests often fail to match geologic maps of the same area. This is true for many of the Blue Ridge forests. One exception occurs along outcrops of pure quartz sandstone and quartzite that contain few nutrients. Because they are drought tolerant, adapted to fire, and better defended chemically against herbivores, pine trees are among the few trees that can survive on these rocks and the soils produced from them (Fig. 4-11). You will notice this connection in many places throughout Shenandoah National Park (Ch. 6E).

Studies of plants have long been used in the exploration for metals. Some metallic deposits contain elements that are toxic for plants. For example, most plants will not grow where the soil contains more than 14% copper, high levels of iron, platinum, or borates. Certain plants may prosper in the presence of specific metals, often taking up the metals from the soil; so, geochemical analysis of the plant may reveal the presence of the metal. In the western states, remote sensing via satellite or airplane has been used to detect the presence of unusual indicator plants.

Soil Acidity

The acidity of soil also determines what types of plants may grow or prosper in any given area. Acidity is described in terms of its pH (a measure of the hydrogen ions present). On this scale pH 7 is neutral, pH 0 is extremely acidic, and pH 14 is extremely alkaline. Most southeastern soils in the U.S. have a pH ranging from 4 to 8 (Table 4-2). Soil acidity is important for plants because they pick up nutrients from solutions and acidity

Extreme Acidity				Neutral				Extreme Alkalinity
pH 0	pH 4	pH 5	pH 6	pH 7	pH 8	pH 9	pH 10	pH 14
Battery acid	Tomatoes, Acid Rain	Coffee, 'Soft' water	Urine, Saliva	Distilled water	Sea water	Baking soda	Great Salt Lake, Milk of Magnesia	Liquid drain cleaner

← Rhododendron →
← Loblolly Pine →
← White Pine →
← Spruce →
← Hemlock →
← Red & White Oaks →
← Birch →
← Dogwood →
← Maples →
← Poplar →
← Black Walnut →

Table 4-2. Tree preferences for acidic or alkaline soils. (Source: Virginia Cooperative Extension of Virginia Tech)

affects the solubility of minerals and organic compounds from which nutrients are derived. High acidity also has negative effects on microorganisms such as bacteria that play an important role in the decomposition of organic matter.

Fig. 4-12. Mountain ash

The presence of certain chemical elements in soil depends largely on pH. For example, phosphorus, one of the most important plant nutrients is more abundant in soils with a lower acidity of about 6.5. A number of common trees including Fraser firs, maples, serviceberry, sweet gum, magnolias, red oaks, and mountain ash (Fig. 4-12) grow best in soils with a pH of 5.0 or above. Rhododendron (Fig. 4-13), most conifers, dogwoods, cedars, pin oaks, beech, and river birch, on the other hand, all grow well in acidic soils. These soils commonly have high concentrations of aluminum, iron, and manganese. This is notable because many of these elements compose the rocks that are widespread in the Southern Blue Ridge.

Fig. 4-13. Rhododendron

Classifying Natural Environments

Fig. 4-14. Fall foliage variations indicate differences in exposure to sunlight and water infiltration.

The conditions that control physical environments vary, often gradually, across any large area as diverse as a mountain range such as the Blue Ridge. These variations give rise to microclimates that are related to such things as slope, rainfall, and the orientation of the ground surface relative to sun and wind (Fig. 4-14). These differences, along with the overlapping growth patterns of plants that survive in various types of locations make the job of defining environments on the basis of the local plants quite difficult. Consequently, forests are usually named after the dominant trees forming the canopy of that forest (Figs. 4-15, 4-16). Some classifications subdivide these forest types, and in a few instances, the subdivisions no longer contain the dominant tree for which the main category was originally named. Many such plant-based environmental definitions rely on noting the close associations of trees that form the upper-story or canopy, with undergrowth made up largely of saplings and shrubs as well as a ground cover composed largely of herbaceous seed plants like grasses and wildflowers.

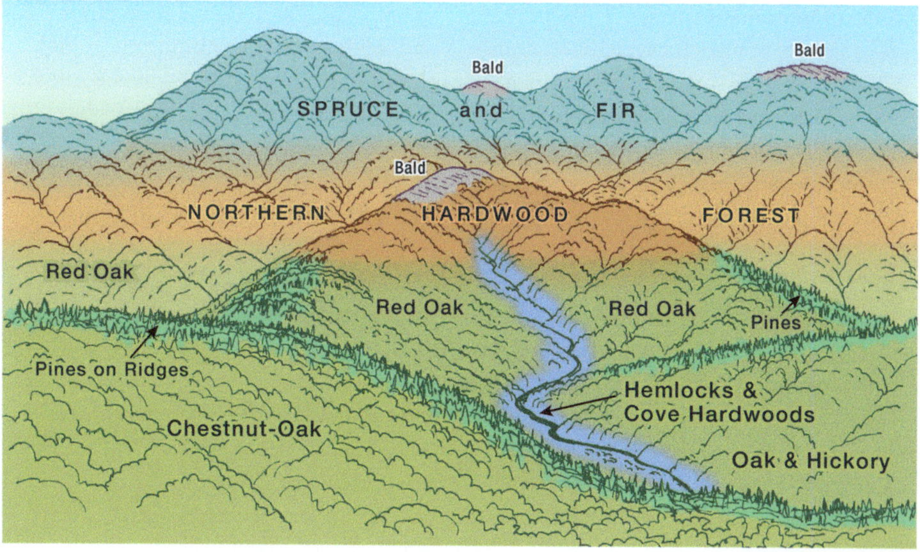

Fig. 4-15. Drawing showing forest types and conditions.

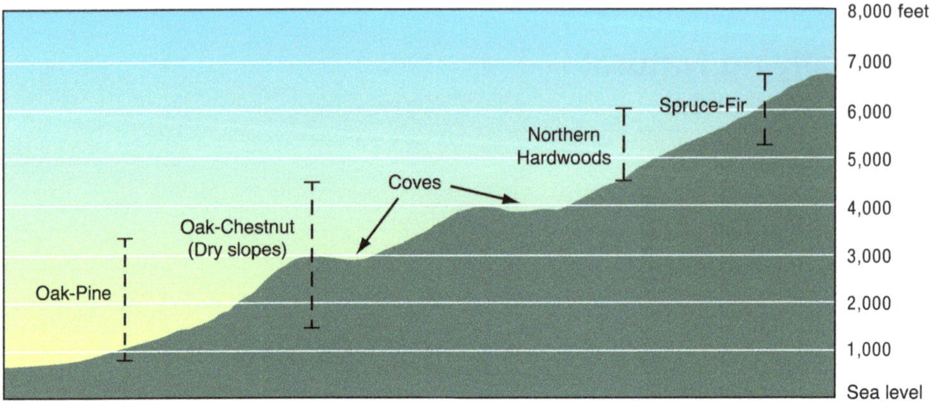

Fig. 4-16. Mountain cross-section showing forest types commonly found at different elevations. Coves may occur at a greater variety of elevations than other forest-types.

Classical Forest-types in the Blue Ridge

Almost all of the Blue Ridge was originally classified as an oak-chestnut forest with a number of subdivisions. American chestnut trees were one of the most abundant trees found throughout this region until a fungal blight (Endothia parasitica) that was accidentally imported along with some Asian chestnut trees eradicated virtually all of them in the early 1900s. Since then other trees that were present, but less abundant, have replaced chestnut as the predominant species. Tulip-poplar and a variety of oaks have assumed dominance among the trees on well-drained slopes. Other trees prosper in the moister areas such as mountain coves. Still others are dominant at elevations above 4,500 ft. and in places where the soil is dry and lacking nutrients. These differences have led to the widespread recognition of a number of categories. Dry slope and ridge forests, oak-pine forests, northern hardwood Forests, cove forests, spruce-fir forests, and balds are commonly recognized designations. The greatest diversity of subdivisions exists in the southern parts of the Blue Ridge (south of the Roanoke River).

Oak-chestnut forests (also referred to as northern oak forests) are named for the trees that once formed the canopy, especially on dry slopes. Oak-chestnut forests once covered vast areas in the elevation range of 1,500 to 4,500 ft. Today they are almost gone, the accessible ones having been logged long ago and the remaining few having been decimated by a fungal blight that was introduced to the country in the early 20th century. Of the remaining trees, those on exposed sites are stunted, gnarled in stature, and reflect the influences of high winds and frequent ice damage. Pre-settlement chestnut and mixed oak-chestnut forests at high elevation also experienced fires ignited by lightning strikes. These took place every few decades. Today fires are often suppressed, and competing vegetation in the understory contributes to poor oak regeneration. This is accompanied by the invasion of sugar maples. Gypsy moth infestations, which have led to repeated defoliation and widespread tree mortality in the Northern Blue Ridge pose another serious threat to these oak-dominated communities in the south.

The role in the canopy played by chestnuts has been taken over by red oak, chestnut oak, white oak, and hickory. Tulip-poplar, yellow birch, red maples, and hemlocks are common at low to mid elevation while yellow birch, red maple, and a few spruce appear at higher elevations. The soils found here contain less organic material, tend to be infertile, and do not support many plant varieties. Wild cherry and witch hazel are often abundant in the understory, and shrubs including rhododendron, azalea, and mountain laurel inhabit the forest floor. These forests abound near Mt. Mitchell, Grandfather Mtn., and along many stretches of the Appalachian Trail.

Northern Hardwood Forests contain an assortment of hardwood trees commonly found in New England and eastern Canada. These are present at higher elevations, typically between 4,500 and 5,500 ft., where climatic conditions resemble those found much farther north, especially in the Southern Blue Ridge Mtns. Sugar maple, yellow birch, American beech, hobblebush, and striped maple are the common key indicator tree and shrub species in the northern hardwood forest. Hemlock and white pine may also be present. All of these plants prosper on north-facing slopes above 3,600 ft., and many of them represent new growth that has become established following the removal of the original plant communities as a result of logging and forest fires in the early 1900s. Maples and beech will likely become the dominant species in the future.

A second community type is typically associated with convex, southerly slopes and ridge spurs which receive a lot of sunlight and often have infertile soils. Beech trees dominate this canopy along with some yellow birch, Fraser magnolia, hemlock, red spruce, and sugar maple. Young beech and various maples make up the understory and hobblebush is the dominant shrub. Many kinds of flowers and ferns may be present.

Herbs (the non-woody species that cover the forest floor) that are common to northern hardwood forests include wintergreen, wild sarsaparilla, and wood sorrel. Birds and animals include the black-capped chickadee, white-throated sparrow, cedar waxwing, porcupine, snowshoe hare, white-tailed deer, and American red squirrel.

Oak-pine Forests: Along the western margin of the Blue Ridge, many of the oak-chestnut forests contain large numbers of pine trees, especially at lower elevations. Because Table Mtn., shortleaf, and pitch pines require periodic fires to regenerate, they are commonly found in drier, more fire-prone sites. Pines survive in well-drained soil that is poor in nutrients, a condition that exists along many of the ridges that are held up by rocks composed primarily of nutrient-poor quartz, such as sandstone.

Fig. 4-17. An assortment of forest types and conditions, including coves and ridges in Shenandoah National Park.

Cove Forests: Coves are bowl-shaped valleys with relatively flat valley floors. These environments commonly extend downslope from the head of a valley where a bowl shape is best developed (Fig. 4-17). The shape of the coves tends to protect them from high winds. Coves generally contain streams that flow down the low slopes found on the valley floor. Rich, fertile, and damp soils are often present on these valley floors. Hence, water-loving rhododendron and mountain laurel thickets commonly line the creeks in coves. Because the valley shape provides protection from long hours of direct sunlight the valleys are cooler than the open slopes. The resulting forests have great tree diversity that may include dozens of different trees and a great variety of shrubs and herbs (Fig. 4-18). A mixture of five or six hardwoods such as buckeye, basswood, sugar maple, tulip poplar, beech, birch, white ash, wild cherry, cucumber magnolia, red maple, red oak, and hickory may constitute the majority of any particular cove forest. Unfortunately, as a result of insect attacks, hemlocks are rapidly disappearing from coves where they were once a prominent presence. Many herbs, notably trillium, cutleaf toothwort, dogtooth violet, bleeding heart, and fairy bells may be present on the floor of cove forests, but the layer of herbs here is usually sparse. The diversity of herbs is most prominent in the few remaining old-growth forests that have escaped logging.

Fig. 4-18. Plants on the forest floor vary with the maturity of the trees above them and with other environmental factors such as elevation, rainfall, slope, and exposure to sunlight.

The cove forest is home to many creatures. Amphibians found here include wood frogs, Fowler's toad, red salamander, black-bellied salamander, Ocoee salamanders, Jordan's salamander, two-lined salamander, and spotted salamanders. Among the reptiles we find eastern box turtles, garter snakes, copperhead snakes, water snakes, rat snakes, and ringneck snakes. Golden mice, deer mice, red-backed voles, shrews, chipmunks, flying squirrels, gray foxes, deer, bears, and bats are among the most commonly seen mammals.

Spruce-Fir Forests: At high elevations (above 4,500 ft.) the oak forests give way to yellow birch, beech, and sugar maples, along with trees that are prominent in northern hardwood forests such as white ash, wild black cherry, cucumber magnolia, and basswood. At even higher elevations, above 5,000 ft., red spruce trees tend to dominate the canopy accompanied by Fraser firs that often occupy some of the highest peaks and their northerly slopes (Fig. 4-19). These high level forests have very severe winters and cool summers accompanied by high rainfall and frequent fogs. The soils are often acidic, and depending on the type of bedrock may be of the type described below as Amphibolite Mtns. Red spruce and Fraser fir trees make up large parts of the forest near mountaintops. These are often so dense that smaller trees are

Fig. 4-19. Spruce forest on a ridge overlooking a northern hardwood forest below.

Fig. 4-20. Bald on a mountaintop near Craggy Gardens, N.C.

shaded out. Near ground level ferns and a dense layer of herbs are abundant. Damage by the woolly adelgid (Fig. 4-8), acid rain, and effects of air pollution are common in these forests. Dense thickets of rhododendron commonly grow beneath the red spruces. Closer to the summits, carpets of mosses, liverworts, and ferns cover the forest floor.

Balds: Treeless mountaintops known as **balds** are present in the Southern Blue Ridge (Fig. 4-20). Grasses dominate many balds that may also be dotted with ferns and other herbs. Other balds are dominated by shrubs such as rhododendrons and blueberries. The few trees that grow around the edges of balds are typically gnarled and stunted, largely a result of strong winds.

The American Chestnut Foundation (established in 1983) is working to restore the American chestnut tree using multiple scientific methods of research and evaluation. Once a foundation species in the eastern U.S., the American chestnut is now functionally extinct due to the importation of a fungal pathogen know as chestnut blight. The blight fungus was first introduced into the U.S. in 1904, arriving on imported Asian chestnut trees. After twenty-two years of patient field work by a largely volunteer force of 5,000 members from Maine to the Carolinas, the Foundation harvested its first potentially blight-resistant chestnuts in 2005. By using a unique method called "backcross breeding," they arrived at a hybrid containing approximately 15/16 of the genes of the original American chestnut plus that essential 1/16 of the blight-resistant Chinese chestnut genes carrying the immunity. The new "Restoration Chestnut 1.0" is currently undergoing rigorous testing in orchards and forests to confirm its viability. Ultimately, the Foundation hopes to return these trees to the forests where they once flourished. To learn more about The American Chestnut Foundation visit their website at: **www.acf.org**

Chestnut bur from the ACF's Meadowview Research Farms in Virginia (Photo: The American Chestnut Foundation)

Other Forest Classifications

The **National Fish and Wildlife Foundation** distinguishes between six forest types in the Southern Blue Ridge. Of these, oak-hickory forests and oak-pine forests are the most widespread. On maps, these forests appear as parts of a patchwork pattern with forests and tree stands distributed throughout the region. Loblolly-shortleaf pine forests are prominent in South Carolina with smaller patches in North Carolina and Tennessee. Forests containing white pine, red pine, and jack pine, spruce-fir, and maple-beech-birch trees cover much smaller areas.

The **Natural Heritage Program** of the Virginia Department of Conservation and Recreation is developing an alternative ecological community classification based mainly on plant life. Their "terrestrial" system is subdivided into classes based on climate, geography, and soil conditions. Within each class, ecological community groups are identified on the basis of topography, soil, and plant-life. These communities are further subdivided into types that share more specific environmental, structural, and floristic similarities. For details, refer to the website for Virginia Natural Heritage. You will find lists of plants and characteristics of each of the identified groups. ❖

Natural Heritage Program Classifications

Oak/Heath Forest: Forests that were formerly populated by large chestnut oaks. The few remaining chestnut oaks are now stunted. Oak/heath forests are located on dry, infertile rocky slopes and ridges up to about 3,200 ft. in elevation. The soils are usually acidic soils derived from sandstone or shale containing little calcium or magnesium but lots of iron and aluminum. These forests are prone to forest fires. A dense understory of heath-family shrubs such as black huckleberry and laurel are present.

Basic Oak-Hickory Forest: Located at elevations below 2,500 ft. in the northern part of the Blue Ridge. They form on soils that contain high levels of iron and magnesium such as metabasalt and amphibolite. A great variety of herbs are present at ground level including grasses, milkweed, goldenrods, yellow pimpernel, feverwort and violets. White oaks, northern red oak, black oak, chestnut oak, hickory, white ash, and tulip poplar form the canopy with dogwood and red bud are prominent in the understory.

Southern Appalachian Northern Hardwood Forest: These areas, located up to elevations of 3,600 ft., contain a highly diverse understory including wood nettle, blue cohosh, wood fern, and snakeroot herbs. Maples and mountain holly make up much of the understory with sugar maple, American beech, yellow birch, and yellow buckeye forming the canopy.

Montane-Oak-Hickory Forest: Located between 2,500 and 4,000 ft. in elevation and commonly developed on fertile, high pH soils derived from granitic and metavolcanic rocks. The understory consists of black bugbane and leatherleaf meadowrue. Red oaks, white oaks, hickories, white ash, and striped maples form the canopy.

Northern Red Oak Forest: Located on upper slopes between 3,000 and 4,200 ft. Most are on infertile soil derived from granitic rocks. Red oaks dominate this forest.

Part 2:
Field Guides

Fig. 5-1. Digital elevation map of the Northern Blue Ridge and parts of the Piedmont and Valley and Ridge. A white dashed line shows the boundaries of the Blue Ridge Mountains. (The National Atlas, USGS, with modifications)

Chapter 5

The Northern Blue Ridge

Defining the Blue Ridge

The Blue Ridge as a Topographic Feature: The term "Blue Ridge" is used in two ways. The first is based on topography, especially elevations, and includes a ridge in the north, a high plateau south of Roanoke, Va. and a region of high peaks in North Carolina and Tennessee. This topographic Blue Ridge starts as a prominent ridge that begins south of Harrisburg, Pa. and continues to Roanoke Va., where the Roanoke River cuts across it (Fig. 5-1). South of the Roanoke River, the prominent ridge found to the north becomes less distinct and the core of the Blue Ridge becomes a plateau, referred to here as the Upland Plateau (Fig. 7-1). The eastern edge of this plateau is marked by a sharp escarpment—a steep slope where the land surface rapidly falls nearly 2,000 ft. over a short distance. The western edge of the plateau occurs where sandstones of the Chilhowee Group are thrust faulted onto limestones of the Valley and Ridge. Near the southern border of Virginia, the plateau surface gradually changes to a large region of much higher mountains, the Blue Ridge Highlands. Many of these peaks rise above 5,000 ft. and in some cases to over 6,000 ft. This region includes the Mt. Rogers area in Virginia and the Great Smoky Mtns. in Tennessee, and it continues into northern Georgia where the Highlands give way to a more subdued landscape.

The Blue Ridge as a Geological Province: The second way of defining the Blue Ridge grew out of mapping done by geologists in the northern part of the Blue Ridge. They discovered that the northern end of the Blue Ridge has a folded anticlinal structure somewhat like the simplified structure shown in Fig. 5-2. Compare this image with the overall Northern Blue Ridge shown in Fig. 5-1 and with the more detailed cross-section in Fig. 5-3. In Pennsylvania and Maryland the Blue Ridge anticline plunges to the

Fig. 5-2. Schematic sketch of an anticlinal fold that plunges to the north and has a core composed of igneous and metamorphic rocks.

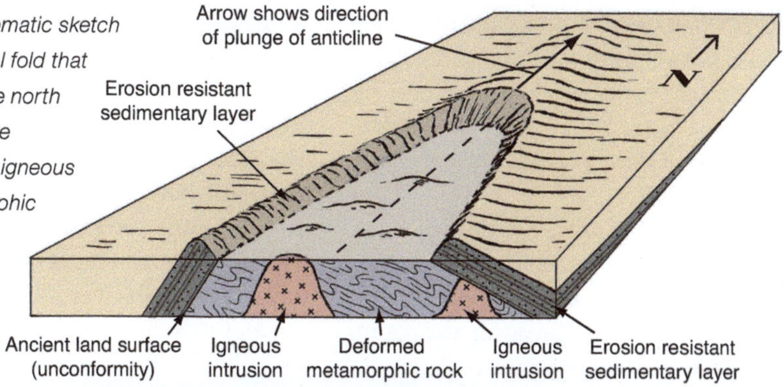

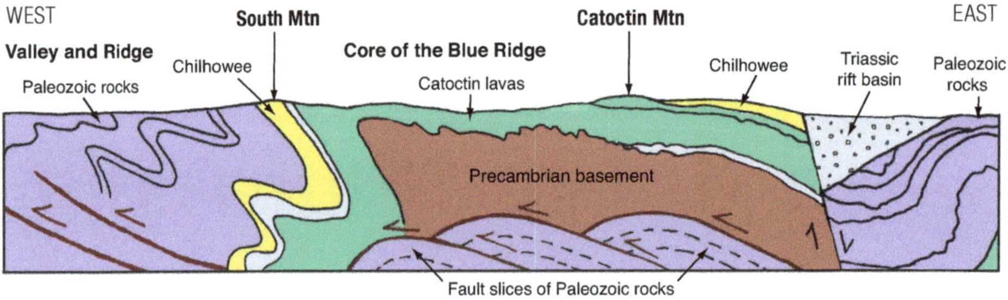

Fig. 5-3. Cross-section of the Blue Ridge in Maryland with a hypothetical interpretation of the structure beneath the core of the Blue Ridge. (Modified after the Geologic Map of Maryland, 1968)

north. Catoctin Mtn. forms the eastern limb of this anticline and South Mtn. and Blue Ridge Mtn. form the western limb. From a geological perspective both limbs and the core between them are all part of the same major structure, referred to as the **Blue Ridge Geological Province** even though much of the structure is not ridge-like. The limbs of the fold are mountain ridges held up by the Chilhowee and Catoctin rock units (Fig. 5-3). The central part, known as the core of the anticline, is underlain by Precambrian basement rocks. Thrust faults that carry the Chilhowee onto rocks of the Valley and Ridge lie along the western edge of the Blue Ridge Province.

Faults also define the eastern side of the Province. In central Virginia a steeply inclined normal fault separates a Triassic-age (200 Ma) basin from the Blue Ridge Province. This fault formed after the Alleghanian mountain building took place when Gondwana started to break apart. Other, much older faults notably the Brevard fault (Fig. 7-1), continue along the eastern edge of the Blue Ridge Province all the way to the coastal plain.

In this book, the name **Blue Ridge** is used in place of Blue Ridge Geological Province. This name is applied to the anticlinal feature in the north, the region between the Chilhowee Group rocks on the west and the zone of thrust faults that continue farther south. It includes the Northern Blue Ridge covered in this chapter, the Upland Plateau region south of Roanoke, and the Highlands region farther south (Chs. 7 and 8).

Geologic Setting
of the Northern Blue Ridge

The northern end of the Blue Ridge begins in Pennsylvania and rises as it continues southward as two roughly parallel ridges (Fig. 5-1). In this northern section, the western ridge is known as South Mtn., and the eastern ridge is known as Catoctin Mtn. The two ridges define the margins of the Blue Ridge in Maryland and northern Virginia. They also form the western and eastern limbs of the large anticlinal structure that plunges to the northeast (Figs. 5-1, 5-2). The crest of the anticline is preserved north of Catoctin Mtn. National Park. Farther south stream erosion has removed the crest of the anticline exposing much older rocks in the core of the Blue Ridge (Figs. 5-2, 5-3). Differences in the resistance of the rocks to weathering and erosion are major factors in the development of the landscape. Millions of years of erosion have lowered the elevation of large parts of this core

Fig. 5-4. View from Shenandoah National Park toward the east across the core of the Northern Blue Ridge. Most of the relatively low relief area located in this part of the core is underlain by igneous and metamorphic rocks that are over a billion years old.

as well as the eastern margin of the Blue Ridge south of Catoctin Mtn. Despite this long-term erosion, the western flank of the Blue Ridge (which includes Shenandoah National Park and the crest of the anticline near South Mtn.) is still prominent in the landscape (Fig. 5-1). It is high because rocks that are resistant to erosion (e.g. quartzites, sandstones, phyllites, lava flows, and the igneous and metamorphic rocks of the Blue Ridge core) form this ridge along the western margin of the Blue Ridge all the way to the Roanoke River. It also appears that erosion with its lowering of the landscape is moving progressively from east to west, as described in Ch. 3.

The Core
of the Northern Blue Ridge

Rocks that were once part of the ancient supercontinent of Rodinia (Fig. 2-10) are exposed in the core of the Blue Ridge. These rocks, referred to as the **Blue Ridge basement complex** include those that were part of the Grenville Mtns. The building of those mountains took place 1.3 to 1.1 billion years ago during the assembly of Rodinia. Detailed mapping and radiometric dating is gradually making it possible to identify the sequences of igneous and metamorphic events involved in that Precambrian evolution of the core of the Blue Ridge. Some of the igneous intrusions in this core (such as the Robertson River intrusions) are of Neoproterozoic age (about 700 Ma), much younger than the Grenville Orogeny, but most of the core is made up of much older, igneous and metamorphic rocks, the remains of two or more mountain belts that formed earlier in the Neoproterozoic.

Much remains to be learned about the age and history of the rocks in the core of the Blue Ridge. Many of the older rocks in the core are similar to granite in composition, but differ in that they contain a pyroxene mineral (hypersthene) not commonly found in granite. Pyroxene is found in rocks that crystallized under

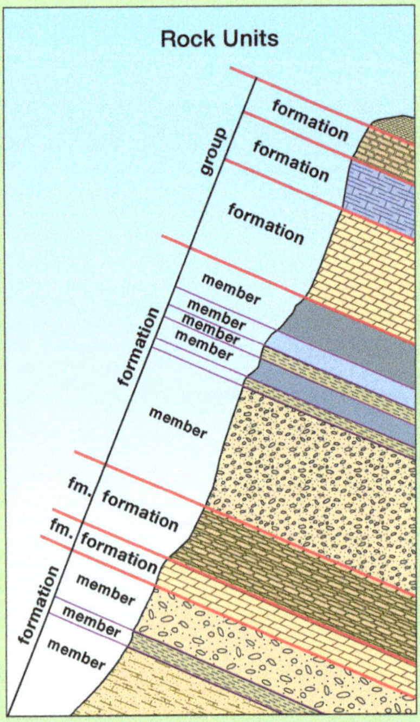

Fig. 5-5. Subdivision of rock units from groups, to formations, to members.

Reading the Rock Record of the Blue Ridge

In describing the geology of an area, geologists map locations where rocks of different types occur at or near the surface of the ground. Two kinds of maps are often produced. One shows the distribution of materials found at the surface. These might include stream gravels, deposits from landslides, lakebed sediments—all referred to as surficial materials. The second type of map, called a bedrock map, shows where the rocks buried beneath those surficial materials are located. They may be igneous, sedimentary, or metamorphic (Ch. 9).

Sedimentary rocks usually occur as layers (also called strata or beds), and can be identified either on the basis of their composition, the environment in which they were deposited, or their position in a sequence of layers. These layers of similar or related rocks are called rock units. If the rock unit consists of a single layer in a number of beds that can be identified and distinguished from other rock units that occur beneath or above them then it is commonly called a formation (Fm.). Some formations are only a few tens of feet thick; others may be thousands of feet thick. For more detailed study, formations may be subdivided into smaller units called members. Where a number of formations are similar or closely related in the way they originated, they may be referred to as a group. Formations, members, and groups are the rock units shown on geologic maps and are used in the study of the rock record (Fig. 5-5).

Different formation names may be applied to sedimentary and volcanic rocks of the same age. This happens because the conditions of sedimentation and the types of sediment deposited at different places over a large area, such as that covered by the Blue Ridge, were quite varied. Names of rock units used in different parts of the Blue Ridge also vary because the geologists who initially assigned names in one area were uncertain about the application of names applied to rocks in the same sequence of units by geologists working in other areas. Thus a rock unit named in Maryland may have a different name from an equivalent rock unit of similar composition and position in a sequence named in Tennessee. Many of the rock units in the Appalachians have multiple names because early geologists tended to work in limited areas. As geologists who had been working in different areas came together they discovered that different names had been given to equivalent rock units. Consequently, two names are often attached to the same rock unit.

extremely high pressure from very dry magmas, such as those produced at great depth (6-12 miles) below the surface of the Earth. Obviously these rocks in the Blue Ridge core must have been deeply buried at some point in their history. This gives us a hint about how much uplift took place during the Grenville Orogeny and how deeply those uplifted rocks were eroded before Rodinia began to split apart. We can only imagine how these mountains looked before they were eroded away long before the most primitive invertebrates appeared on Earth. Images of the Alpine-Himalayan chain come to mind.

Rock Units along the Western Margin *of the Northern Blue Ridge*

Sedimentary and volcanic rock units exposed along the western flank of the Blue Ridge during the late Precambrian and early Cambrian times following the breakup of Rodinia (Figs. 2-10, 2-11) were derived from the Grenville Mtns. that had been formed when Rodinia was assembled. They were deposited along the continental margin of North America (Laurentia) (Table 5-1).

Following the breakup of Rodinia about 560 Ma, stream erosion wore away the high Grenville Mtns. until only a relatively low-lying landscape remained. As Laurentia (North America) moved away from Gondwana (Africa) the Iapetus Ocean flooded the margins of the two continents and sediments accumulated. Initially the edge of the continent broke apart and displacements dropped the eastern side down, forming rift basins (Fig. 5-6). Sediments eroded from the adjacent landmasses then settled into these basins forming what is now known as the Swift Run Fm. Faults along the edges of these rift basins opened paths for lava to rise from deep in the crust to the surface. As the continents moved apart the weight of the continental crust was removed, reducing the pressure at depth and making it easier for the rocks there to melt. Volcanic activity began to take place in the rift basins. This led to the formation of massive flows of lava interlayered with sediment. These rock units are known as the Catoctin Fm. Another volcanic center that formed about 200 million years earlier at Mt. Rogers in southern Virginia is described in Ch. 8F.

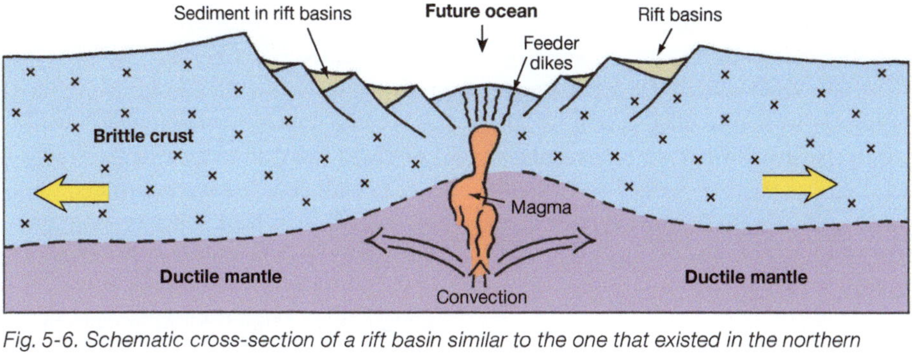

Fig. 5-6. Schematic cross-section of a rift basin similar to the one that existed in the northern Blue Ridge near the end of the Precambrian Era. Similar rift basins formed as Pangaea broke apart and the Atlantic Ocean formed.

Swift Run Formation

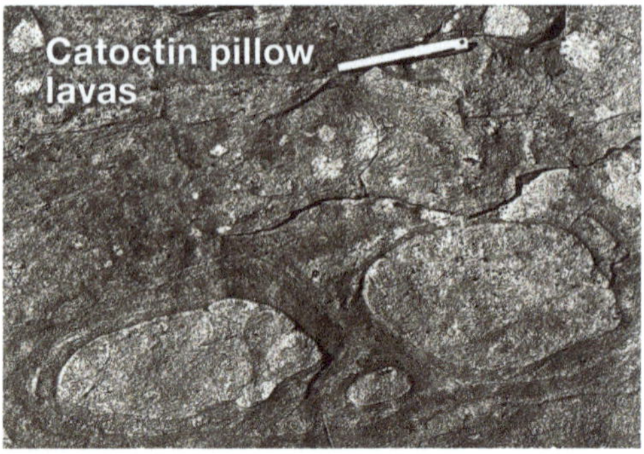

Pillow lavas in the Catoctin Formation.

The Catoctin and Swift Run Fms.

Geologists named the lavas found in the northern part of the Blue Ridge after Catoctin Mtn. where they were first discovered. Imagine red hot lava, like that now seen in Hawaii, pouring out of long fracture zones, spreading across the Grenville topography, and slowly building up lava beds over a thousand feet thick in some parts of the rift basins.

The margin of Laurentia subsided as Laurentia moved farther away from Gondwana and the Iapetus Ocean deepened. Eventually thousands of feet of sediment accumulated as the edge of the continent sank into water. At first the water was very shallow, but it became deeper as the continental crust pulled apart and the Iapetus Ocean grew. In the area that is now the Northern Blue Ridge, lavas flowed out on top of the Swift Run Fm. In the Shenandoah National Park, some of these lava flows poured out over a rough landscape with over 1,000 ft. of relief. In places the lava flowed over an irregular land surface that had formed as the landscape of the Grenville Mtns. was eroded. The lava buried hills and ridges that had been as much as a thousand feet high.

Like many modern lava flows the Catoctin lavas contained bubbles that became filled with minerals such as quartz, epidote, chlorite, or calcite (Fig. 6E-18). Other features seen in modern lavas such as columnar joints and surfaces of blocky lava are present in the Catoctin lava flows beautifully exposed at Catoctin Mtn. Natl. Park, along the Skyline Drive in Shenandoah National Park, and along the Parkway as far south as the James River. Many millions of years later, during the Paleozoic Era, the Catoctin lavas were covered by thick layers of sediment, sank, and were subjected to high temperatures that transformed the basaltic lava into a rock known as greenstone. The green color comes from the presence of the minerals chlorite, epidote, and actinolite that formed during the metamorphism.

The Chilhowee Group

When the extrusion of the Catoctin lavas stopped, the eastern margin of Laurentia (North America), the area now known as the Blue Ridge, continued to subside. The eastern margin of the continent was pulled apart as the Iapetus Ocean widened. Streams continued to carry sediments from the continent into the eastern continental margin. One of the rock units composed of these sediments and that is found only in the northern part of the region is named after Loudoun County, Va.

Major Rock Units - Western Margin of the Northern Blue Ridge	
Rock Unit	**Description**
Chilhowee Group	(See Figs. 5-7, 5-8.)
Antietam (Erwin) **Fm**	Quartzite and quartzite schist.
Harpers (Hampton) **Fm**	Dark greenish gray phyllite and schist with thin layers of quartzite grading into shales and sandstones farther south.
Weverton (Unicoi) **Fm**	Gray to purplish gray feldspathic quartzite and conglomerate.
Loudoun Fm	Phyllite overlain by conglomerate that is cemented by a mixture of greywacke.
Catoctin Fm	Metabasalt (greenstone), metarhyolite, and greenstone schist.
Swift Run Fm	Metamorphosed quartzite, sandstone, conglomerate, and schist.
~~ An erosion surface – a time when the basement complex was being eroded. ~~~~~~~	
Blue Ridge basement Complex:	This name is an informal way of referring to undifferentiated igneous and metamorphic rocks. Most of them are granulite facies gneisses of Precambrian-age, but a number of other rocks, some of Paleozoic-age, may be included.

Table 5-1. Major rock units exposed along the western margin of the Blue Ridge. Tthe youngest units are at the top. Limestone, dolomite, and shaly units of Cambrian-age are found in the Valley and Ridge Province above the Chilhowee Group.

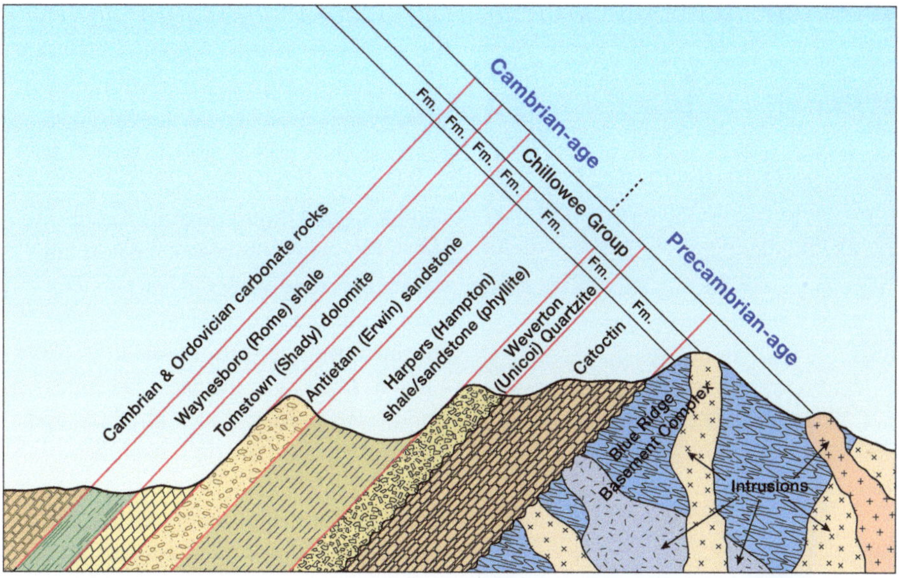

Fig. 5-7. Stratigraphic units found along the western margin of the Northern Blue Ridge. A thin unit, the Loudoun Fm. is only found in northern Virginia and Maryland. The Swift Run is found to the south in Shenandoah National Park, and the Catoctin Fm. continues as far south as central Virginia. The Chilhowee continues to the southern end of the Blue Ridge. The Blue Ridge basement complex and the younger igneous intrusions in it are located in the core of the Blue Ridge.

Fig. 5-8. Representative rocks found in the Chilhowee Group, Catoctin Fm., and Precambrian basement.

Other, much younger layers of sediment covered the continental margin from Penn. to Tenn. These units, called the **Chilhowee Group** are prominent all along the western flank of the Blue Ridge (Fig. 7-1). They are sedimentary rocks, mainly conglomerates, sandstones, shales, and some volcanic ash, along with their slightly metamorphosed equivalents. All the sediments making up the Chilhowee Group were deposited on the Laurentian continental margin. (Note: in Table 5-1, the names listed are used in the northern part of the Blue Ridge; their southern equivalents are shown in parenthesis).

Weverton Fm: (Unicoi Fm.) The oldest and hence lowest part of the Chilhowee Group, the Weverton, was formed from sediments carried by streams flowing across a land surface that was gradually being submerged. In general, this lowest unit consists of conglomerates with pebble- and granule-sized fragments. It was named for exposures located near a town of that name located near the Potomac River where it cuts across the Blue Ridge.

Harpers Fm: (Hampton Fm.) As subsidence of the area continued, most of the sediments composing the Harpers Fm. were deposited in deeper water on the continental slope. This middle unit, (over 1,000 ft. thick in some places) includes impure sandstones, shales, phyllites, and some nearly pure quartz sandstones.

Antietam Fm: (Erwin Fm.) Eventually subsidence stopped and sand destined to become the Antietam was deposited on beaches that formed along the edge of the Iapetus Ocean. As sea levels rose the ocean invaded Laurentia and occupied the eastern part of it for many millions of years. These deposits are still present as far inland as Wisconsin. This upper unit is composed of nearly pure quartz sandstones in the north and more shaly sandstones farther south and is named for exposures located near the town of Antietam, site of a major Civil War battle.

Blue Ridge Basement Complex

There is much that remains to be discovered about the history of the rocks in the core of the Blue Ridge. Most of these rocks are over a billion years old (Mesoproterozoic-age). They are referred to as the Blue Ridge basement complex, and most were part of a mountain chain that formed along the margin of what is now North America between 1.3 and 1.1 billion years ago. This was the time when the primitive supercontinent Rodinia (Fig. 2-10) was forming as older continents and continental fragments coalesced.

Detailed mapping and radiometric dating is gradually making it possible to identify the sequence of events in the Precambrian evolution of what is now the Blue Ridge core. The oldest groups of rocks in the core are highly foliated gneisses that were metamorphosed during deformation that occurred when mountain building took place during what is known in Canada as the Shawinigan Orogeny. This mountain building was followed by the intrusion of large granitic plutons that were emplaced, then metamorphosed during the Grenville Orogeny (a name often applied to both of these Mesoproterozoic orogenies). The Grenville Mtns. rose between 1,144-1,028 million yrs. ago. After millions of years of erosion what remains of them is called the Shenandoah Massif.

Much later in the evolution of the Blue Ridge core, 700 to 760 million years ago, Rodinia began to break apart. A number of granite plutons known as the Robertson River plutons intruded the core complex around that time. Most of these plutons lie between the older gneisses and the somewhat younger metamorphosed granitic intrusions associated with the Grenville Orogeny. This suggests that the Robertson River plutons may have formed close to the zone where Rodinia was beginning to split apart. They formed soon after the crust of Rodinia began to stretch. A short time later the continental crust began to break apart, huge quantities of lava (the Catoctin lavas) rose from deep in the crust or perhaps from the upper mantle. These lavas, now seen on the margins of the Blue Ridge, broke through to the surface and flowed out over much of the Northern Blue Ridge. They are well exposed at Catoctin Mtn. in Maryland (Ch. 6A). Some may still be found as far south as the James River.

The youngest rock unit found in the basement complex crops out along the Mechum River, for which it is named. These rocks are metasediments that are similar to those in the Lynchburg Group, named for the city near where they were found. They are exposed for long distances along the eastern margin of the core complex. Following the Grenville Orogeny, erosion gradually lowered the mountains over a period of several hundred million years. Eventually a land surface (an erosion surface) existed at low elevation, perhaps near sea level, and sediments of the Mechum River Fm. and the Lynchburg Group were deposited. Thus the Mechum River sediments are likely remnants of sediments that were deposited above an unconformity that covered much of the northern part of the core

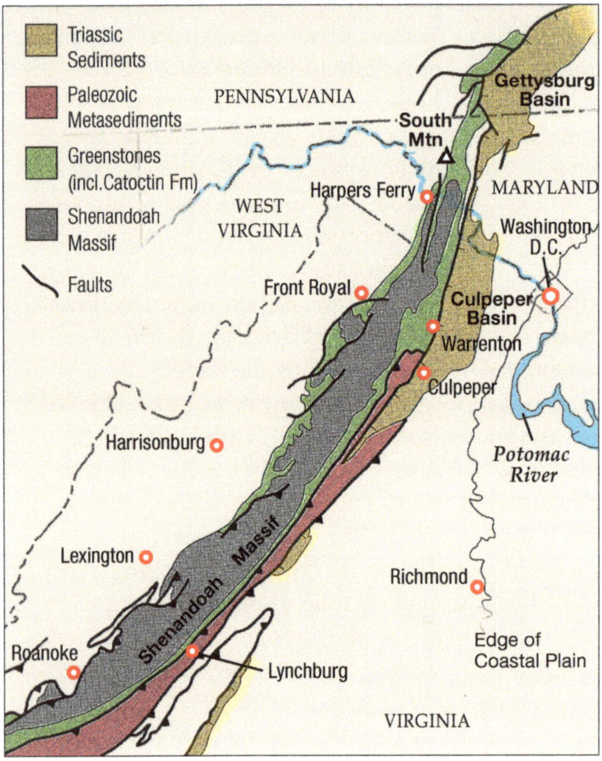

Fig. 5-9. The Shenandoah Massif (gray) is composed of igneous and metamorphic rock, most of which are middle Precambrian-age. They were part of Laurentia. They are covered by Catoctin and Chilhowee units (green). Zones of major thrust faults separate the Chilhowee units from Neoproterozoic and Paleozoic-age metamorphosed sediments and lavas (red) that were accreted to Laurentia during the Paleozoic Era. The Triassic-age (Mesozoic) sediments (gold) were deposited in rift basins following the last major episode of mountain building. They are separated from the other units by steeply inclined faults in which the east side is lower relative to the west side. (Modified after Rankin,1993)

complex. The metasediments we see today were preserved in a long, relatively narrow, down-dropped basin that runs northeast to southwest in the core complex.

The Eastern Margin
of the Northern Blue Ridge

Today, Catoctin lavas and lower parts of the Chilhowee Group are present on the eastern limb of the large complex anticlinal structure that is best developed from Catoctin Mtn. in Maryland to Warrenton, Va. Near Warrenton, the eastern limb of the anticline is cut out by a fault that forms the western boundary of the Culpeper Basin (Fig. 5-9), one of the rift basins created as the Atlantic Ocean began to form in the Triassic Period following the mountain building of the Alleghanian Orogeny (Table 5-2).

What happened at this time was similar to what had happened about several hundred million years earlier when Rodinia began to break up, but this time it was Pangaea, the supercontinent that had formed as the Appalachian Mtns. were uplifted, that began to split apart. Fault-bounded basins emerged as the continental crust pulled apart. Sediments from the older mountains filled in the basins and basaltic lavas came to the surface. Today we see these Triassic-age rift basins filled with sediment derived from the high mountains all the way from the Connecticut Valley to Florida where the rift basins have been covered by the younger sediments of the coastal plain. Lavas, Triassic-age counterparts of the Catoctin lavas that had formed close to 300 million years earlier, are now exposed in these rift basins named for towns that lie within the basins, such as Newark, Gettysburg, and Culpeper. One of these intrusions forms the famous Palisades along the Hudson River at New York City.

The Culpeper Basin (Fig. 5-9) forms the eastern margin of the Blue Ridge for a long distance to the south starting at Warrenton, Va. Rocks of the Chilhowee Group are not present along the edge of the core complex. Instead we find metamorphosed sedimentary rocks and greenstones thrust onto the edge of the Blue Ridge basement complex. They, as well as part of the basement complex, are cut into long strips by numerous thrust faults. Some geologists think these rocks formed in a rift system along the margin of Laurentia fairly close to the present edge of the core complex. Others think they formed much farther to the east and were separated from North America by the Iapetus Ocean until they were accreted onto the eastern margin of Laurentia in the orogenies that took place during the middle to late Paleozoic Era.
In any case, the rocks found on the eastern margin of the Blue Ridge consist of metamorphosed sediments, volcanic ash, and lava flows that were subsequently moved to the northwest where we find them today along the eastern side of the Blue Ridge basement complex (Fig. 5-9). The absence of rock units that are highly resistant to erosion is partially responsible for the low topography we see along the eastern margin of the Blue Ridge. This margin has also been reduced in level as the headwaters of the streams that flow into the Atlantic Ocean cut into the land to the west,leaving the high ridge of the Blue Ridge along its western margin.

Lynchburg Group

Near Lynchburg, Va. Hundreds, possibly thousands of feet of late Precambrian to early Cambrian-age rocks are present and overlie the Grenville and pre-Grenville Blue Ridge basement complex.

Major Rock Units - Eastern Margin of the Northern Blue Ridge	
Rock Units	**Description**
Triassic rift basin	In central Virginia a normal fault separates Triassic sediments from the Blue Ridge basement.
Evington Group (Candler Fm)	(A Cambrian-age phyllite and schists altered from clay and silt) is present south of central Virginia.
Chilhowee & Catoctin	Is present from Catoctin Mtn. to Warrenton, Va.
Lynchburg Group (Rockfish conglomerate)	Metasedimentary rock – silt, sand, greenstones & conglomerate.
~~ Major erosion surface – much of the record has been eroded away ~~~~~~~~~~	
Blue Ridge basement Complex	The Grenville-age basement complex labeled Shenandoah Massif on Fig. 5-9 lies in the central and western part of the Blue Ridge Province. Igneous and metamorphic rocks (red on Fig. 5-9) exposed along the eastern side of the Blue Ridge are late Precambrian to Paleozoic-age in age and have been thrust onto the Shenandoah Massif during Paleozoic mountain building.

Table 5-2: Major Rock Units along the eastern margin of the Northern Blue Ridge.

Fig. 5-10. Rockfish conglomerate at Rockfish, Va., (above) containing rounded fragments of the Lovingston Fm. is located near outcrops of the Lovingston Fm. (below) which is part of the Precambrian-age basement complex. Fragments of the basement in the Rockfish conglomerate prove that the conglomerate is younger than the basement, and that the basement was exposed nearby at the time the conglomerate was being formed.

The contact between these rock units is exposed here. These rocks, known as the Lynchburg Group, were originally dark shales, sandstones and conglomerates. Some were originally volcanic ash and basaltic intrusions. The conglomerates and sandstones that are prominent near the base of the Lynchburg Group are known as the Rockfish conglomerate (Fig. 5-10). It is probably the oldest rock of clearly sedimentary origin found on top of the crystalline core complex on the eastern side of the Blue Ridge core. Some parts of the Lynchburg group have been metamorphosed into phyllite, mica schist, amphibolite, and gneiss, but most are meta-conglomerate and quartzite. Originally the amphibolites were probably volcanic ash or basaltic rock.

The Lynchburg Group is thought to be equivalent to rock units known as the Ashe and Alligator Back Fms. farther south in Virginia and North Carolina. The Ashe Fm. is composed of fine-grained, thinly layered sulfidic biotite-muscovite gneiss, mica schist, phyllite, gneissic conglomerate containing granule to pebble-size quartz and feldspar, amphibolite, and garnet amphibolite. Layered marble and impure marble is rarely present, and pyrite-chalcopyrite-calcite veins are common near the top of the Ashe Fm. Radiometric ages for the Ashe Fm. are in the range of 550-800 Ma). These are described in more detail in Chs. 7 and 8. The Lynchburg Group is overlain by a Cambrian-age phyllites and schists known as the Candler Fm. It occurs in the fault slices found along the eastern margin of the Blue Ridge.

Evington Group

The rock units above the Catoctin Fm. near Lynchburg are composed of layers of sedimentary rocks altered by metamorphism to form schist and marble. Although they are totally different in composition, these units, called the Evington Group, are the same age as the sandstones, shales, and conglomerates of the Chilhowee Group found on the western side of the Blue Ridge. One member of the Evington Group, the Candler Fm., composed of phyllite and schist, found along the eastern Blue Ridge margin has been traced to much

The Eastern Limb of the Blue Ridge Anticlinorium

The contact between the crystalline core of the Blue Ridge and the metasedimentary rocks that make up the eastern margin of the Blue Ridge is located about a mile northwest of Lynchburg near the small community of Reusens. Near Reusens the basement consists of an unusual mixture of gneiss and injections of granitic veins. A metamorphosed conglomerate known as the Rockfish conglomerate lies over the basement. It is probably the oldest rock of clearly sedimentary origin found on the crystalline complex on the eastern side of the Blue Ridge Province. It is the lower part of the Lynchburg Fm., a rock unit that also includes biotite-quartz gneiss, quartz-mica schist, and amphibolite layers that may have been volcanic or intrusions of basaltic rock that were later altered to amphibolite. These rocks are exposed along the river in Lynchburg which is the type locality for the formation. Layers of greenstone, probably equivalent to the Catoctin Fm. found in the northern part of the Blue Ridge are present above the Lynchburg Fm. The Catoctin lavas have been dated as 670 Ma. The next units above are composed of layers of sedimentary rocks altered by metamorphism to form schist and marble. Although they are totally different in composition, these units, called the Evington Group, appear to be equivalent in age to the sandstones, shales, and conglomerates of the Chilhowee Group found on the western side of the Blue Ridge.

The rocks along the southeastern edge of the Blue Ridge are highly deformed and much more altered by metamorphism than are their equivalents on the northwestern edge. This is true all along the eastern edge of the Blue Ridge. In Virginia, most of the faults present along this edge are southeast-dipping thrust faults. See a description of this zone in the Philpott Reservoir area (Ch. 8C). Farther south, the thrust faults have a strike-slip component (Fig. 3-2). The Brevard fault is the best known of these (Fig. 2-12). In these cases, the east side of the fault moved southwest parallel to the fault as well as up toward the northwest.

farther south (Fig. 8C-17).

The rocks along the southeastern edge of the Blue Ridge are highly deformed and much more altered by metamorphism than are their equivalents, the Chilhowee Group, on the northwestern edge. This is true all along the eastern edge of the Blue Ridge. In Virginia, most of the faults present along this edge are southeast-dipping thrust faults indicating the rocks were moved from southeast to the northwest. From central Virginia to the southwest, major thrusts faults such as the Brevard fault, define the eastern margin of the Blue Ridge (Fig. 2-12). Although these are thrust faults that have moved the southeastern side to the northwest (Fig. 7-10), they also have a strike-slip component of movement (Fig. 3-2). The eastern side of these faults has moved obliquely (in part toward the southwest parallel to the fault as well as up and toward the northwest). These faults formed as the Eurasian plate collided with the Laurentian continent (North America) giving rise to the Appalachian Mtns. of today.

Ecological Setting
of the Northern Blue Ridge

Localized climatic factors, microclimates, and elevation as well as the character of the soil and underlying bedrock all affect plant growth and hence shape the forests of the Northern Blue Ridge. Most bedrock in the Blue Ridge falls into three major categories: granitic rocks of the Precambrian basement, metabasaltic rocks of the Catoctin Fm., and metasedimentary rocks that form the Chilhowee Group. Each weathers to form distinctive soil types. Gathright (1976) captures this complexity in his description of the soils in Shenandoah National Park: "Fertile and locally deep but stony soils have formed on the more soluble plutonic and volcanic rocks, but only very thin sandy or shaly soils are present on the metasedimentary rocks. The difference in these soils is directly expressed in the size and species of trees growing on them…The difference is most noticeable in October when the fall colors are at their peak…Then, the entire forest area underlain by the metasedimentary rocks takes on a drab yellow-brown hue while the colors in the forest on the volcanic and plutonic rocks are bright reds, oranges, and yellows with many trees still partly or wholly green. The sharp color contrast probably reflects differences in soil moisture content as much or more than species differences between trees in the two rock types."

This spring-fed pond is located on the Blue Ridge Upland Plateau near one of the oldest cabins found along the Parkway.

Soils

Soils are generally produced by chemical weathering of underlying rock (when they are not the transported deposits of streams or downslope movements). Water plays a central role in these processes. It dissolves the components of the bedrock and soil, moving dissolved material to greater depth in the soil and into the circulation of groundwater. It also enters into chemical reactions with minerals in the bedrock. For these reasons soil chemistry shows a strong correlation between acidity—the quantities of base cations (ions, such as calcium and magnesium, that are positively charged) present in soil moisture—and the underlying bedrock. In turn, these characteristics of soil chemistry help determine which plants prosper in the soil.

In general, soils formed from metasedimentary rocks and granitic rocks rich in silica have low pH and are not fertile. As the amount of clay and feldspar in the soil increases, the soil pH rises and the soil becomes more fertile. In contrast, soils formed from rocks rich in magnesium and iron, such as greenstones and gneisses containing pyroxene, tend to be deep, well drained, and relatively fertile. Local vegetation, in turn, influences soil chemistry. For example, where spruce and fir dominate coniferous forests in the Blue Ridge Highlands, the soil is a combination of organic matter derived from conifers and sandy soils formed from the weathering of the granitic gneisses and sandstones. Most of these high areas also have heavy rainfall that causes leaching of the soil. All these factors produce an acidic and infertile soil.

Oak-hickory forest

Ecological Communities

The classification of ecological communities by the The **Natural Heritage Program** of the Virginia Department of Conservation and Recreation recognizes five major types of deciduous tree forests in the Northern Blue Ridge. This classification is similar to the classical descriptions described in Ch. 4 but contains additional subdivisions. Basic oak-hickory, oak/heath, Southern Appalachian northern hardwood, montane oak-hickory forests, and northern red oak are the five main categories referred to as **matrix forests** in the Natural Heritage classification. Each of these covers large areas and contains specialized patches of vegetation referred to as “large patch” and “small patch” communities.

Matrix Forests

The forests throughout the lower elevations of the Blue Ridge underwent drastic changes toward the end of the 1800s when blight killed most of the large chestnut trees that along with oaks had covered most of these areas. Today only a few saplings and stunted, older chestnuts survive. The larger chestnut trees have been replaced by oaks, hickories, and tulip poplars. These **basic oak-hickory forests** form at low elevation on fertile soils that are alkaline. This habitat has a high diversity of species that includes redbuds and hop hornbeam in the understory, along with spring rue-anemone, star chickweed, cut-leaved toothwort, spring beauty and several varieties of grasses.

At elevations above 2,500 ft. where the bedrock is composed of metasedimentary rocks that produce a dry, infertile, acidic soil **oak-heath forests** form. These contain an overstory of oaks underlain by a dense layer of heath-family plants such as black huckleberry, mountain laurel, pink ladyslipper, and dwarf iris. These are the main survivors in a habitat that supports a low diversity of species. This environment is subject to forest fires and gypsy moths and is a difficult place for even invasive plants to gain a foothold.

Southern Appalachian northern hardwood forests can be found at elevations up to 3,600 ft. Their canopies are dominated by sugar maple, American beech, yellow birch, and yellow buckeye, with maples and mountain holly in the understory.

At higher elevations, **montane oak-hickory forests** develop where the bedrock is composed of greenstone and basement gneisses that produce alkaline soils. The resulting forests contain white oak, white ash, and striped maple trees, with white snakeroot, horse-balm, leatherleaf meadowrue, and black bug bane growing near the ground.

Red oak forests share these higher elevations and tend to thrive in parts of the mountains where winter temperatures are very cold, where winds are high, and where the soil is less fertile. Mountain holly, azaleas, fly poison, and hayscented fern are abundant in the understory here.

Large Patch Communities

Boulderfields, pine-oak-heath forests, and coves are the largest of the patch communities in the Northern Blue Ridge.

River running through a cove forest on Mt. Rogers. Coves tend to have moist, fertile, and often acidic soil. They are usually sheltered from winds and sun. Most cove forests have a great diversity of species of large trees and many shrubs in the understory.

Boulderfields form where rock fragments and blocks break off of cliffs or bedrock exposures above steep slopes. The fragments accumulate as talus that takes the form of sheets or cones that form as the loose fragments move downslope. These surfaces have very little soil and are thus difficult places for most plants to grow, especially above 3,000 ft. Here lichen mats may be the main survivors. Conditions are more favorable at lower elevations where plant communities of Virginia creeper, birch trees, and lichens may be present. Which plant populations develop depends to some extent on the composition of the rocks, making it possible to distinguish acidic boulderfields from basic boulderfields on the basis of the diversity of the plant population.

Pine-oak-heath forests contain a mixture of pitch and table-mountain pines along with bear-oaks and stunted chestnuts. This forest type grows along ridges and steep southwestern slopes where metasedimentary rocks weather to form a dry, acidic, infertile soil. This combination makes them vulnerable to forest fires which are needed by the pines and bear-oaks for regeneration.

Coves are known for their great diversity of plants. Forests that form in coves and deep valleys that are moist and protected from long hours of sunlight are ideal for many plants. Those that form on bedrock that is rich in iron and magnesium, notably greenstones and gneisses containing pyroxene minerals, are classed as **rich cove forests.** A great variety of trees grow here, including maples, basswood, tulip-poplar, white ash, birches, hickories, and red oaks. Many wildflowers including trillium and flowering forbs are part of the lush herb layer found here.

The lake at Peaks of Otter has been enlarged by the construction of a dam, but originally it was a natural pond.

Basic mesic forests can be found at lower elevations where conditions are somewhat similar to those in rich coves. They are distinguished from other forest types by the presence of paw-paw trees, twinleaf, and toadshade trillium.

Small Patch Communities

Highly specialized environmental conditions of limited extent are responsible for the development of small patch communities. Basic woodlands, high and low elevation basic outcrop barrens, as well as wetlands and seepage swamps (as described by Fleming, 2003), show the variable character of these communities.

Basic woodlands occur on dry, rocky slopes. Despite this setting, the vegetation here includes a large number of species. The soil, containing high levels of calcium, magnesium, and manganese, is largely responsible for this diversity. Usually it is derived from the weathering of greenstone, but occasionally it is found on pyroxene-bearing gneisses, phyllite, and metasiltstone. Stunted white ash and pignut hickory can be found in open stands. Drought-tolerant grasses and sedges are also present.

Low elevation basic outcrop barrens, a community closely related to the basic woodlands, occurs on dry and sunny, south to west-facing slopes. These resemble the shale barrens found in the Valley and Ridge Province. They are present on slopes underlain by siltstone and phyllites that are common in the lower parts of the Chilhowee Group (Keener, 1970). These soils support scattered shrubs, stunted trees, moss mats, and patches of herbs that grow on ledges with little soil. Red cedars are a characteristic tree in this community. In basic outcrop barrens at high elevations (that are subject to severe winter temperatures, ice, and strong winds), the soil is a thin veneer of gravel, silt, and organic matter. Only lichens, sparse shrubs, and hardy herbs are present here.

Wetlands, including swamps formed where water seeps out of surrounding fractured rocks, occur in some valleys and on a few high level areas of low relief. Water seeps to the surface along cliffs and near the base of sheets of talus. Where seeps occur at high elevation plants must be resistant to the cold winter air. These high altitude swamps often contain hummocks that are covered by bryophyte mats composed of mosses, liverwort, and hornworts. American water-carpet and whorled wood aster are among the common herbs in these swamps.

More on the ecological communities of the Northern Blue Ridge can be found online at **http://www.dcr.virginia.gov/**. Search on "community ecology." ❖

Fig. 6-1. Map of the Northern Blue Ridge showing the Skyline Drive and the northern section of the Blue Ridge Parkway. (The National Map, USGS, with additions)

Chapter 6

Places of Special Interest: Northern Blue Ridge

From Catoctin Mountain to Roanoke

The northern end of the Blue Ridge rises from an elevation of about 200 ft. a few miles southwest of Harrisburg, Pa. to a peak of about 1,900 ft. at Catoctin Mtn. in Maryland. (Fig. 6-2). Several parks are located on the northwestern part of the Blue Ridge in Maryland where the Appalachian Trail passes through Greenbrier and Washington Monument state parks and Harpers Ferry National Historical Park. Catoctin Mtn. National Park, and Cunningham Falls and Gambrill state parks lie on the eastern side of the Blue Ridge (Fig. 6-1). All of these parks provide a refuge for plants and animals that are threatened by development in surrounding areas. Because the environment is much less disturbed in these parks, they are good places to hike, study the ecology and geology of the region, and to view plants and wildlife.

In the following sections, we will examine Catoctin Mtn. National Park and Harpers Ferry in more detail. Both are excellent places to study the northern part of the Blue Ridge. Catoctin Mtn. is situated near the crest of the Blue Ridge and Harpers Ferry is located where the Potomac River cuts across it.

For more information:
Look for details about the state parks of Pennsylvania and Maryland on their respective state park websites.

For geological field localities in Pennsylvania see "Geology of the South Mountain Area, Pennsylvania" by W.D. Sevon and Noel Potter, Jr. which is available from the Field Conference of Pennsylvania Geologists, Inc., P.O. Box 1124, Harrisburg, Pa. 17108.

Tourist Sites: The Civil War battlefield at Gettysburg, Pa. is located a short distance east of the parks near the northern end of the Blue Ridge. ❖

Places of Special Interest: PAGE

6A. Catoctin Mtn. National Park & Cunningham Falls State Park 93

6B. Gambrill State Park 99

6C. Washington Monument State Park...... 102

6D. Harpers Ferry & The Potomac Water Gap 104

6E. Shenandoah National Park 113

6F. Blue Ridge Parkway – Waynesboro to the James River......... 142

6G. James River Traverse across the Blue Ridge 154

6H. Blue Ridge Parkway – James River to Roanoke 174

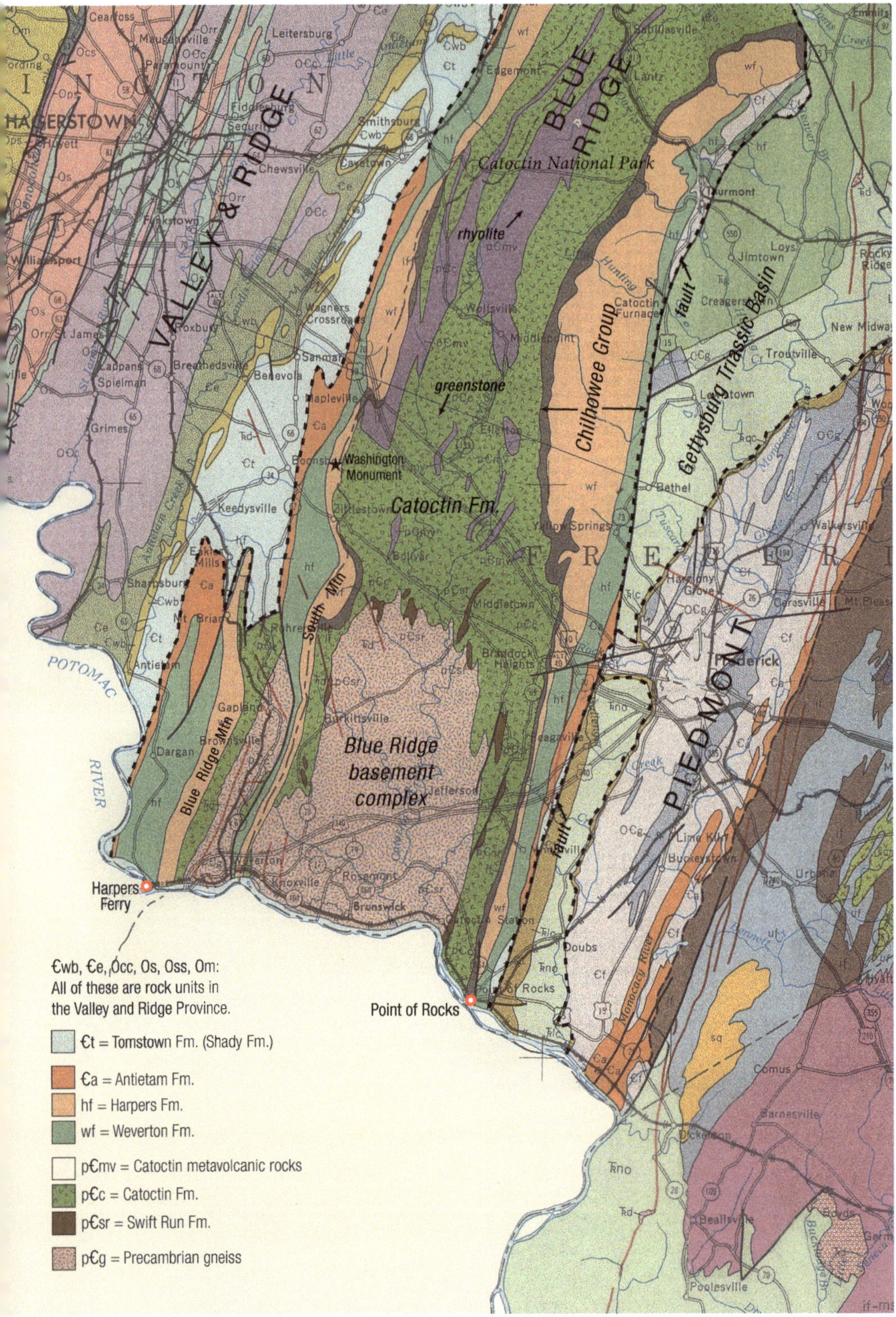

Fig. 6-2. Geologic map of the Blue Ridge in Maryland. (Md. Geological Survey, Baltimore, Md., 1968)

Chapter 6A

Catoctin Mountain National Park & Cunningham Falls State Park

History

The name "Catoctin" came from an American Indian tribe, the Kittoctons, who lived near the Potomac River and hunted along South Mtn. Early European settlers, most of whom entered the country from ports near Philadelphia, arrived in the region in 1732. Some of them became loggers or made charcoal for the Catoctin Iron Furnace, which produced pig iron from ores that came from the valley east of the furnace. The furnace was located in what is now Cunningham Falls State Park near wood that was needed for fuel. Other settlers established farms in the high valleys, or supplied oak and chestnut bark (rich sources of tannin), to tanneries established in the Monocacy Valley. Catoctin Mtn. is also the site of the presidential retreat established by President Eisenhower that he named "Camp David," after his son.

Above: View of the cascades located at Cunningham Falls State Park.

Location: SR 77 in Maryland, which connects Thurmont with Smithsburg and Hagerstown, passes along the border between Catoctin Mountain National Park and Cunningham Falls State Park. Many of the points of interest in Catoctin Park are along or near Park Central Road (Figs. 6A-1, -5).

Accommodations: Campgrounds including group camping, cabins, and picnic areas are located in both parks and along Route 77 close to Thurmont. Groceries and fuel must be purchased outside the parks. Motel accommodations are available at Thurmont.

Trails: Maps of trails in Catoctin Mtn. Natl. Park are available at the park entrance on SR 77.

Outstanding Features: The parks are excellent places to see the Catoctin Fm., obtain views over the Triassic lowland east of the park, and study birds and plants along the trails.

Climate: Precipitation is evenly spread over the four seasons with averages of 44 in. per year. Snowfall, which varies considerably from year to year, averages about 35 in. Summer temperature average about 80°F; winters average 30°F. Fall and spring are mild seasons.

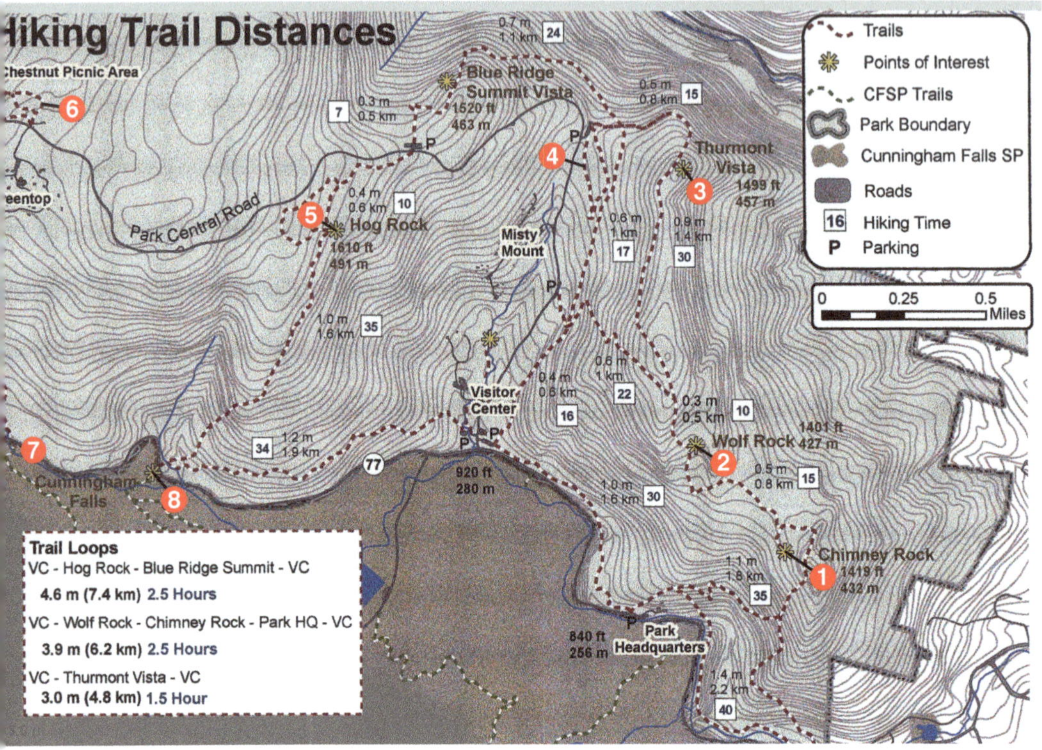

Fig. 6A-1. Points of Interest (pp. 97-98) in Catoctin Mtn. National Park, Md. (NPS, with additions)

Ecological Setting

Most of the parklands belong to forest, field, or wetland ecosystems (Ch. 4). Mixed hardwood forests that contain chestnut oak, hickory, and black birch cover most of the northern part of the Blue Ridge (Fig. 6A-2). White oak, hemlock, ash, birch, poplars, beech, and red oak grow in moist areas. Most of the forests located on Catoctin Mtn. near the old furnace are covered with second growth forests that formed following the extensive logging that took place in the 18th and 19th centuries. Much of the wood was used to make charcoal needed in the nearby iron furnaces.

Soil formed by weathering of the Catoctin lava flows covers large parts of the forest floor. Some areas were cleared of forest and cultivated during early settlement. Few of these farms remain intact. They were abandoned because the soils were not very productive and the steep slopes made the area difficult to cultivate. Most the old fields are now grown up with

Fig. 6A-2. Northern hardwood forests have developed on ancient Catoctin lava flows.

black locust, wild cherry, sassafras, and yellow poplar, but a few were set aside in 1933 to be used to demonstrate the rehabilitation of "sub-marginal" farmland.

Interesting wetlands are present along several of the creeks in the parks. Skunk cabbage (Fig. 6A-3) prospers near Spicebush Trail, near Misty Mount Cabins, and west of the Chestnut Picnic Area. They are also excellent places to look for wildflowers and other plants in the spring and summer.

Fig. 6A-3. Skunk-cabbage prospers in a swampy area near Chestnut Picnic Area.

Of the animals present in the park, you are most likely to see one or more of the four types of squirrels that inhabit the forest. They are fox, grey, red, and flying squirrels. The flying squirrels actually glide rather than fly. Many of the streams that flow down the steep slopes on the flanks of South Mtn. support brown and brook trout.

Birds on Catoctin Mtn.

Catoctin Mtn. is an excellent place for birding. 170 bird species have been seen in the park. Birds most likely to be seen are the American kestrel, Cooper's hawk, red-tailed hawk, sharp-shinned hawk, Carolina chickadee, tufted titmouse, wild turkey, mourning dove, downy and pileated woodpecker, northern flicker, yellow-bellied sapsucker, white-breasted nuthatch, barred owl, Carolina wren, eastern wood-pewee, brown-headed cowbird, common grackle, European starling, American crow, blue jay, gray catbird, brown thrasher (Fig. 6A-4), northern mockingbird, northern cardinal, rufous-sided towhee, American goldfinch, dark-eyed junco, purple finch, American robin, and the eastern bluebird..

Fig. 6A-4. Brown thrasher

Many of these birds are described in Ch. 11. The NPS provides a bird checklist for this area on its website at **nps.gov/cato/naturalscience/index.html**. This list includes migratory species and residents during the winter months.

Geologic Setting

Catoctin Mtn. is located near the crest and on the eastern limb of the Blue Ridge anticline. (Figs. 5-1, 6-1). Starting at Thurmont, Md., you will find yourself situated in one of the large Triassic basins formed when North America began to rift apart from Africa. At this time the initial stages in the formation of the Atlantic Ocean were taking place (Fig. 5-6). Major, steeply inclined faults developed as the basin developed. Sediments that filled this basin lie beneath Thurmont and extend northeast and southwest along the boundary between the Blue Ridge and the basin. They are reddish or brown sandstones that you may see in roadcuts and new excavations. Heading west on US 77, you will drive across the fault.

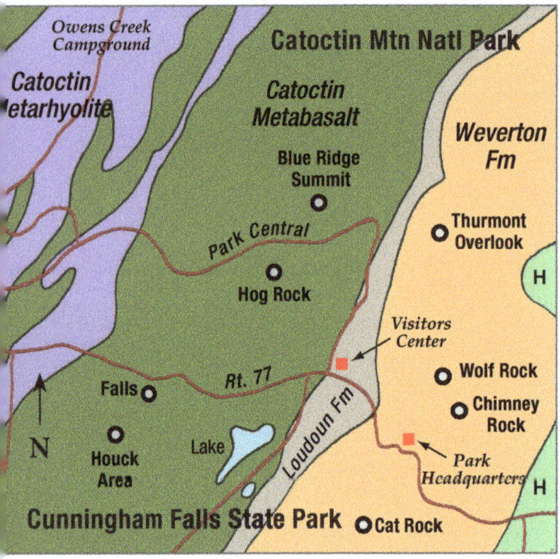

Fig. 6A-5. Catoctin Mtn. N.P. and Cunningham Falls S.P. (After Fauth, 1977, and Gene Slatick)

Unfortunately, this fault and most of the rocks close to it in the basin are buried under gravel and broken rocks eroded from Catoctin Mtn. and soils that developed on the sediments in the basin during the last few million years. After you cross the fault, you will drive by east-dipping (inclined) layers of the Weverton Fm. (p. 80). The quartzites in the Weverton Fm. are exposed as ledges along the road. They are very resistant to erosion, even in this humid climate. The Weverton, Loudoun, and Catoctin fms., are exposed in the park. They were formed toward the end of the Precambrian Era and in the earliest parts of the Cambrian Period. The lava flows of the Catoctin have an aggregate thickness of more than a thousand feet in some places. The flows once covered a vast area and are still found on both sides of the Blue Ridge anticline and as far south as the James River. Most of the flows have the composition of basalt, similar to those found today in Hawaii. Other flows have the composition of rhyolite. This indi-

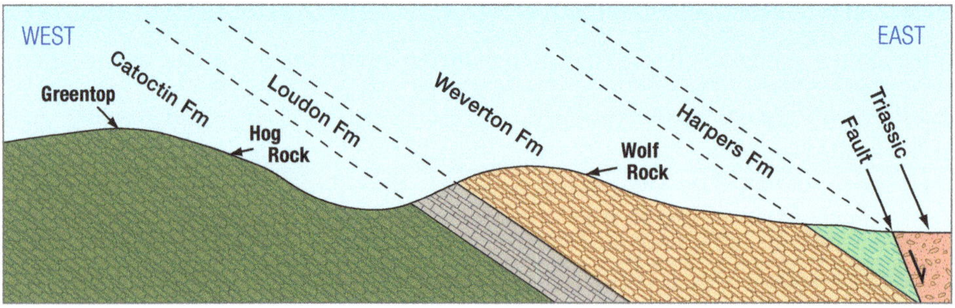

Fig. 6A-6. This schematic cross-section across Catoctin Mtn. Natl. Park is oriented from west to east and extends farther to the east than the map above in Fig. 6A-5. At the eastern end of the cross-section a steeply inclined fault cuts through the Harpers Fm. and brings the Harpers Fm. into contact with Triassic-age sedimentary rocks. (After Fauth, 1977, and Gene Slatick)

Fig. 6A-7. Japanese stiltgrass (left). Stiltgrass forming a groundcover (right).

Fig. 6A-8. Spicebush blossoms in spring at the Chestnut Picnic Area.

cates that the molten rock (magma) from which they formed was derived in part from continental crust, which contains much more silica than the lavas that rose from deeper in the crust.

All of the formations found along US 77 are also exposed in the park, a much safer place to view them. Take the road to the right at the park entrance and proceed along Park Central Rd. Most of the following points of interest may be reached from this road.

Points of Interest

in Catoctin Mtn. Natl. Park and Cunningham Falls St. Pk.

Several of the trails in Catoctin Mtn. National Park (Spicebush Nature Trail at Chestnut Picnic Ground, Charcoal Trail, Hog Rock Nature Trail, and Falls Nature Trail) have signs explaining natural features. See Figs. 6A-1, -5 for site locations.

Stops 1 & 2: **Chimney Rock and Wolf Rock:** These Stops are located along a trail that ascends about 500 ft. from the visitor center. Both sites are outcrops of the quartzite that makes up the Weverton Fm. (p. 80).

Stop 3: **Thurmont Vista** is located farther along the trail. This site provides excellent views of the relatively flat and narrow strip of land that connects the Gettysburg and Culpeper Triassic basins located east of Catoctin Mtn. Farther east you can see the rolling landscape of the Piedmont, which is underlain by lower Paleozoic rocks similar in age to those found in the Valley and Ridge Province.

Stop 4: **Charcoal Trail** departs from a parking lot about a mile north of the visitor center and ties into the Chimney Rock Trail. The Loudoun Fm. is exposed near the parking lot.

Stop 5: **Hog Rock Trail** passes over outcrops of the greenstone (metamorphosed basalt) which is the most distinctive feature of the Catoctin Fm.

Stop 6: **Spicebush Nature Trail** is an excellent place to see the extent to which the invasive stiltgrass (Fig. 6A-7) has displaced natural ground cover. Spicebushes

Fig. 6A-9. Two types of Catoctin lava. The darker one is a porphyritic rhyolite (p. 315).

(Fig. 6A-8) along the trail display beautiful blooms in April. Two varieties of the rhyolite lavas found in the Catoctin Fm. are present along the trail, and outcrops occur off the trail. One of these lavas has a porphyritic texture (Fig. 6A-9). Note the larger crystals of feldspar that formed slowly when the magma was buried deep in the Earth. They are surrounded by a fine-grained, silica-rich matrix that cooled and crystallized quickly when the magma reached the surface. The other rhyolite, a light colored and uniformly fine-grained lava, exhibits patterns formed by flowage of the molten lava.

Stop 7: **Outcrops along US 77:** Excellent exposures of the Catoctin greenstones occur along US 77 but it is difficult to park and traffic is often heavy.

Stop 8: **Cunningham Falls State Park:** This park is named for a waterfall that drops about 70 ft. where it crosses Catoctin greenstone (Fig. 6A-10). The falls are located a short distance off of US 77. Parking is available for handicapped in a small parking lot along the highway. Others can reach the falls by following a trail through the park. The entrance to the park is located on the north side of Catoctin Hollow Road and on the western side of Hunting Creek Lake (Fig. 6A-11). Swimming and boating facilities are located along the lakeshore. ❖

Fig. 6A-10. Cunningham Falls located adjacent to US 77 in Cunningham Falls State Park.

Fig. 6A-11. Artificial beach in the recreation area at Hunting Creek Lake in Cunningham Falls State Park.

Chapter 6B

Gambrill State Park

Geologic Setting

The introduction to the Northern Blue Ridge (Ch. 5) contains general information about the topography and geologic setting of this region. Gambrill State Park is located on Catoctin Mtn., which is the eastern limb of the regional scale anticline that is prominent at the northern end of the Blue Ridge (Fig. 5-1). The rocks exposed in Gambrill State Park (Fig. 6B-1) include the metamorphosed basaltic lava flows of the Catoctin Fm. They were extruded into rift basins as Rodinia began to break apart late the Precambrian and the Iapetus Ocean began to form. As most of the volcanism ceased mud, gravel, and sandy sediments that now appear as dark gray conglomerate and black, tuffaceous (containing ash) phyllite, known as the Loudoun Fm. covered the lava flows. Streams carrying coarse sand and pebbles now known as the Weverton Fm. then deposited sediments that covered the region. The Weverton Fm. is subdivided into three members. The Owens Creek Member is a medium-to-dark gray, medium-bedded, pebbly, ferruginous (iron bearing) conglomerate, and conglomeratic sandstone altered to quartzite. The Maryland Heights Member is a dark-greenish gray to medium gray metagraywacke, quartzite, and phyllitic siltstone, silty sandstone with pebbly conglomerate layers. The Buzzard Knob Member is a light gray to medium gray quartzite with dark gray argillaceous (clay rich) layers separating quartzite beds.

Location and Access: Gambrill State Park is located in Maryland on Catoctin Mtn. northwest of Frederick, Md. and northeast of US 40. If driving from Frederick, take US 40 (not US 40-Alt). Small rural roads connect Gambrill State Park and Catoctin Mtn. Natl. Park. If traveling from Gambrill to Catoctin Mtn., take Gambrill Park Road, Mink Farm Road, Catoctin Hollow Road, and US 77 (east) to the entrance to Catoctin Mtn. Natl. Park (Fig. 6A-1).

Accommodations: Campgrounds, cabins, shelters, and trailer parking are available. Bring food supplies.

Trails: A number of trails are marked through the park and interpretative programs are offered during summer months.

Outstanding Features: The overlooks at High Knob provide excellent views of the valley, the core of the Blue Ridge between Catoctin Mtn. and South Mtn., and the southern continuation of Catoctin Mtn. Outcrops of the Weverton and Loudoun formations are located near the road.

Above: View to the southwest from Gambrill State Park, with the Blue Ridge in the distance.

As you drive into the park from Phelps (lower left on geologic map (Fig. 6B-1) you will drive across the Catoctin greenstones. They are exposed in outcrops along the road at Phelps. Most of the road passes across the Loudoun Fm. The top of High Knob where the tower is located is underlain by the Buzzard Knob quartzite. The fault that lies along the western edge of the Culpeper Triassic basin crosses the map along its eastern margin. The area to the right of the fault dropped down as Pangaea began to split apart in the early part of the Mesozoic Era about 200 million yrs. ago. Africa was then beginning to drift away from North America, an event that would lead to the development of the Atlantic Ocean.

Note on the geological map (Fig. 6B-1), that the traces of the contacts between the Weverton Fm. members have a v-shaped pattern. The v's point to the east and formed where streams cut valleys down the eastern slope of the park. You see this pattern at many places in the Blue Ridge. The v-shaped pattern indicates the direction of downward inclination of the layers (called dip). Also note that the Buzzard Knob quartzite is nearly horizontal where it crops out at the top of the ridge where the tower is located. The rock units are exposed along the trails and at the parking area on High Knob.

Points of Interest

From the town of Cherry Hill, Md., the road leading into the park crosses low slopes covered by colluvium (a mixture of stream deposits, soil, and material moving downslope). The road crosses the major normal fault that separates the Blue Ridge Province from the western edge of the Culpeper and Gettysburg Triassic basins. Farther up the slope the road crosses the Weverton and Loudoun formations. (Fig. 6B-1). This section is similar to the one between Thurmont, Md. and Catoctin Mtn. NP. The Buzzard Knob member of the Weverton is exposed at the

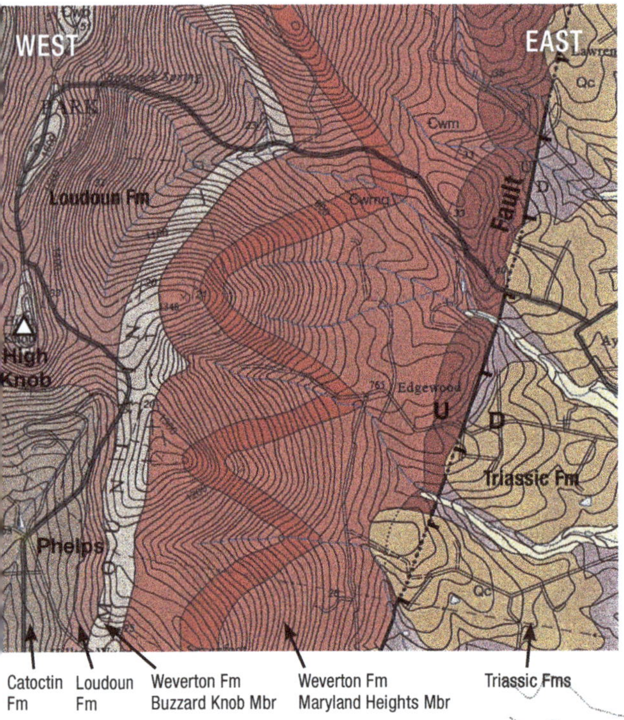

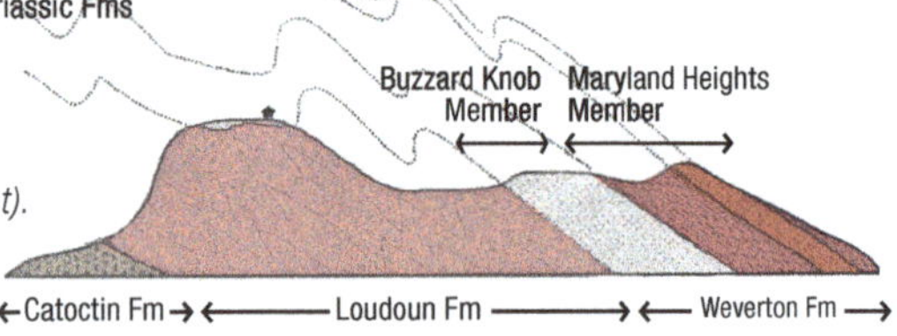

Fig. 6B-1. Geologic sketch map (above) and cross-section of part of Gambrill State Park (right). The section passes through High Knob and ends at the normal fault. (After Brezinski, 2004)

Fig. 6B-2a. Pegmatite and quartz veins in Loudoun Fm.

Fig. 6B-2b. Phyllite with quartz veins in the Loudoun Fm.

Fig. 6B-2c. Kinks in the Loudoun phyllites.

High Knob parking area. It contains quartzite and pebble conglomerates. Downslope, at the viewpoint, phyllites and quartz veins, parts of the Loudoun Fm., crop out (Figs. 6B-2a-b). Note the small-scale kinks in the phyllites (Fig. 6B-2c).

Looking south and east from the overlook at High Knob, Catoctin Mtn. (Fig. 6B-3a) which was formed by lava flows, continues to the south on the eastern limb of the Blue Ridge. Precambrian-age metamorphic rocks underlie the valley southwest of Gambrill State Park. Because these rocks contain a large amount of the mineral feldspar, they are more susceptible to weathering in the humid climate of this region than is the quartz-rich Weverton Fm. Erosion of these ancient rocks has resulted in the formation of this broad valley (Fig. 6B-3b). On a clear day, the western

Fig. 6B-3a. View from High Knob Overlook to the south. To the left is a Triassic basin. In the center is Catoctin Mtn., a ridge held up by Catoctin lavas on the eastern limb of the Blue Ridge anticlinorium.

Fig. 6B-3b. View from High Knob Overlook to the southwest. Catoctin Mtn. is on the left, the Blue Ridge basement is in the center, and the western limb of the Blue Ridge anticlinorium is on the right.

limb of the Blue Ridge anticlinorium is visible, and you may be able to make out two ridges. Both are part of the western limb of the Blue Ridge Geological Province. The more distant one is named Blue Ridge Mtn. and the closer one is called South Mtn. The Weverton quartzites hold up both ridges. They are separated by a fault that is responsible for the duplication of the rock layers. This will be more obvious at Harpers Ferry (Fig. 6D-4). ❖

Chapter 6C

Washington Monument State Park

Location: This park is located on South Mtn. in Maryland, a few miles northeast of US 40-Alternate and southeast of Boonsboro, Md. Although South Mtn. is part of the much larger area known as the Blue Ridge Geological Province, South Mtn. and Blue Ridge Mtn. are locally distinguished as separate ridges (Fig. 6-1).

Accommodations: Picnic areas, shelters, and a playground are available.

Outstanding Features: In 1827, the citizens of Boonsboro built a stone tower to honor George Washington. The tower (Fig. 6C-1) is a superb place to observe birds and is the site of annual bird counts of migrating hawks and eagles.

Above: Fig. 6C-1. The Washington Monument, one of the outstanding viewpoints for spotting migrating raptors.

Geologic Setting

Looking west from the Washington Monument, you see the Valley and Ridge Province (Fig. 6C-2). South Mtn. continues to the southwest as far as the Potomac River (Fig. 6-1), but ends a short distance south of the river. The northern end of the ridge, named Blue Ridge Mtn., is visible to the southwest and is slightly offset to the west from South Mtn. Both are held up by the Weverton Fm., (Fig. 6C-3) which is repeated near Harpers Ferry as a result of faulting.

Quartzites of the Weverton Fm. are the most resistant rocks in the Chilhowee Group. They are exposed in the outcrops you walk by along the trail from the parking lot to the tower. Fragments dislodged from outcrops of the Weverton at the top of the ridge form a talus slope on the western side of the monument (Fig. 6C-4). The Catoctin Fm. underlies most of the area to the east of the tower. The view to the southeast from the tower looks across the low topography that characterizes this part of the central core of the Blue Ridge. The rocks in this core

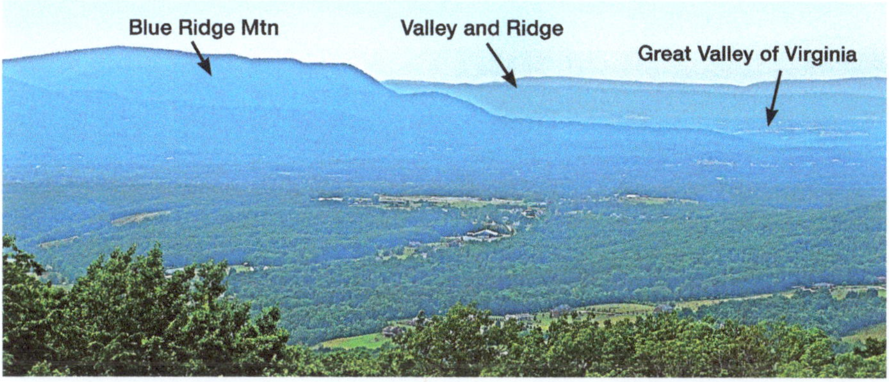

Fig. 6C-2: View toward the west of Blue Ridge Mtn. Past the mountain and beyond the Great Valley of Virginia, the Valley and Ridge Province can be seen in the far distance.

Fig. 6C-3. The Weverton quartzite showing cross-bedding that formed in a stream.

Fig. 6C-4. Talus slope located downslope from the monument.

are Precambrian-age igneous and metamorphic rocks. On the eastern horizon you can see the eastern limb of the Blue Ridge anticlinorium, which is held up by Catoctin greenstones.

Migrating Raptors

In recent years many raptors have been seen at the park. Broad-winged hawks, red-tailed hawks, osprey, and sharp-skinned hawks have topped the list, with a lower number of bald eagles, northern harriers, Cooper's hawks, and American kestrels. ❖

Fig. 6C-5. Silhouettes of three raptors that can be seen at the Washington Monument: a) bald eagle, b) osprey and c) turkey vulture.

Chapter 6D

Harpers Ferry and The Potomac Water Gap

Location and Access: The small town of Harpers Ferry is located at the confluence of the Potomac and Shenandoah rivers at the edge of the Great Valley at a point where the borders of West Virginia, Virginia, and Maryland join. US 340, from Charles Town, W. Va. to Frederick, Md. passes through the Potomac River Water Gap (Fig. 6D-1). Start your visit to this area at the Harpers Ferry National Historical Park visitor center located off of US 340 on the western side of Harpers Ferry (Fig. 6D-2) The park brochure contains information about historic sites. A few parking spaces are located at the west end of Shenandoah St. and at the train station in Harpers Ferry; otherwise parking in town is very difficult. A shuttle bus runs from the visitor center to town about every 15 min. until 6pm.

Accommodations: Motels are present along US 340 to the east and west of Harpers Ferry, and several hotels are located in the town of Harpers Ferry.

Trails: Maps showing trails in the park are available at the park visitor center.

Tourist Sites: The old town is a popular tourist center. See their website or inquire at the National Historical Park visitor center for details.

History

"The passage of the Patowmac through the Blue Ridge is perhaps one of the most stupendous scenes in Nature." In these words Thomas Jefferson described his impression of the Potomac water gap, which is one of three places where major streams flow across the northern part of the Blue Ridge. The Potomac water gap and the town of Harpers Ferry are historically significant for a number of reasons. It provided a passage for American Indians and early settlers across the mountains. The first railroad connecting the east coast with the region west of the Blue Ridge passes through the gap. This connection remains an important link in the national railroad system. Harpers Ferry was a significant site during the Civil War. It was here that John Brown attacked a federal armory in his effort to obtain arms he planned to use in an effort

Above: View of the Potomac River near Harpers Ferry.

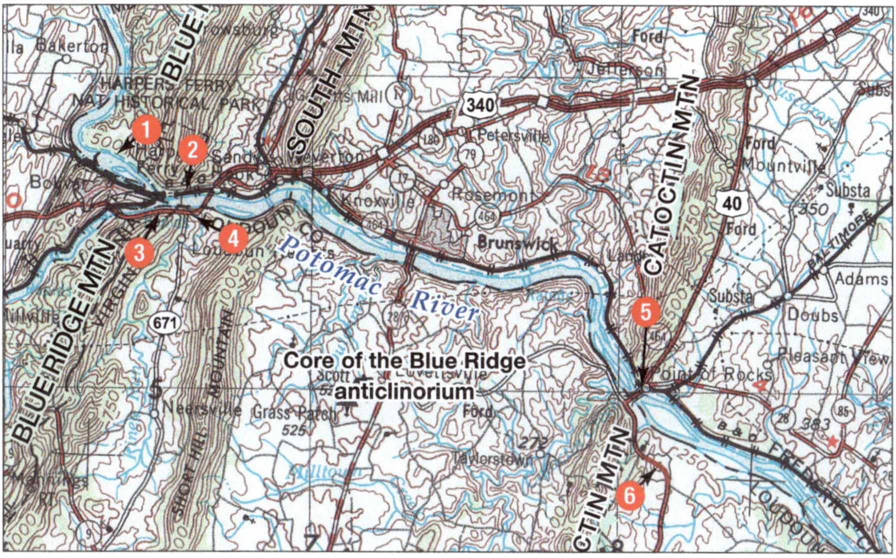

Fig. 6D-1. Stops (red dots) along the Potomac River Water Gap (see p. 110). The low topography of the core of the Blue Ridge anticlinorium separates the two limbs of the Blue Ridge Geological Province. Blue Ridge Mtn. and South Mtn. form the western limb. Catoctin Mtn., where the duplicated Chilhowee Group rocks hold up the ridge long, forms the eastern limb. (NPS map, with additions)

to end slavery. He seized the armory but was captured there by Federal troops and hung in nearby Charles Town, W. Va. Much more information about the town's history is available in the park visitor center and in the town.

Ecological Setting

The NPS reports that 170 bird species and 30 mammal species have been seen in Harpers Ferry NHP. Many of these birds will be found along the shore of the Shenandoah River. From the viewpoints on Maryland Heights you may see bald eagles, red-tailed hawks and peregrine falcons that were reintroduced to the area in 2001. You will find many of these birds along the river at the River Vista near the Murphy-Chambers House located close to the visitor center off US 340. Others can be found in town. For access to the river in Harpers Ferry follow the Virginius Trail that starts near the bus stop located in the center of the town.

White-tailed deer, southern flying squirrels, ground hogs, rabbits, foxes, and an occasional black bear are among the mammals found in the park.

Eastern deciduous forests cover most of the park area but the makeup of the forest reflects the particular environment present. Red maples, flowering dogwood, spicebush, mountain laurel, and pawpaw

Canadian geese at Harpers Ferry.

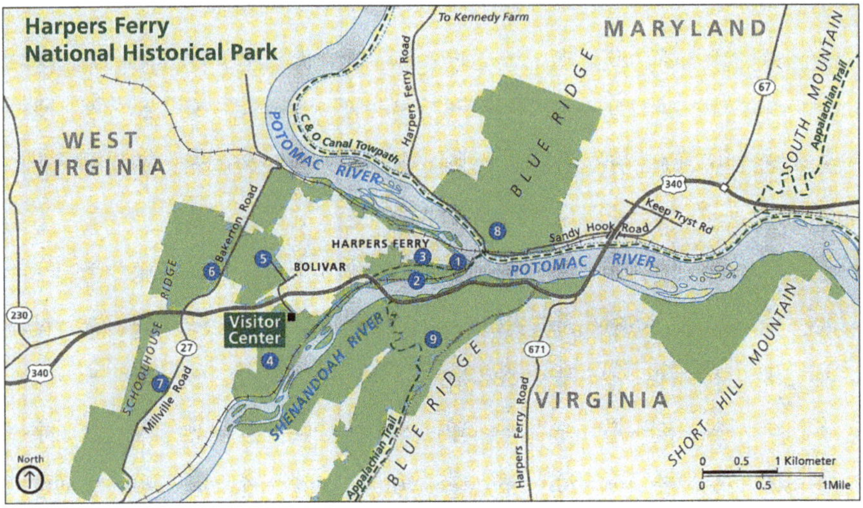

Fig. 6D-2. Map of Harpers Ferry National Historical Park. Note the location of the visitor center. Short Hill Mtn. is a southern continuation of the ridge known as South Mtn. (NPS)

make up much of the understory in most of the forest in the park. Black huckleberry, Blue Ridge blueberries, deerberry, and maple leaf viburnum are common shrubs, and many varieties of wildflowers grow in the park.

Soil composition, the character of the underlying bedrock, elevation, slope, and exposure to sunlight are all-important factors determining which plant species thrive. Upland, lowland, and floodplain vegetation communities are distinguished by particular groups of plants. Chestnut oak and tulip poplar form much of the canopy on the highest rocky soils such as those found on Maryland Heights. Black oak occurs more frequently on south, west, and east-facing slopes while northern red oak and chestnut oak are frequently present on the north-facing slopes. A somewhat different group of plants populate the residual soils formed on the Catoctin greenstones and Harpers phyllite. These are usually at lower elevations and on lower slopes. The trees here include red oak, white ash, sugar maple, basswood, hackberry, bitternut hickory, elm, and tulip poplar. Spicebush, hornbeam, and pawpaw are common shrubs.

Silver maple, sycamore, green ash, and cottonwood trees are present on portions of the floodplains most of which are inundated every year or two. The soils of floodplains are composed of unconsolidated sand, silt, and gravel that have been transported in—not formed in place from the underlying bedrock. Higher, less frequently flooded portions of these floodplains and terraces support a much more diverse group of trees including sycamore, ash, poplar, hickory, sugar maple, black walnut, and a species of southern red oak known as the Shumard oak, which is rarely found in the Blue Ridge. The understory contains many of the same shrubs found at high elevation.

Geologic Setting

The Potomac River is one of three large streams that flow completely across the northern section of the Blue Ridge

(Fig. 6-1). It crosses through in a short distance through a gorge that is less than a mile wide. Rock exposures along its course provide an exceptional opportunity to examine the geology across this important element of the Appalachians.

The Potomac (Fig. 6D-3) drains a large part of the Valley and Ridge Province between central Virginia and southern Pennsylvania. The drainage pattern is one often seen in the Valley and Ridge where the larger streams tend to flow in valleys cut into limestone and shale rock units while the small tributaries follow short courses down the slopes of ridges formed by sandstones and quartzites. The overall pattern resembles the shape of a trellis and is given that name. At a few places the larger streams turn and flow almost at right angles across the ridges, producing the water gaps described earlier (Ch. 3). Many geologists attribute these gaps to stream piracy (p. 48). They also favor this explanation for the origin of the Potomac Water Gap that begins at Harpers Ferry. Harpers Ferry is located on the western limb of a complex anticlinal structure that is most clearly developed at the northern end of the Blue Ridge (Fig. 6D-4). You can get a great view of this limb of the anticline from the trail up to Jefferson Rock or to Maryland Heights. This is an excellent place to see the juncture of the Potomac and Shenandoah rivers as the enlarged Potomac starts its journey across the Blue Ridge. To the west, the Valley and Ridge Province is visible. Though they are not obvious from a distance, the early Paleozoic limestones, shales, and sandstones that underlie the valley are folded and cut by thrust faults that displaced parts of the sequence of rock units over great distances.

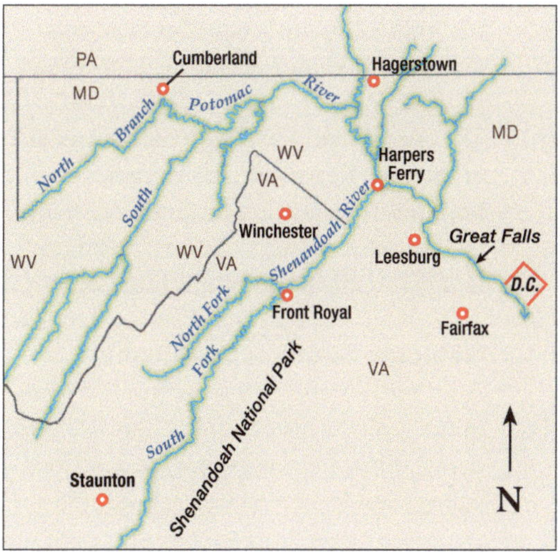

Fig. 6D-3. Drainage basin of the Potomac River. The pattern formed by the North and South forks of the Shenandoah River is known as a trellis pattern.

Faults also break the western limb of the Blue Ridge anticline that is exposed in the water gap. One major fault is responsible for the duplication of the Precambrian basement rocks and the overlying Catoctin greenstones and Chilhowee units seen near Harpers Ferry (Fig. 6D-4). This major fault has undergone two episodes of movement. The first one took place when the continent of Rodinia split apart forming rift basins in the area of what is now the Blue Ridge. Displacement on the fault at that time was of the type we call **normal** with the eastern side moving down (Fig. 3-2). Nearly 300 million years later when the Alleghanian Orogeny took place the Blue Ridge moved to the west and the earlier faults were reactivated as a thrust fault with the east side moving up and to the west. As a result of this faulting, the western limb of the Blue Ridge anticlinorium has two ridges: South Mtn. and Elk Ridge (also known locally as Blue Ridge Mtn.). South Mtn. dies out to the southwest. Blue

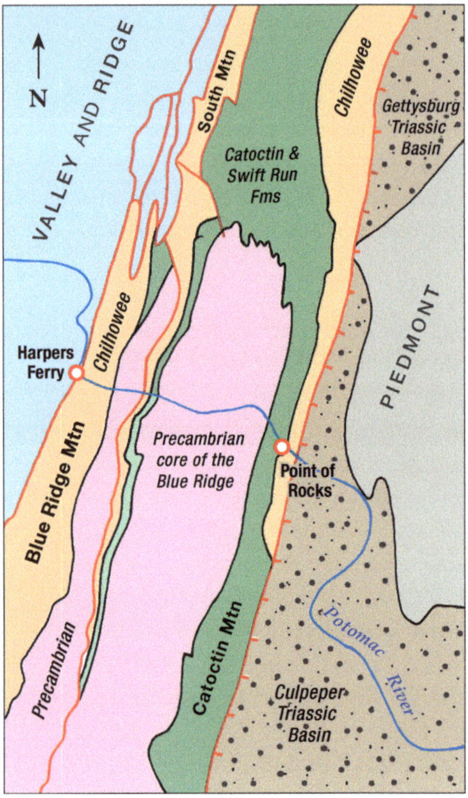

Fig. 6D-4. Geologic sketch map of the Blue Ridge anticlinorium near the Potomac River. (After Southworth and Brezinhski, 1996)

Ridge Mtn. continues to the southwest. These two ridges stand out because they are composed of the Chilhowee and Catoctin units. They are part of what was once a large fold formed during the Alleghanian Orogeny (Fig. 6D-4). The shape of the fold suggests that it was formed as a result of compression of the crust as it was pushed from the southeast toward the northwest.

Proceeding east across the water gap you will go through the core of the Blue Ridge anticlinorium. In this area you will find Precambrian igneous and metamorphic rocks, many of which were altered during the Grenville Orogeny (p. 23). The feldspar content of these rocks makes them much more susceptible to chemical weathering than the quartz-rich rocks that make up the Weverton Fm. For this reason the landscape is much more subdued in the core than it is on either limb of the fold.

The eastern limb of the Blue Ridge anticlinorium is exposed at Point of Rocks where the Potomac cuts across the greenstones and Weverton quartzites that make up Catoctin Mtn. The Catoctin greenstones are exposed at Point of Rocks and the Weverton is present along the mountainside to the northeast. If you continue east from Point of Rocks you will cross a high angle normal fault like the ones described earlier at Catoctin Mtn. Natl. Pk. and Gambrill State Pk. This fault was formed in the Triassic about 200 million years ago when Pangaea started to split apart. This fault cut across most of the rock layers in the Chilhowee Group and brought Triassic-age conglomerates into contact with the Catoctin Fm. along the eastern limb (Figs. 6D-4, -5). Since the Triassic ended over 200 million years ago, erosion has removed the crest of the Blue Ridge anticlinorium exposing Precambrian igneous and metamorphic rocks in the core of the anticlinorium. The elevation of the region on both sides of the Blue Ridge anticlinorium has also been reduced, leaving the more resistant Chilhowee and Catoctin units as ridges that define the limbs of this huge fold.

Geologic Sequence of Events

A recently established sequence of events in the geologic history of the Blue Ridge in the vicinity of Harpers Ferry is described below (Southworth and Brezinski, 1996):

1) About a billion years ago a major mountain belt, the Grenville Mtns.,

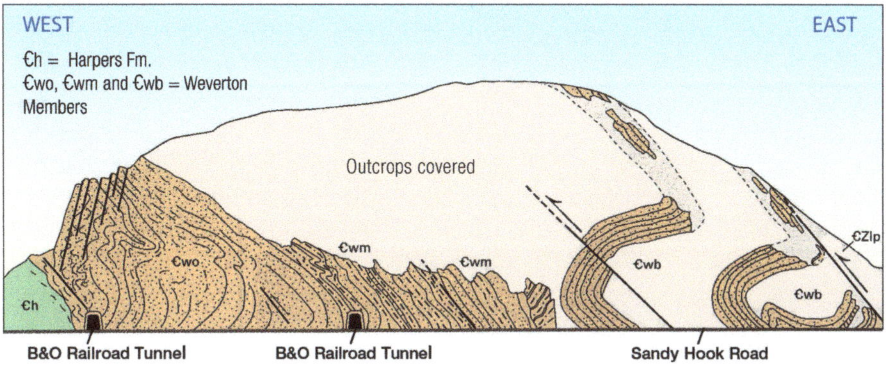

Fig. 6D-5. Cross-section of the north side of the Potomac near Harpers Ferry. This section is across the northwestern flank of the Blue Ridge. (From Southworth and Brezinhski, 1996)

were situated along what is now the east coast of North America. At the time it was formed the mountains lay in the interior of Rodinia just as the Himalayas now lie between the subcontinents of India and Eurasia. Stream erosion immediately began to reduce the high elevations of this mountain belt and after several hundred million years produced a subdued landscape. We can imagine one with hills or perhaps one with relief as high as parts of the modern Appalachians.

2) About 700 million years ago, convection in Earth's mantle began to exert forces on the base of Rodinia, causing the crust to stretch and pull apart. As extension took place normal faults (see p. 36 and Fig. 2-11) formed in the continental crust and blocks of the crust began to subside, creating large fault-bounded basins.

3) Sediments, sands, and gravels now known as the Swift Run Fm. began to fill in the basins. As crustal extension continued, molten rock (magma) rose from the mantle as dikes and poured out on to the surface, covering the rift basins and gradually the land surface as far south as the James River in Virginia. These lavas are now exposed along both flanks of the Blue Ridge. We don't know how far the lava flows extended to the east and west of the Blue Ridge. They may not have extended very far—they may have been removed by erosion or they may lie beneath other rock units that were thrust on top of them.

4) It appears that crustal extension continued until some time in the Cambrian Period (about 450 million yrs. ago). Great thicknesses (measured in thousands of feet) of limestone, muds, and sands were deposited on the continental margin west of the modern Blue Ridge. Because this section contains evidence of having been deposited in shallow water, the continental margin must have continued to subside during much of the Paleozoic Era. Today we find very little of the Paleozoic-age sedimentary record in the Blue Ridge. Either it was not deposited because the rocks now seen in the Blue Ridge were too far to the east or erosion removed most of the sedimentary rocks that were deposited in the Blue Ridge area.

5) The culmination of mountain building took place late during the Alleghanian Orogeny in the Paleozoic Era. At that time Gondwana (Africa) approached and collided with Laurasia (North America and Eurasia) creating the Appalachian

Fig. 6D-6. Outcrops of the Harpers Fm. seen across the Potomac River from Harpers Ferry (Stop 1 on Fig. 6D-2). This old railroad bridge now supports a walkway.

Mtns. within a huge continent named Pangaea (Ch. 2). During this collision the sedimentary layers (Swift Run, Weverton, etc.) above the Catoctin lavas became detached from the underlying rocks, and were folded and stacked up as they slid in thrust sheets toward the northwest.

6) Pangaea began to break apart during the early part of the Mesozoic Era. Magmas again rose into the extended parts of the crust; normal faults, like those that formed half a billion years earlier when Rodinia broke apart, developed; sedimentary basins formed; and eventually the continents were separated—enough for an ocean, the Atlantic, to form.

7) Since the Mesozoic, erosion has reduced the elevation of the Appalachians to their present form.

Geologic Traverse

This traverse along US 340 between Harpers Ferry at the western end and Point of Rocks (Fig. 6D-1) at the eastern end crosses almost continuous exposures across the western limb of the Blue Ridge. Note that traffic along this highway is heavy and no space is available to pull off the road when traveling eastward. A few pullouts are present on the westbound side of US 340, but it is not safe to walk across this highway. Visit points of interest off the highway and use the pullouts for viewing the river and mountain slopes across the river. You can also drive or walk on trails along Sandy Hook Road on the north side of the river.

Points of Interest

Stop 1. **In the town of Harpers Ferry**: You can obtain excellent views of the junction of the Potomac and Shenandoah rivers and the western edge of Blue Ridge from the point where High Street and Shenandoah Street intersect in Harpers Ferry (Fig. 6D-6). A walkway crosses the Potomac at this point.

The Harpers Fm., middle unit in the Chilhowee Group, was named for the excellent exposures found in a long section of outcrops located adjacent to Shenandoah Street. A few parking spaces are located at the western end of Shenandoah Street near the connection with US 340. Most of the Harpers Fm. exposed here is a phyllite, formed by the metamorphism of shale. This shale originated as muds deposited in marine waters on the margin of the North American continent in the early Cambrian Period. As the thickness of the sediment increased the muds became compacted, water was driven out, and the sediment became shale. Later the shale was metamorphosed as it was covered by thousands of feet of sediment, heated up and was deformed during Paleozoic-age mountain building. Signs of this deformation, notably rock cleavage and folding, are evident in many of the exposures located

Fig. 6D-7. Subhorizontal rock cleavage (aligned clay metamorphosed to micaceous minerals) formed as a result of compression during deformation of the Blue Ridge is a common feature of the Harpers Fm. The steeply inclined surfaces mark bedding in the fm.

Fig. 6D-8. Sharp folds developed in the cleavage that is exposed along Shenandoah Road leading into downtown Harpers Ferry.

here. **Rock cleavage** consists of closely spaced planes formed by the alignment of micaceous minerals (Figs. 6D-7, -8, -9). Slippage has taken place on many of these planes. This cleavage formed as a result of the compression that affected the continental margin of North America during Paleozoic orogenies. The original bedding of the shale may be identified by color changes, grain size variation, and textures that are parallel to the bedding. Both the bedding and cleavage are folded and vary in inclination (dip) from subhorizontal to vertical. Most of the layers exposed here have been deformed so much that they are upside down. The rock cleavage is inclined at a low angle to the southeast.

Stop 2. **The Weverton Fm. at Harpers Ferry:** Exposures of the Harpers and Weverton Fms. are present along Sandy Hook Road on the north side of the Potomac. Access this locality by walking across the Potomac walkway from the lower part of town in Harpers Ferry or by driving on Sandy Hook Road (this road can be reached by turning east off of US 340 onto Keep Tryst Road a short distance from the north end of the bridge over the Potomac River). The road is narrow and has little space for parking. The entrance to the Maryland Heights Trail is located along Harpers Ferry Road (an extension of Sandy Hook Road) about half a mile north of the end of the

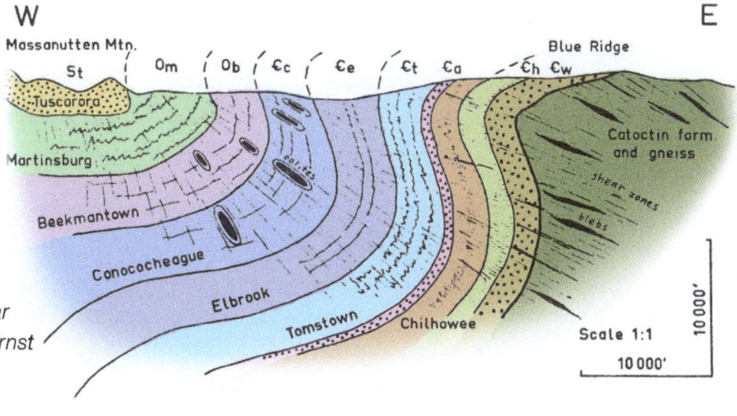

Fig. 6D-9. Schematic representation of cleavage along the western flank of the Blue Ridge. The section is near Luray, Va. (After Ernst Cloos, 1958)

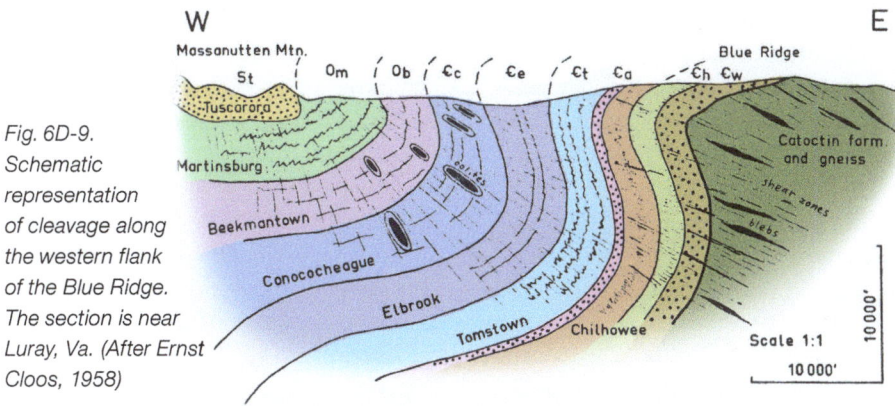

Potomac walkway. This trail ascends the cliff about 1,000 ft. to a superb overlook of the confluence of the Potomac and Shenandoah rivers, offering excellent views of the Harpers and Weverton Fms.

Stop 3. **Views while traveling west on the south side of the Potomac River along US 340:** This is a good place to see the large-scale folds present in the Weverton Fm. When the water level is low, layers in the Weverton create turbulence. The layers of conglomeratic quartzite form several large and many small asymmetric folds. Note that the original layers of the rock on the lower limbs of these folds are upside down. Folds of this type are referred to as overturned asymmetric folds. They formed as a result of stress in the Earth's crust acting from the southeast toward the northwest that was caused by the collision of North America (Laurentia) with Africa (Gondwana). Small-scale folds are visible in roadcuts through the Weverton Fm. along US 340.

Stop 4. **At the intersection of US 340 and SR 671:** Parking is available a short distance west of the intersection. The rock outcrop here is part of the Precambrian metamorphic complex that underlies the younger sedimentary rock layers exposed in the gap. This rock is granitic in composition (contains both potassium and plagioclase feldspars, quartz, and garnet) and has a well-defined alignment of crystals, called **foliation**, as a result of the strong deformation it has undergone. Quartz, feldspar, and garnet are visible in this exposure. Good exposures of the Precambrian basement also occur along Catoctin Creek on the north side of SR 464.

Stop 5. **Point of Rocks:** The Catoctin Fm. is exposed on both sides of the bridge where US 15 crosses the Potomac, but heavy traffic makes this Stop dangerous. It's much safer to go to Gambrill, Catoctin, or Shenanadoah parks to see it.

Fig. 6D-10. Conglomerate of this type is commonly found along the Triassic fault shown in Fig. 6D-4. (Photo: Bob Root)

Stop 6. **Triassic Conglomerates along US 40 and US 15:** A large, Triassic-age basin that extends north and south along the eastern side of the Blue Ridge lies east of the Point of Rocks. A fault formed here when Pangaea split apart. It separates the Triassic basin from the Blue Ridge. As the area east of the fault began to drop down, streams carried sediments including gravels from the high mountains west of the fault into the basin (Fig. 6D-10). The gravels include rock fragments of limestone derived from rocks now exposed many miles to the west. During the Triassic these limestones may have covered the rocks now exposed in the Blue Ridge. Today these conglomerates are exposed in roadcuts and in fields located along US 15. Most rock outcrops along US 15 north of Leesburg, Va. are composed of these conglomerates. One such outcrop is located northeast of Leesburg, approx. 500 yards southwest of the intersection of US 15 and the US 15 Bypass (in the Leesburg Quadrangle). ❖

Chapter 6E

Shenandoah National Park

Location and Access: Virginia's Skyline Drive, which runs through the park, has four entrances. From the north, enter by US 340, south of Front Royal. From Luray or Sperryville use US 211. The Swift Run Gap entrance is located where US 33 crosses the Skyline Drive. The southern entrance is located at Rockfish Gap where I-64 and US 250, running between Waynesboro and Charlottesville, cross the Skyline Drive and Blue Ridge Parkway. A fee is charged to enter the park but there is no charge for use of the Blue Ridge Parkway to the south. During winter months, access to both the Parkway and the Skyline Drive depends on weather conditions.

Accommodations: Rooms, cabins, camping, dining facilities, and fuel are available in the park. Reservations for summer and fall months should be made well in advance. For information about accommodations and availability of food at visitor centers along the Skyline Drive, contact Aramark at 1-800-999-4714, visit **www.nps.gov/shen**, or inquire in person at the entry stations. Some of these facilities may be closed even when the Skyline Drive is open. Tourist facilities are available close to the park at Front Royal, Luray, and Waynesboro.

Trails: Trail maps can be obtained from the NPS website **www.nps.gov/shen** or from the information desks at Dickey Ridge Visitor Center near the north end of the park, the Byrd Visitor Center at Big Meadows, or the Loft Mtn. Information Center near the southern end of the park.

Outstanding Features: The park area is one of the most intensively studied sections of the Blue Ridge. Its ecology and geology continue to be carefully examined by many scientists and a large number of technical papers are available through libraries and websites. A search of the homepage for Shenandoah National Park will lead you to many of these resources. A few specific sources are listed in the reference section.

Nearby Attractions: The many fine caverns near Luray, Va.; Mary Baldwin College, the Shakespearean Blackfriars Playhouse, and the home of Woodrow Wilson in Staunton; James Madison University in Harrisonburg; and the important Civil War battlefield at New Market are all located in the Shenandoah Valley west of the park. The University of Virginia and Jefferson's home, Monticello, are located east of the southern end of the park

Above: View to the northwest from the Skyline Drive in Shenandoah National Park.

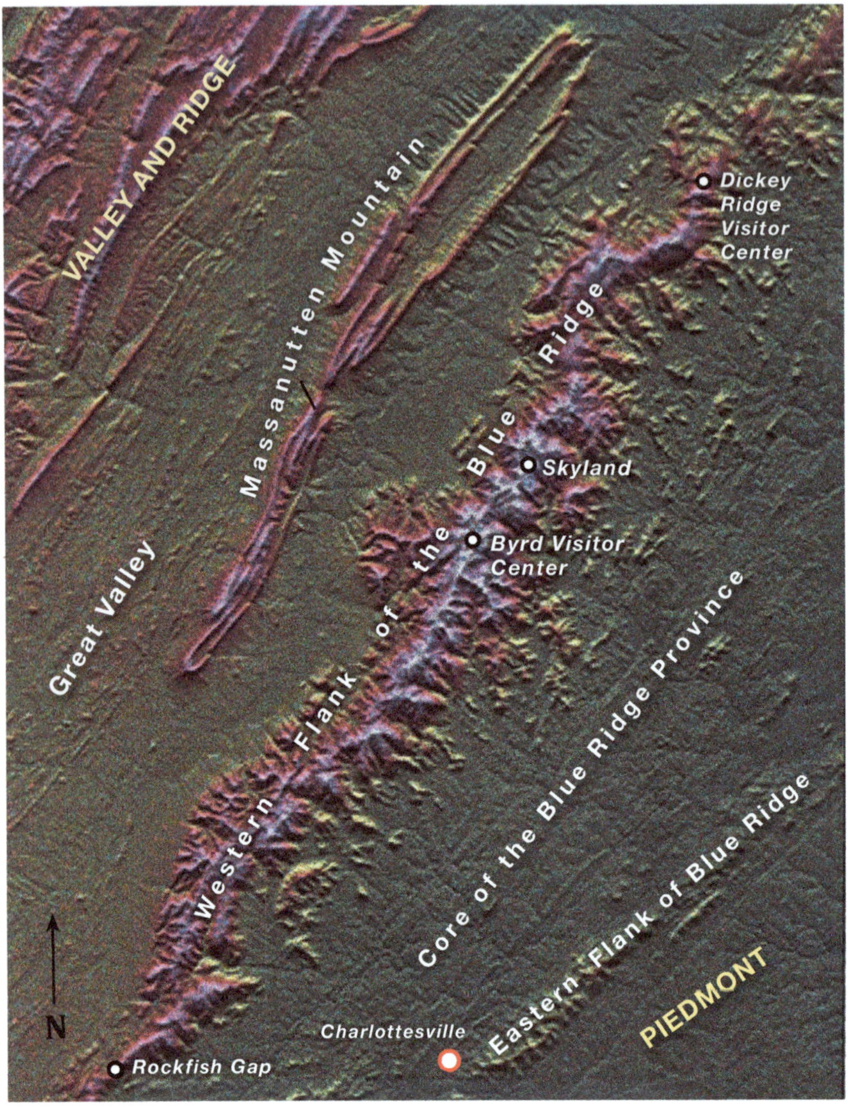

Fig. 6E-1. Shenandoah Natl. Park and surroundings. (USGS, Open file report 04-1321)

History

The Blue Ridge (Fig. 6E-1) was sparsely settled during the early part of the 20th century. A few roads crossed the mountains but most agricultural development was confined to the lower valleys. Logging became widespread and the virgin forests composed of giant chestnut trees were cut down. As people from the Washington, D.C. area took an interest in the mountains a few vacation resorts opened. Encouraged by Va. Gov. Harry Byrd, an early Shenandoah Natl. Park proponent, Herbert Hoover built his summer retreat here at the Rapidan River headwaters. As the park concept grew more popular Hoover authorized drought relief funds in 1931 to finance road construction into the mountains starting at Front Royal, Va. This project later became the Civilian Conservation Corps, a gov-

Autumn view from Calf Mtn. toward the west, across the Great Valley. North Mtn., the first ridge of the Valley and Ridge Province is faintly visible in the distance.

ernment program providing employment during the Depression. After cobbling together lands through purchases and eminent domain, the park was finally established by Franklin Roosevelt in 1935. Soon thereafter the road called the Skyline Drive and tourist facilities were built along the crest of the ridge from Front Royal to Waynesboro, Va. The Shenandoah Natl. Park soon became one of the prime areas for study of ecological communities in eastern North America.

Fig. 6E-2. View at Mooreman Overlook. The gray band of trees in the middle distance were killed in a forest fire.

Ecological Setting

Except for acid rain and the construction of road and tourist facilities, the environment in the park has been little disturbed by humans since 1935. A natural succession of the 1,600 species of plants found in the park has taken place as it has aged. However, fires, outbreaks of gypsy moth caterpillars that defoliate trees (esp. oaks), and various plant diseases have altered the natural succession (Fig. 6E-2). Despite the varied landforms and high diversity of plant species here, the dominant forest has remained an oak-hickory forest. In 1940, 85% of the park area was forested, 72% of which was chestnut oaks and northern red oaks. Today forests cover nearly 95% of the park, but oaks account for less than 60%. In 1940, no yellow poplar stands were present; today they cover more than 16% of the park.

Animal populations in the park are closely connected with the food supply. Deer, bear, turkey, squirrel, and many other animals can be found in oak forests that provide a major food source—acorns. Population sizes often vary with the density of the oak forest.

Early spring in the Shenandoah Natl. Park brings flowering serviceberry, hepatica and red maples, followed later by trillium, azaleas, columbine and mountain laurel. At this time migrating birds begin to pass through the park, among them a variety of brightly colored warblers, thrushes, vireos, grosbeaks, sparrows, finches, geese, ducks, and many owls and hawks. Many of these birds are then claiming territories in the park for breeding. As the summer arrives male songbirds are found defending these territories with their songs while they and their mates work to feed their hungry chicks. Ovenbirds, Acadian flycatchers, scarlet tanagers, wood thrushes and black-throated, blue, as

Northern red oak acorns are a favorite food of many animals in the Blue Ridge, including white-tailed deer, gray squirrels, and black bears (above).

Male gypsy moth (left); females are much lighter in color. Gypsy moth caterpillar (right). They were accidentaly released in 1869 by an amateur entomologist near Boston, Mass. who imported them from Europe to breed with silkworms. Since then they have traveled south, roughly 13 miles per year, decimating forest canapies in their wake. Fond of eating the leaves of acorn-producing oak trees, they have diminished the food supply of many animals in the Blue Ridge. (Moth photo: © entomart; Caterpillar photo: J.E. Appleby, USF&WS)

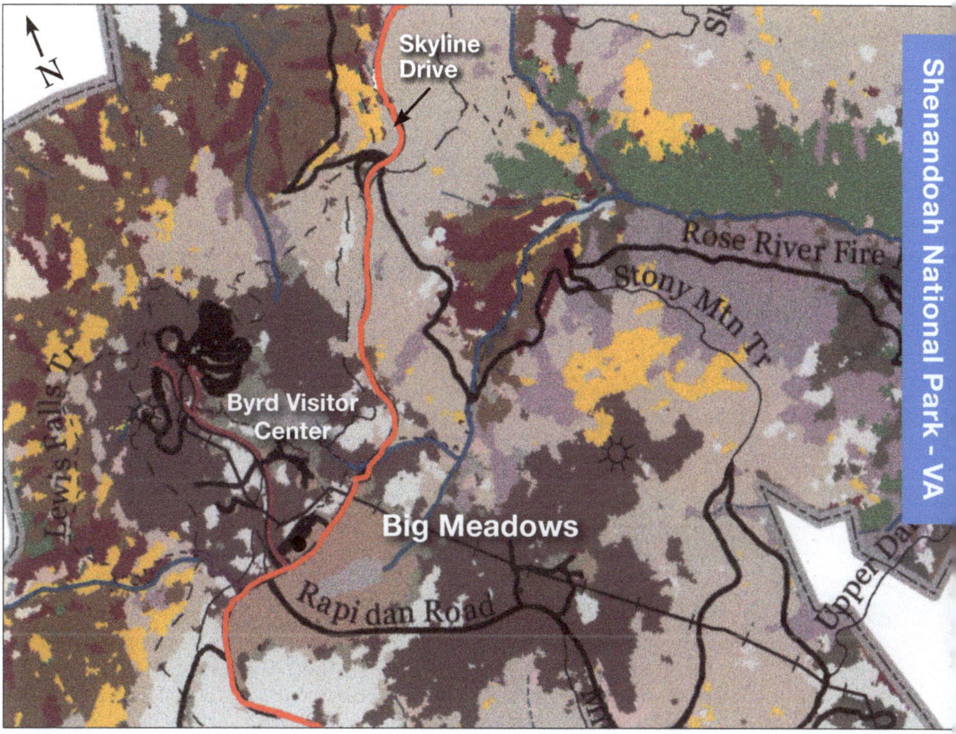

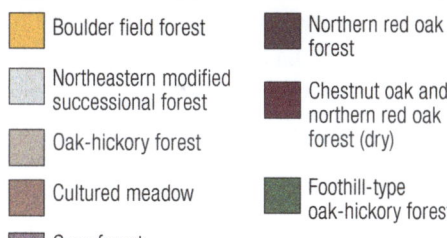

Fig. 6E-3. Patchwork of forest and other land covers in Shenandoah Natl. Park near Big Meadows. The patterns of this patchwork vary depending on a number of variables including: sun exposure, amount of moisture, length of growing season, temperature range, soil type, fires, and other disturbances. (Modified after Young, 2009)

well as hooded warblers can be found among the hardwood trees in the forest while indigo buntings, eastern towhees, and chestnut-sided warblers frequent the low underbrush. In the fall you'll find an array of goldenrods and asters amidst the fall foliage along with waves of migrating birds that are now quieter and cloaked in duller colored plumage. Titmice, woodpeckers and chickadees will stay here through the winter.

A vegetation-mapping program was completed in 2009 by biologist John Young. The map that accompanies that report subdivides the park into 35 vegetative communities. A portion of this map reveals the patchwork pattern formed by these communities, some of which are labeled on the map (Fig. 6E-3). The amount of moisture present during the growing season, the length of the growing season, average and extremes in temperature, soil types, disturbances caused by humans (e.g. road building), and natural processes such as fires are the basic elements that determine which species dominate an area.

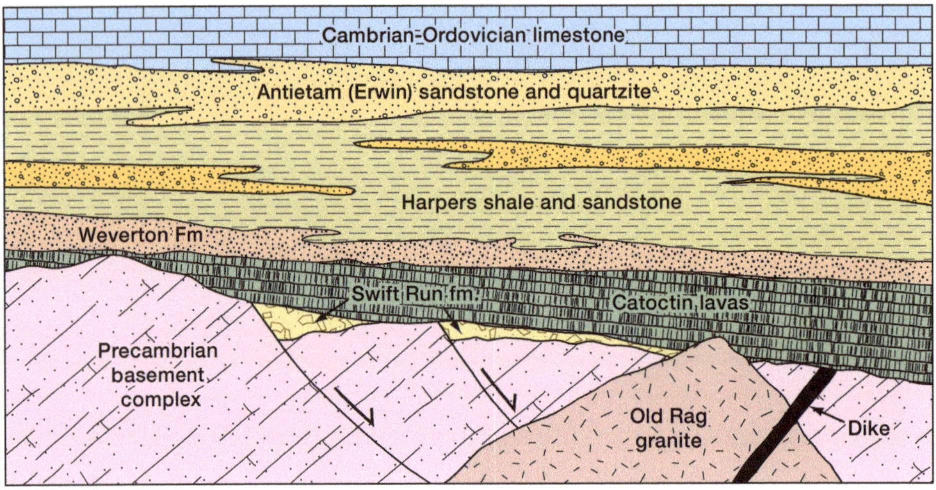

Fig. 6E-4. Cross-section showing rock layers in Shenandoah National Park. (After Gathright, 1976)

Geologic Setting

Shenandoah Natl. Park is located on the western limb of the large complex structure known as the Blue Ridge anticlinorium (Fig. 6D-4). Representative rocks of the **Precambrian basement complex**, the overlying Catoctin lavas, and sedimentary rocks of the Chilhowee Group that formed along the edge of the Iapetus Ocean near the end of Precambrian time and at the beginning of the Paleozoic Era, are exposed in the park (Fig. 6E-4). The Precambrian-age basement rocks and covering units were thrust to the northwest late in the Paleozoic Era. Thrust faults exposed along the western edge of the Blue Ridge (Fig. 6E-5) and along the western edge of the Great Valley of Virginia are surface expressions of the huge amount of movement involved as rocks of the Blue Ridge were uplifted and moved to the northwest. Geologists think rocks that make up the Blue Ridge were moved as a huge block many miles—perhaps more than 100 miles to the northwest. Although the amount varies, enormous displacements occurred on a few of the thrust faults. One such thrust fault can be traced to the east near the entrance to the park at Front Royal, Va. (Fig. 6E-6). It caused the western edge of the Blue Ridge to swing to the east. As movement took place on these thrust faults the thick layers of sedimentary rocks that lay in the Valley and Ridge were compressed. The stronger layers of dolomite and sandstone buckled and yielded by folding into anticlines and synclines such as the ones so beautifully displayed at Massanutten Mtn. (Fig. 6E-1). The weaker rock layers composed mainly of shale became zones of slippage. Ultimately, the whole package of sedimentary rocks were compressed by folding and thrust-faulting.

Landscape Views

The many overlooks along the Skyline Drive provide spectacular views of the landscape both in the park and beyond its borders. Streams and downslope movement of weathered and loose rock are the dominant forces in erosion here. Although few of the rocks in the Blue Ridge dissolve easily in water, solution of carbonate rocks has played an important role in lowering the land surface, produc-

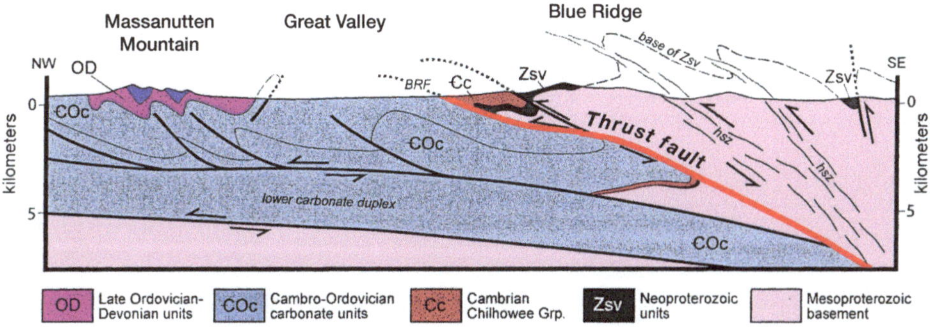

Fig. 6E-5. Cross-section of the western flank of the Blue Ridge, the Great Valley, and Massanutten Mtn. (Simplified after Southworth, et al., 2009)

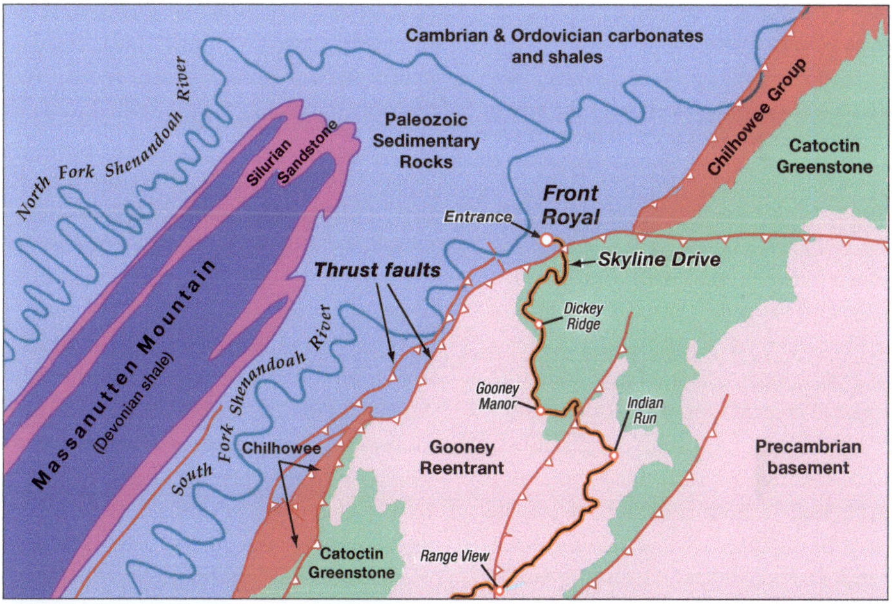

Fig. 6E-6. Geologic sketch map of the area between the park entrance and Range View Overlook. Shown are the ridge-forming Silurian sandstones of Massanutten Mtn. (deep pink); Paleozoic sedimentary rocks, mostly shales and limestones of the Great Valley (light blue-gray); the Catoctin Fm., largely greenstones (green); Precambrian gneisses and granitic rocks (pale pink). Precambrian-age basement rocks lie within Gooney reentrant. Catoctin rocks that are highly resistant to erosion underlie most of the high ground. (Simplified after Southworth, et al., 2009)

ing the Great Valley of Virginia west of the Blue Ridge. Continental glaciers did not reach the Blue Ridge during the major advances of ice down into what is now the United States in the Pleistocene, but ice did contribute to erosion on the higher parts of the ridges in the park. The distinct difference between the landscapes to the west and to the east of the Skyline Drive is largely a result of the composition and structure of the underlying rocks. Stratified rocks, with layers that vary greatly in their resistance to erosion, are present to the west. Because igneous and metamorphic rocks, with less pronounced differences, lie to the east, the eastern landscape reveals less about the shape of the underlying rock.

Fig. 6E-7. View from Hogback Overlook toward the west across the valley of the South Fork of the Shenandoah River. The Great Valley of Va., and North Mtn. are visible in the distance.

Views of the Landscape West of the Park

Distinctly different geological provinces lie to the west of the Skyline Drive from those you see to the east. To the west, you look across the Valley and Ridge Province. To the east is the core of the Blue Ridge Geological Province. The contact between the Blue Ridge and the Valley and Ridge is located at the foot of the Blue Ridge. A portion of the Valley and Ridge, known as the Great Valley of Virginia separates the Blue Ridge from a long, straight, mountain ridge known as Massanutten Mtn. Its northern end can be seen from the northern overlooks along the Skyline Drive, and it continues 47 miles to the southwest, ending near Harrisonburg, Va. From maps and models of the region it is clear that Massanutten Mtn. lies within the Great Valley of Virginia, which can be seen on both the eastern and western sides of Massanutten (Fig. 6E-1). The Great Valley continues north into Maryland and Pennsylvania and south, nearly to Roanoke, Va.

The Great Valley of Virginia is underlain by limestones, dolomite (a calcium-magnesium carbonate), and shales that are of Cambrian and Ordovician-age (550-440 million years old). They have an aggregate thickness of many thousands of feet and were deposited while the eastern margin of North America was covered by shallow marine waters. This margin continued to subside during most of this time, allowing the great thickness of sediment to accumulate. This subsidence ceased in the early Silurian as uplift took place in the Appalachian region when mountains rose along the eastern edge of the continent and sands began to spread along the margin of the sea. Much later, near the end of the Paleozoic Era, as the final mountain building took place in the Appalachians, the carbonates in the Great Valley and the sandstones that form the ridges in most of the Valley and Ridge were folded and pushed to the northwest. Erosion over the last 250 million years has etched out the valleys of the Valley and Ridge and left the sandstones to form ridges. These sandstone ridges are evident at Massanutten Mtn. and are present at the surface along the western edge of the Great Valley. When the air is clear you can see these ridges in the far distance, as much as 30 miles to the west.

Massanutten Mtn. is much more complex than a simple ridge. The rocks exposed here have been folded and broken by thrust faults. The mountain is sometimes referred to as a "canoe-shaped synclinal mountain," but smaller anticlines are present on both limbs, and two synclinal structures separated by an

Field group on an outcrop of Silurian sandstones located on North Mtn.

anticline make up the southern half of the structure. The central valleys between the ridges are underlain by easily eroded Devonian-age shale. The two limbs of the Massanutten syncline are visible from many overlooks. At the northern and southern ends of the park you can see that the Great Valley is much larger than the valley between the Blue Ridge and the eastern limb of Massanutten Mtn. The South Fork of the Shenandoah River flows through this smaller valley. The North Fork of the Shenandoah River is located on the western side of Massanutten Mtn.

Both the South and North Forks of the Shenandoah exhibit remarkable meandering patterns (Figs. 6E-1, -6). The regularity of the meanders suggests that the river is following a pattern of fractures that are present in the underlying shale unit. The width of the meander belt is determined by the presence of resistant rock units on either side of the belt. From the Skyline Drive you can catch glimpses of the South Fork, but the high ridges that make up Massanutten Mtn. cut off views of the North Fork of the Shenandoah River.

Views of the Landscape East of the Park

Views to the east from the Skyline Drive are dramatically different from those to the west. Many large, deep valleys open to the east of the high mountainous ridge on which the park is situated. Beyond these mountains the landscape is more subdued (Fig. 6E-8). Numerous isolated mound-like mountains rise in this area, which is underlain by a variety of Precambrian-age igneous and metamorphic rocks, most of which are between 1-1.3 billion years old. These form the core of the Blue Ridge anticlinorium. No distinct ridges like Massanutten Mtn. are present, but if the air is clear you can see a low ridge in the far distance. This ridge composed of Catoctin lavas that are resistant to erosion is located along the eastern edge of the Blue Ridge Province.

Fig. 6E-8. View to the east from the northern part of Shenandoah National Park. Note the air pollution that appears as a dark band in the lower part of the atmosphere.

Points of Interest
in Shenandoah National Park from North to South

Shenandoah Valley Overlook:
Skyline Drive-milepost 3 (SD-mp 3)
This viewpoint, located on outcrops of Catoctin greenstone overlooks the South Fork of the Shenandoah River. The river follows a meandering pattern as it flows in a valley underlain by Ordovician-age (about 450 million yrs. old) shale of the Martinsburg Fm. Folded, Silurian-age (about 430 million yrs. old) sandstones form the ridges of Massanutten Mtn. Limestones and shales of the Great Valley of Virginia underlie the valley north and west of Massanutten Mtn.

Dickey Ridge Visitor Center: (SD-mp 4.6) Here you will find good information about the park, trail maps, books, nature exhibits, ranger program information, restrooms, and a picnic area—but no food or gas. Several trails have entrances near the center. Fox Hollow Trail is an easy self-guided 1.3-mile loop through an old farm, now covered with oaks, hickories, and poplars.

Gooney Reentrant: (SD-mp 0-20)
From its entrance, the Skyline Drive follows a great crescent-shaped indentation in the mountain front, called a **reentrant** (a low-lying landform between two hill spurs) that is visible for 20 miles from the first overlook to Hogback Overlook. It shows up clearly on the 3D model of the park (Fig. 6E-9) and in a geologic sketch map of the area (Fig. 6E-6). Much of the topography in the reentrant has low relief and is much lower in elevation than the crest of the Blue Ridge.

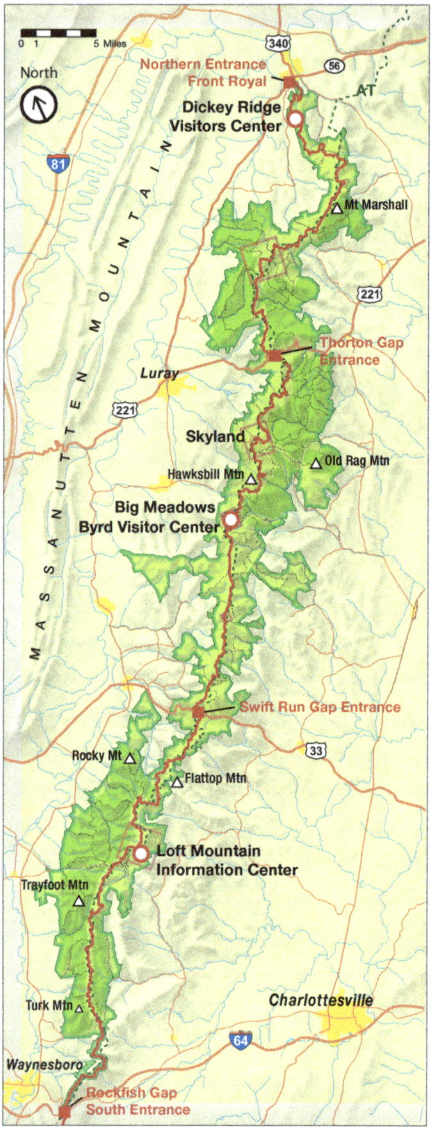

Overview of Shenandoah National Park. (Modified from NPS map)

NOTE:
- "SD-mp" = Skyline Drive milepost number.
- Plant and rock collection in the park is illegal.
- Easier trails are often near visitor centers. Free maps can be found at some trailheads. For trails off the Skyline Drive, obtain maps from a visitor center or buy a good commercial map (such as one by Natl. Geographic).
- **Rust-colored subheads** = Stops with facilities, picnic areas, camp stores, etc.

Fig. 6E-9. 3-D model at Dickey Ridge Visitor Center showing the northern end of Shenandoah NP, the South Fork of the Shenandoah River, and Massanutten Mtn. (NPS map, with additions)

Farther south along the Skyline Drive, ridges composed of resistant quartzites of the Antietam and Weverton fms. are in the foreground of many of these views. V-shaped notches mark places where streams have cut through the thick layers of beach sands once deposited along the continental margin before being cemented together as hard rock. Uplift during mountain building that elevated the Blue Ridge has tilted and in some places folded these quartzites.

The Skyline Drive is built on a high ridge formed largely by Catoctin lava flows that were altered from basalts, like those found in Hawaii, into metamorphic rocks, called greenstones. They are exposed in many of the roadcuts along the Drive. Their resistance to erosion accounts for the high level on which the Drive is constructed. Precambrian-age granitic rocks, the same rocks that make up the low topography in the core of the Blue Ridge Province are present in the subdued topographic area of the reentrant. The thrust fault on which the Blue Ridge has moved to the west lies at the western edge of this outcrop belt (Fig. 6E-5). It carries Precambrian granitic rocks onto the much younger shales (Ordovician-age Martinsburg shale) of the Great Valley of Virginia.

A thrust fault, inclined to the southeast and passing under the Blue Ridge, is present along much of the outcrop belt of the Antietam Fm., carrying it onto the Cambrian-age carbonate and shale rocks exposed in the valley at the foot of the Blue Ridge. The South Fork of the Shenandoah River has shaped the valley in which Front Royal, Luray, Stanley, Shenandoah, and Elkins are located. Beautifully developed and regularly spaced meanders give both the South Fork and North Fork distinctive patterns (Figs. 6E-1, -9). Both lie within belts underlain by a thick shale known as the Martinsburg Fm.

Fig. 6E-10. View from Hogback Overlook across the crescent-shaped arc of Gooney reentrant.

Shenandoah National Park - VA

Fig. 6E-11. Catoctin greenstone with columnar joints in this outcrop of metabasalt.

Fig. 6E-12. Chemical weathering of the surface of Catoctin greenstone causes color variations.

Views from Signal Knob (SD-mp 5), **Gooney Run** (SD-mp 6.8), **Gooney Manor** (SD-mp 7.2), **and Hogback** (SD-mp 20.7) **overlooks:** These overlooks provide good views of the Great Valley, the northern end of Massanutten Mtn., meanders in the Shenandoah River forks, and a distant North Mtn. One of the major thrust faults formed during the Alleghanian Orogeny in the late Paleozoic Era (Fig. 6E-5) crops out along the base of North Mtn. This fault lies below the Great Valley and the Blue Ridge. It is most likely that the rocks now exposed in the Blue Ridge moved many miles to the northwest over this fault plane.

At all of these overlooks you will be standing on exposures of Catoctin greenstone lava flows that are about 570 million yrs. old. These rocks are exposed at many overlooks between Signal Knob Overlook and the Thornton Gap Entrance. Because of their resistance to erosion, the Catoctin lava flows that spread across the landscape about 570 million yrs. ago, stand higher than the billion-plus yrs. old Precambrian rocks on top of which they were formed. These Precambrian rocks are the bedrock under a bowl-shaped valley, the Gooney reentrant, in the foreground (Fig. 6E-10). A few hills and ridges stand at the western edge of this valley. Cambrian-age sandstones and quartzites of the Antietam Fm. hold up the hills and ridges along the western edge of the Blue Ridge. Beyond that the meandering South Fork of the Shenandoah River flows along a valley underlain by easily eroded shales of the Martinsburg Fm. To the west of this valley Massanutten Mtn. rises. The rocks that form this mountain, like the Antietam Fm., were originally deposited as beach sands at the edge of an ocean that invaded the North American continent. However, they were formed during the Silurian Period and are nearly 100 million yrs. younger than the Antietam Fm. Massanutten Mtn. extends to the southwest almost to the south end of Shenandoah NP. The north end of the mountain is visible in the near distance. Beyond that is the Great Valley which extends to a ridge that is visible on clear days. The same Silurian sandstones that form the ridges on Massanutten Mtn. hold up this distant ridge, which is known as North Mtn.

Indian Run Overlook: (SD-mp 10.5) **Columnar joints** in the Catoctin lavas that formed during the cooling and contraction of the lava (Fig. 6E-11), are best exposed at the north end of the pullout.

Striped skunk (Photo: Tom Friedel†)

Notice that the cracks in the greenstone are filled with secondary minerals such as quartz and epidote (green). Chemical weathering has caused these multi-colored rims (Fig. 6E-12). To the east, you see isolated low mountains and large areas of low-relief land typical of this part of the core of the Blue Ridge Province. In summer months, the roadside and cliff at this Stop contains a remarkable assemblage of plants, wildflowers, vines, trees, and shrubs.

Jenkins Gap Overlook: (SD-mp 12) Views to east. Note the large v-shaped valley in the foreground. It has a profile characteristic of valleys cut by stream erosion. Much of the material (rocks, sand, and soil) carried in such streams gets into the stream by moving down the valley sides into the stream channel where the water carries it downstream.

Hogwallow Flats Overlook: (SD-mp 13.6) In the far distance, you can see the eastern limb of the Blue Ridge. It is composed of Catoctin lava flows like those in Shenandoah NP.

Range View Overlook: (SD-mp 17) This is one of the few overlooks where you can obtain excellent views to the east across the core of the Blue Ridge, to the south across the high peaks in the park (Figs. 6E-13a-b), and to the west of the Valley and Ridge Province. The folded

Fig. 6E-13a. View to the south across the highest peaks in Shenandoah National Park.

Fig. 6E-13b. View to the southwest, including the Great Valley and the distant Valley and Ridge.

Fig. 6E-14. Dikes such as this one cut through underlying gneiss and brought magma from depth to the surface where it spread out as lava flows.

sedimentary rocks of the Valley and Ridge form long, straight ridges. In contrast, the igneous and metamorphic rocks of the basement complex exposed in the core to the east and south of the overlook have irregular, mound-shaped forms. Rocks (mainly gneisses) from the Blue Ridge basement complex were used to construct the retaining wall around this overlook. Some of these rocks are also exposed in the outcrops across the road.

Gimlet Ridge Overlook: (SD-mp 18.5) Good views of the north end of Massanutten Mtn. When the air is clear you can see distant ridges in the Valley and Ridge Province.

Mt. Marshall Overlook: (SD-mp 19) Looking to the northeast you see two mountains. The one closest to the Skyline Drive, Mt. Marshall, is composed of Catoctin greenstone. The mountain to the right of Mt. Marshall, called "The Peak" is composed of Precambrian-age gneisses. The rocks that make up The Peak have been thrust to the west and lie on top of the younger greenstones. Thus the lava flows of the Catoctin Fm., which once poured out and covered the Precambrian gneisses, now lie beneath them along this fault.

Little Hogback Overlook: (SD-mp 19.4) Good view of the north end of Massanutten Mtn. The retaining wall here is built of metamorphic rocks that are part of the Blue Ridge basement complex.

Little Devils Stairs Overlook: (SD-mp 20) Views to the east. The rock exposures in the cliff on the west side of the Skyline Drive are part of the Blue Ridge basement complex. Several metabasaltic dikes cut across the basement rocks in outcrops located about 50-100 yards north of the parking area (Fig. 6E-14). When the rocks are wet, the dikes may be hard to see because water flowing down the cliff gives the whole cliff a dark color.

Hogback Overlook: (SD-mp 20.7) This overlook provides an excellent broad view of the reentrant along the western Blue Ridge front. Outcrops here are composed of Precambrian basement gneisses that contain pyroxene.

Mathews Arm: (SD-mp 22) A campground is located here.

Elkwallow: (SD-mp 24) A camp store, cafe, and gas station are located here.

Jeremys Run Overlook: (SD-mp 26.5) As you look down the v-shaped valley, you can trace the western edge of the Blue Ridge into the valley of the South Fork of the Shenandoah River. Massanutten Mtn. rises on the river's western side.

Pass Mountain Overlook: (SD-mp 30) View to west. Chilhowee Group sandstones and quartzites of the Cambrian-age Antietam Fm. hold up the ridge along the mountain front. The town of Luray, home of the well-known caverns, is visible as is the four-lane highway that leads west through a wind gap and across Massanutten Mtn. into the Great Valley to the west.

Thornton Gap Entrance Station: (SD-mp 31.5) US 211, connecting Luray, Va. with Sperryville, Va. crosses the Skyline Drive here. Note the abrupt change from outcrops of Catoctin greenstone on the north side of the gap to metamorphic rocks of the Blue Ridge basement on the south side. A thrust fault is responsible for the gap. Rocks cut by the fault were broken and have been eroded much more rapidly than those outside the fault zone. Blue Ridge basement rocks are thrust onto the Catoctin greenstones at this gap.

Tunnel Parking Overlook: (SD-mp 32.5) View to east across the core of the Blue Ridge Province. Excellent exposure of Precambrian (1.16 Ga) basement rocks. These include gneisses that show a well-defined alignment of minerals and some one-inch long feldspar crystals (Fig. 6E-15). If the outcrop is well lit by sunlight, you can see bright reflections of light from cleavage planes in the crystals.

Fig. 6E-15. Gneiss at Tunnel Parking Overlook.

Hazel Mountain Overlook: (SD-mp 33) Excellent view to the east. There is an exceptionally good exposure of the Precambrian basement here. The rock, a gneiss, contains small crystals of purple garnet (Fig. 6E-16a) and has a strongly defined alignment of crystals in a plane (a foliation) (Fig. 6E-16b) that lends a linear texture to i. If you look at the flat side of some of these rock masses, you can see the aligned minerals. These formed during the deformation of the Blue Ridge during the Grenville Orogeny in the Precambrian.

Fig. 6E-16a. The garnets in this gneiss indicate the grade (temperature and pressure) the rock was subjected to during metamorphism (p. 319).

Fig. 6E-16b. Foliation is the alignment of minerals (common in micaceous minerals) that often form in rocks under high pressure during metamorphism.

Fig. 6E-17. Views of the Skyline Drive and the park's high peaks as seen from Stony Mtn. Trail.

Pinnacles Overlook: (SD-mp 35) Looking east across the Blue Ridge basement core, you obtain a clear view of Old Rag Mtn. It is composed of a Precambrian-age granite that was intruded into the surrounding gneisses. Dikes composed of greenstone cut through the granite producing staircase-like passages.

Pinnacles Picnic Ground: (SD-mp 36.7) Excellent for large groups, it also has a shelter and restrooms. During construction of the Skyline Drive a CCC (Civilian Conservation Corp.) camp was located near this site.

Hughes River Gap: (SD-mp 38.8) Views here are to the west toward the town of Luray. Blue Ridge basement rocks are visible in the retaining wall and outcrops across the road from the overlook.

Stony Man Overlook: (SD-mp 38.5) From here you can see the rugged profile of a human head that gave Stony Man Mtn. (4,011 ft.) its name. The mountain is composed of Catoctin greenstone. Weathering along contacts between lava flows contributed to the shaping of the mountain. Red oaks, ash, paradise trees, maples, locust, and mountain laurel are present at the overlook.

Little Stony Man Trail: (SD-mp 39.1) This popular trail is an excellent place to look for plants and birds, especially hawks. It leads you on an easy hike to the second highest peak in the park. A trail guide is available at the trailhead near the parking lot. Along the trail you will pass witch hazel, northern red oak, red maple, birch, spruce, chestnut oak, white oak, mountain laurel, and mountain ash. At the top you will find superb views of the Valley and Ridge as well as along the Blue Ridge mountain front (Fig. 6E-17) where rocks of the Chilhowee Group hold up the ridge. A thrust fault that carried the Blue Ridge to the northwest lies along the base of the ridge. Most of the low slopes west of the mountain front are underlain by stream deposits originally deposited as ancient alluvial fans. Outcrops at the top of the trail are Catoctin Fm. lavas. In the 1800s, a copper mine was located near the top in the Catoctin Fm. Blue and green stains, typically associated with copper carbonate minerals (malachite and azurite) are still present at the mine site.

Thorofare Mountain Overlook: (SD-mp 40.5) In the far distance to the east, you can see the southern end of a ridge that lies along the eastern margin of the Blue Ridge Province. The mound-shaped low mountains are in the core of the Blue Ridge. Old Rag Mtn. lies to the south in the middle distance. Exposures of Precambrian basement rock occur

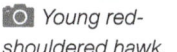

Young red-shouldered hawk.

Cones on the branch of an eastern hemlock.

Common thistle (non-native).

in the parking area, and the curbstones provide fresh samples of these rocks. Outcrops of the Catoctin Fm. are exposed along the road north of this overlook. Take a look at the large block of greenstone located about 100 ft. down the slope. The rock contains two greenish minerals: chlorite and epidote. The chlorite is disseminated throughout the rock and is responsible for its greenish color. Chlorite, epidote, and quartz fill small spherical cavities that initially formed as gas bubbles in the molten lava. These minerals filled the cavities as the lava cooled and became solid (Fig. 6E-18).

Skyland Resort: (SD-mp 42) This is one of the two largest tourist facilities in the park with cabins, a campground, restaurant, camp store, and gas station. Hemlock trees, now rapidly disappearing in the Blue Ridge due to woolly adelgid infestations, can be seen here and along the Drive to the south. It is also a good place to spot migrating birds.

Old Rag Mountain Trails: (SD-mp 43) With spectacular 360-degree views from the summit and its famously intense, steep rock scrambles near the top through boulderfields and narrow passageways cut into the granite by greenstone dikes, this hike is the most popular in the park. However, the 7-8 hr., 9.3-mile circuit is also notoriously dangerous. ⚠ **Caution.** Pack lightly and compactly and bring 2-3 quarts of water, plenty of food, good hiking shoes, and a headlamp or flashlight for late returns. Never hike it in wet or icy conditions. To avoid the scrambles take a fire road. Hiking permits can be bought at the Old Rag Fee Station. Several trails lead to Old Rag from the Skyline Drive, but the most popular circuit begins from the east, off SR 600.

Limberlost Trail: (SD-mp 43) This easy 1.3-mile loop trail offers the opportunity to see a variety of wildlife, including the pileated woodpecker, solitary vireo, Blackburnian warbler and veery, as well as black bear, white-tailed deer, and rabbits.

Fig. 6E-18. Catoctin greenstone with cavities filled with secondary minerals such as chlorite, epidote, and quartz.

Timber Hollow Overlook: (SD-mp 43) A ridge held up by Antietam quartzite lies along the edge of the Blue Ridge. The Appalachian Trail passes by the overlook. If you follow it to the southwest, you will walk across outcrops of Catoctin greenstone and the Swift Run Fm. and finally across Precambrian basement rocks. Park rangers often use this trail for nature hikes. The outcrop immediately below the retaining wall contains a rock called unakite (Fig. 6E-19). This relatively rare rock contains feldspar (pink), quartz (looks like glass), and epidote (a green mineral) that formed when hot watery solutions altered minerals in the original rock that contained iron and magnesium. Pitch pine, oaks, witch hazel, red oak and black locust grow in the median at this overlook.

The rare Shenandoah salamander lives only in this park. It has two color phases: uniformly dark and striped. It feeds nocturnally on small insects in the soil. (Photo: USGS)

Fig. 6E-19. Unakite exposed just below the retaining wall at Timber Hollow Overlook.

Crescent Rock Overlook: (SD-mp 44) Looking west you will see a low ridge in the middle ground that is held up by sandstones and quartzites of the Antietam Fm. (Fig. 6E-20). The notch you see in Massanutten Mtn. has been interpreted as a place where, much earlier in its history, a stream cut a water gap across the ridge but was unable to cut down fast enough to maintain the gap as the Shenandoah Valley was lowered by erosion caused by the Shenandoah River. After the stream was diverted from this course a gap was left. Gaps formed in this way are called "wind gaps." This process is described in Ch. 3. The AT, which passes nearby, crosses outcrops of Catoctin greenstone exposed along the cliff. White and red oaks and maples are prominent in the forest here.

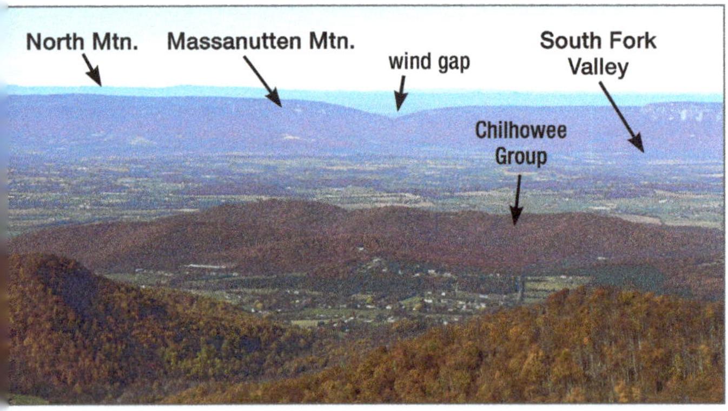

Fig. 6E-20. View of the western edge of Blue Ridge from Crescent Rock and Timber Hollow Overlooks. The ridge in the middle ground is held up by Antietam quartzites (Chilhowee Grp). The ridges of Massanutten Mtn. are also composed of quartz-rich rocks that are much younger than the Antietam. Note the wind gap in Massanutten Mtn.

Fig. 6E-21. Old Rag Mtn. is one of the most prominent landmarks in the park and its most popular hiking destination. The mountain is composed of Precambrian granite that was intruded into the older gneisses of the Blue Ridge basement.

Franklin Cliffs Overlook: (SD-mp 49) View to the west. Note the notch (wind gap) in the ridge on Massanutten Mtn. Trees in this section of the Skyline Drive were severely damaged by ice storms. Catoctin greenstone is exposed here. A trail at the north end of the overlook offers clear views of the cliff over red oaks, witch hazel, and locust trees.

Hawksbill Gap Trail: (SD-mp 45.6) At 4,051 ft., Hawksbill Mtn. is the highest point in Shenandoah NP. This popular hike to the summit takes about two hours along a 2.8-mile loop trail that is steep in places and moderately difficult. From the summit you will find excellent 360-degree vistas. Look for hawks migrating in spring and fall. At the summit are balsam fir and red spruce. Along the nearby Salamander Trail lives the elusive Shenandoah salamander, found only in this park.

Old Rag Overlook: (SD-mp 46.5) Old Rag Mtn. (Fig. 6E-21) is the type locality for the Old Rag granite, a granite that was intruded into the older gneisses of the Blue Ridge Precambrian basement complex about 1,060 Ma while they were still buried thousands of feet below the surface of the Earth.

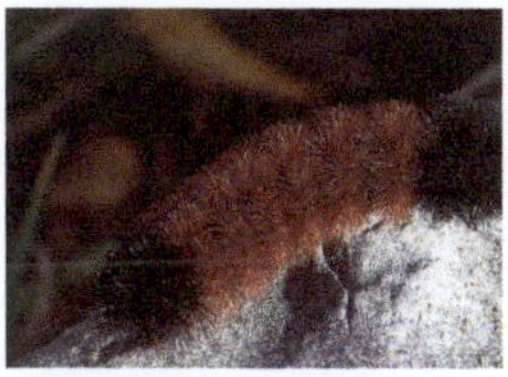

Woolly bear caterpillar

Dark Hollow Falls: (SD-mp 50.7) The parking lot marks the trailhead to Dark Hollow Falls (Fig. 6E-22) which is closer to the Skyline Drive than any other waterfall in the park. The trail is 0.5 miles down to the falls. The falls have formed at the contact between Catoctin greenstone and a Precambrian-age gneiss that has weathered, becoming less resistant to erosion.

Fig. 6E-22. Dark Hollow Falls. (Photo: Michele Fletcher, 2011)

Fig. 6E-23. Big Meadows in autumn. A number of rare grasses grow at Big Meadows in the swampy "fens" areas. Common to Big Meadows are the delicate red top grass and tall oat grass. Invasive Japanese stiltgrass (p. 97) can be seen in the picnic area. Many plants and birds found here are identified in Part 3.

Byrd Visitor Center at Big Meadows: (SD-mp 51) This is the main visitor center for the park. Here you can visit the museum and bookstore, attend lectures, take hikes with rangers, and obtain information about what is happening in the park. A lodge, cabins, and campground are available for overnight stays. Food is available at the lodge or the cafe next to the gas station.

Byrd Visitor Center

Big Meadows: (SD-mp 51) (Fig. 6E-23) Lowbush blueberries are prominent among the many plants that grow at Big Meadows (Fig. 6E-24). Studies of the current flora as well as pollen obtained from cores drilled in the unconsolidated sediments that lie beneath the meadow provide interesting insights to the past climate of this region. The modern trees in the meadow (including balsam fir, red spruce, dogwood, Canadian burnet, and Table Mtn. pine) are usually found at latitudes 4-5° north of this location. The present day cool

Black bear near the Byrd Visitor Center. Bears are attracted to the many berries at Big Meadows.

microclimate and moist soil make this a refuge for flora that prospered over large areas in the Blue Ridge during the last period of glaciation which ended about 11,000 years ago. During the last glacial period peak temperatures in this part of the Blue Ridge were about 10° C (50° F) cooler than present temperatures.

During the cold periods of the Ice Ages the valleys and hills around the Blue Ridge supported alpine tundra and the mountaintops were covered with patches of ice. Today a modern temperate Appalachian oak forest surrounds Big Meadows (Fig. 6E-3). Pollen studies show that this oak forest has gradually replaced a forest that had been composed mainly of conifers a few hundred years earlier. The present flora is not yet completely in equilibrium with modern climate. Vegetation continues to change as it slowly adjusts to modern climatic conditions (see Litwin, et al., 2004). Catoctin greenstone underlies most of Big Meadows.

Fig. 6E-24. Several types of wild berries grow at Big Meadows. Top row: closeup of Blue Ridge lowbush blueberries (Photo: Alan Cressler); lowbush blueberry plants turning flame-red during fall; black huckleberry. Bottom row: maleberry is also called "he-huckleberry" because, though similar to huckleberry, it produces dry capsules rather than juicy fruit (Photo: Fritz Flohr Reynolds†); deerberry (Photo: Ted Bodner†); and red elderberry (Photo: M.J. Richardson†).

Yellow warbler (Photo: Geoff Clarke†)

Wild turkeys climbing atop hay bales.

Virginia Birding & Wildlife Trail: (SD-mp 52.8) A favorite of birders, this 1.1-mile segment of the Appalachian Trail (Milam Gap to Tanners Ridge Fire Rd.) passes through apple groves, an old farm, and fields. The area is also an excellent place to see migrating birds in spring and fall.

Naked Creek Overlook: (SD-mp 53) View to the west. Note that the Great Valley is much wider as you look to the southwest beyond the southern end of Massanutten Mtn. During winter, the ski slopes at a resort located near the south end of Massanutten Mtn. may be visible. On clear days you can see across the Great Valley to the ridge on North Mtn.

Hazeltop Ridge Overlook: (SD-mp 54.5) Most of the retaining wall was constructed out of Blue Ridge basement gneisses but a large block of Antietam quartzite is located to the left of the marker for Powell Trail.

The Point Overlook: (SD-mp 55.5) Excellent views to the west. This is one of the few places where some of the sediment that was deposited between the extrusion of lava flows of the Catoctin Fm. is exposed along the Skyline Drive. Like the lava, these sediments have been meta-

Fig. 6E-25. "The Oaks"

morphosed. The resulting rock is a phyllite characterized by micaceous minerals formed from what was originally shale and silt. Some of the Catoctin greenstone is exposed at the south end of the outcrop across the road from the overlook.

Bearfence Mountain: (SD-mp 56.4) The parking lot marks the entrance to a 0.8-mile moderately difficult trail that climbs 275 ft. to a rocky summit with stunning panoramic views. Some scrambling over rocks is required.

Lewis Mountain Campground: (SD-mp 57.6) This site contains cabins, rooms, a campground, and a picnic area.

The Oaks Overlook: (SD-mp 59) Views to the west. Black, red, and chestnut oaks are prominent in the forest here (Fig. 6E-25). Note that the outcrop of Catoctin greenstones exposed here has a prominent southeast dipping (inclined) cleavage (p. 53). The curbstones contain excellent examples of Precambrian-age crystalline rocks and Catoctin greenstones. Note the large feldspar crystals in the metamorphic rocks.

Swift Run Gap Entrance Station: (SD-mp 62.7) Access is from US 33, which connects Elkton with Stanardsville, Va. The type locality for the Swift Run Fm. is located in this gap, but exposures are poor and the traffic on US 33 makes this dangerous for rock hunting.

South River Picnic Area: (SD-mp 63) The South River Falls Trail is considered one of the best places in the park for observing breeding birds in late spring/early summer. You might see a variety of warblers, Louisiana waterthrush, northern parula, yellow-throated vireo, rose-breasted grosbeak, scarlet tanager, and peregrine falcons

Hensley Hollow Overlook: (SD-mp 63.5) View to the west. Outcrops of the Catoctin greenstones show southeast dipping cleavage (p. 53). A small thrust fault can be seen in the cliff face across the Parkway from the parking area.

Swift Run Overlook: (SD-mp 67) The south end of Massanutten Mtn. is visible in the middle ground, Antietam quartzites form the ridges along the base of the Blue Ridge, and Precambrian gneisses are present in the outcrops beside the Skyline Drive.

Sandy Bottom Overlook: (SD-mp 67.5) During winter ice forms in the water that moves through the cracks in this exposure of granitic basement rock. As the water freezes the ice expands, it puts pressure on the rock and enlarges the cracks. Locust and red oak trees, and Virginia creeper are abundant here.

Bacon Hollow Overlook: (SD-mp 69) View to the east. The valley is flanked by steeply dipping (inclined) faults (Fig. 6E-26). One might think that the sides of the valley had been uplifted but, in fact, the central part of the valley that is underlain by Precambrian basement gneisses has risen relative to the flanks over time, which are composed of Catoctin greenstone (Figs. 6E-11, -27). On clear days, you can see ridges on the eastern flank of the Blue Ridge Province. Note the good

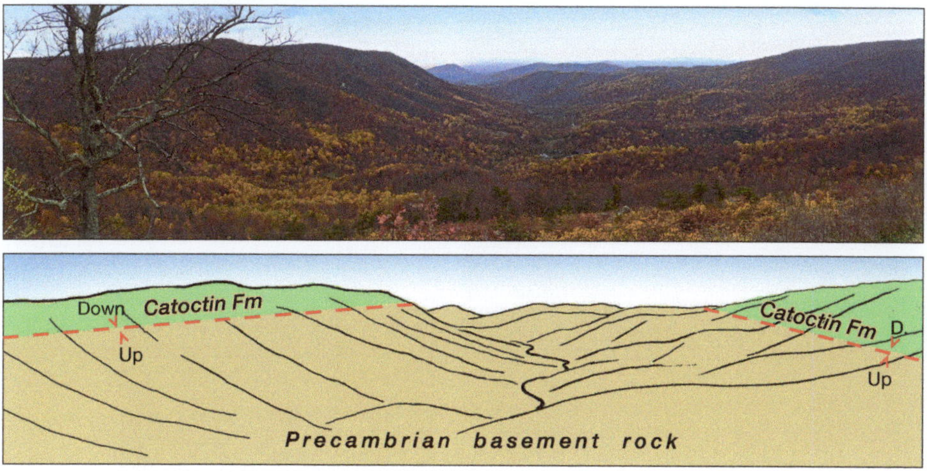

Fig. 6E-26. Photo and schematic drawing of Bacon Hollow showing the uplifted block of basement rock that has been eroded down to form the valley floor. The two faults do not merge in the distance.

Fig. 6E-27. Slicken-fibers in Catoctin greenstone.

Fig. 6E-28. Trails of the scolithus linearus worm are nearly perpendicular to the top of the layer of sand through which the worm was boring for food.

exposures of Blue Ridge basement gneiss at this overlook.

Eaton Hollow Overlook: (SD-mp 70.5) View to the northwest. Rocks exposed here are part of the Catoctin Fm. The view to the southwest is across a wide belt of Chilhowee Group units.

Rocky Mount Overlook: (SD-mp 71.1) Views to the west include the Blue Ridge flank (Chilhowee Group rocks) in the foreground, Massanutten Mtn. in the middle ground, and North Mtn. in the far distance. Good exposures of the Catoctin greenstone are present. The overlook is located in an oak-hickory forest.

Bobcats avoid humans and prefer remote woodlands, rocky ledges, and wetlands.

Loft Mountain Overlook: (SD-mp 74.5) The Catoctin greenstone exposed here contains slicken-fibers (Fig. 6E-27), fibrous minerals that have formed along small zones of slippage in the greenstone. If you run your fingers along the fibers, you will be able to determine the direction of the slippage. Note the red cedar trees on both sides of the Catoctin greenstones. On a clear day, you can see the eastern limb of the Blue Ridge Geological Province.

Brown Mountain Overlook: (SD-mp 77) Excellent views of the Blue Ridge front. The Harpers Fm. (middle unit of the Chilhowee Group) is exposed at this overlook. Note the talus slopes and the anticlinal structure formed by the Antietam Fm. in the distance. Good examples of the Antietam Fm. quartzite showing worm fossils (scolithus linearus) (Fig. 6E-28) are present in many of the curbstones. North of this overlook, the Blue Ridge front bends to the east. This

Eastern cottontail rabbit

is caused by the presence of a major thrust fault known as the Elkton fault. The area to the south has moved farther to the northwest than has the area to the north.

Fig. 6E-29. View from Ivy Creek Overlook of pines planted after a fire.

Ivy Creek Overlook: (SD-mp 77.5) View to the east. Note the change in the forest in the foreground of this view (Fig. 6E-29). The pine forest is present because of plantings made after a forest fire. Otherwise, this would likely be an oak-hickory forest.

Rockytop Mountain Overlook: (SD-mp 78) From this overlook you can see the Antietam Fm. along the Blue Ridge Mtn. front, the Great Valley, North Mtn. (part of the Valley and Ridge), and the south end of Massanutten Mtn. Excellent scolithus linearus fossils are exposed in the curbstone. The cliff across from the Skyline Drive is composed of the Harpers Fm. The rock exposed in this cliff is loose and could fall. ⚠ **Caution:** do not stand at the base of this or any other potentially dangerous cliff.

Loft Mountain Information Center: (SD-mp 79.5) Here you will find an information desk, small bookstore, snack bar, restrooms, campground, picnic area, numerous trails, camp store, and gas station that are open seasonally. There are plans for it to be developed into an educational and recreational destination.

Fig. 6E-30. Southwestern end of Massanutten Mtn. (canoe-shaped, upper right). The mountain is flanked by large valleys on both sides.

Big Run Overlook: (SD-mp 81) View to the west. The southern end of Massanutten Mtn. is said to resemble the end of a canoe (Fig. 6E-30). The valley of the South Fork of the Shenandoah River is visible. The talus slopes developed on the ridge in the foreground are composed of fragments of the Antietam Fm. quartzite. Note the anticline formed in the Antietam Fm. that is exposed near the crest

Loft Mountain

White-tailed deer fawns, born in late spring/early summer, tend to lie still, hiding in the brush for the first 4 weeks.

Raccoon (Photo: Dave Menke, USF&WS)

of the ridge to the northwest (Fig. 6E-31). Many such structures are present along the western margin of the Blue Ridge, but few can be seen as clearly as this one. Big Run Valley is the largest drainage area in the park. It includes most of the area you see from the Skyline Drive to the mountain front. Pitch pines, red oak, and hickory are present in the parking area.

Dundo Overlook: (SD-mp 83.5) From this overlook you can see where Stonewall Jackson led his men over the Blue Ridge through Fishers Gap on his way to Fredericksburg in the eastern Piedmont during his Valley campaign. Jackson, who taught physics at VMI before the war, was one of R.E. Lee's most important generals in the early years of the Civil War. After leading a remarkably successful campaign in the Valley, he was accidentally shot and killed by one of the soldiers guarding the front lines.

Blackrock Summit: (SD-mp 84.8) A short hike from the Blackrock Summit Parking Lot will take you to an excellent example of the downslope movement of large blocks (3-8 yards high) composed of quartz sandstone (part of the Harpers Fm. that became dislodged and moved) (Fig. 6E-32). This and many similar features are interpreted as having formed during the last major phase of glaciation (11,000-plus yrs. ago) At that time, freezing and thawing were more active than they are currently. The blocks may have been encased in snow and ice, forming what are called rock glaciers (Eaton, Hancock, and Lamoreaux, 2009). The forest between Blackrock Parking Lot and Trayfoot Mtn. Overlook has been burned, leaving many dead trees and a thick growth of young plants, especially ferns and laurel.

Fig. 6E-31. Anticline in the Antietam (Erwin) quartzite along the Blue Ridge flank.

Dundo Picnic Grounds: (SD-mp 85) Beautiful stands of mountain laurel grow along this section of the Skyline Drive. A campground, picnic area, and restaurant are located here.

Fig. 6E-32. Rock fragments form a talus slope like rock glaciers do in high mountains where the fragments become frozen into a mass of ice that moves downslope.

Horsehead Overlook: (SD-mp 88.5) View to the west. To the right you see good examples of talus slopes that are light in color on the distant ridge. The soil on these ridges, composed mainly of silica-rich quartzites from the Antietam Fm., contains few nutrients. Pines are among the few trees that survive on these quartzite ridges. In contrast, the southeast-facing slopes in the foreground, underlain by nutrient-rich soil, are covered by open oak forest that is growing on rocks similar to those that form the cliff across the Skyline Drive. It is composed of sandstones and shales of the Harpers Fm. Impurities of mud and silt cause the sandstones to have a dark grayish color (Fig. 6E-33, -34). They provide the nutrients used by the red oaks that abound in the forest. Iron and manganese stains are present on the rocks at the overlook. Fossils of bryozoans (sea lilies) and gastropods (snails) can be seen in the limestone blocks used to build the retaining walls. But these carbonate rocks don't occur in the Blue Ridge; they were brought in from the Great Valley to the west.

Riprap Overlook: (SD-mp 91.5) Riprap Ridge, the prominent ridge in the middle distance is composed of quartzites of the Antietam Fm. Blocks of the quartzite loosened by freezing and thawing in the winter slide and roll down the steep slopes forming blankets of talus. If you follow the trail to the ridge, look for worm trails of the primitive scolithus linearus. A parking area for those taking the trail to Riprap Ridge is located close to milepost 90 north of this overlook.

Fig. 6E-33. The Harpers Fm. The orange-brown discolorations are iron stains on a fracture surface that cuts through the layers of sandstone.

Fig. 6E-34. Harpers Fm. Note the flaky shale above the layers of sandstone.

Shenandoah National Park - VA

Note the limestone and dolomite blocks used in the retaining wall. One at the southern end of the wall contains a set of fractures filled with secondary calcite.

Moorman's River Overlook: (SD-mp 92) The Harpers Fm. is exposed in the outcrop across the Drive. (Fig. 6E-34). Limestone and dolomite from the Great Valley was used for the retaining wall.

Crimora Lake Overlook: (SD-mp 92.5) Views of the Blue Ridge western flank, the Great Valley, and North Mtn. in the distance. The walls around the overlook are composed of limestones and dolomites from quarries in the Great Valley.

Turk Mountain Overlook: (SD-mp 93.5) Views to the west. Quartzites of the Antietam Fm. hold up the ridge in the middle ground. Outcrops of sandstone, siltstone, and shale of the Harpers Fm. are present at this overlook and along the road from here to the Sawmill Run Overlook. The curb wall contains a few blocks of fossil-bearing limestones quarried in the Great Valley.

The eastern chipmunk uses its cheeks as pouches for food and to transport dirt when burrowing. They like the cover provided by rocky areas, old logs, and underbrush.

The eastern redbud is a native understory tree. Its distinctive pink blossoms can be seen in the Blue Ridge in late spring.

Sawmill Run Overlook: (SD-mp 95.3) View to the west. Limestone and dolomite from the Great Valley were used in the curb wall. Dolomite is lighter in color than the limestone and it has a smooth surface. Pitch pines, locust, and maples, are present at the overlook.

Sawmill Ridge Overlook: (SD-mp 95.5) Views to the west. Note the prominent talus slopes on the high ridge to the right. These slopes are composed of fragments of the Antietam Fm. When the leaves are off the trees you can see a paved road coming down a very steep slope in the distance. Runoff from the road has stripped vegetation from the side of the road and down the slope, illustrating the erosive power of runoff. The rock exposures here provide unusually good opportunities to examine outcrops of the Weverton (Unicoi) Fm. (the lowest unit in the Chilhowee Group) (Fig. 6E-35). Siltstones, pebble conglomerates, quartzites broken and cut by quartz veins, quartzites showing manganese-iron stains, and purple volcanic ash are present in these outcrops. Note the echelon pattern of some of the tension gashes (Fig 6E-36) now filled with quartz. This pattern usually forms as a result of deformation caused by a shear indicated by the arrows in the photograph. The fractures throughout this outcrop

Fig. 6E-35. Closeups of the Weverton Fm. (lowest unit in the Chilhowee Group) at Sawmill Ridge Overlook. Note the large isolated grains of quartz in the photo at left.

Fig. 6E-36. Quartz fills fractures in the Weverton (Unicoi) quartzite.

indicate that the rock was very brittle at the time the fractures formed. Short leaf pine, pitch pine, small oaks, and dogwood trees are present. The curbstones here are composed of carbonate rocks, limestones (dark blue gray) and dolomite (lighter gray) quarried in the Great Valley. These carbonate rocks are common in the Great Valley but are not found in this part of the Blue Ridge.

Calf Mountain Overlook: (SD-mp 99) The city of Waynesboro is situated at the base of the Blue Ridge. Views here are across the Great Valley to North Mtn., the first ridge of the Valley and Ridge Province in this area. South of here a large fold in the rocks that make up the Chilhowee Group causes the curve you see in the edge of the Blue Ridge.

Beagle Gap Overlook: (SD-mp 100) Although weathering and erosion have reduced the level of most of the core of the Blue Ridge Geological Province, the mountains you see in the core of the Blue Ridge from this overlook are much higher than those located farther north.

McCormick Gap Overlook: (SD-mp 102.5) The small knolls below this overlook are composed of the Antietam Fm. and constitute the edge of the Blue Ridge. From this point south, the rock outcrops located along the Skyline Drive are all lava flows of the Catoctin Fm. The town of Waynesboro, Va. lies at the foot of the Blue Ridge. The Skyline Drive ends and the Blue Ridge Parkway begins at the intersection of US 250 and I-64. You can collect a sample of the Catoctin greenstone on US 250 east of this junction. ❖

Female eastern swallowtail butterfly

Chapter 6F

Blue Ridge Parkway – from Waynesboro to the James River

Location and Access: The northern entrance to the Blue Ridge Parkway in Virginia from I-64 is at exit 99 for Rockfish Gap, where I-64 crosses the Blue Ridge between Waynesboro and Charlottesville. Access to the Parkway is also available where SR 56 crosses the Parkway between Steeles Tavern and Montebello, and also where US 60 crosses the Parkway between Buena Vista and Amherst. The southern entrances to this section of the Parkway are located where SR 130 and US 501 cross the Parkway near the James River.

Accommodations: Motels and restaurants are located at Waynesboro, Buena Vista, Lexington, Amherst, and Natural Bridge. A NPS run campground and seasonally available restaurant are located on the Parkway at Otter Creek about one mile north of SR 130.

Trails: There are many trails in the George Washington National Forest, which flanks the Parkway along most of this section. U.S. Forest Service topographic maps showing trails are available at the Waynesboro and Buena Vista Forest Service offices. The National Geographic map of the Lexington area also has hiking trails.

Nearby Attractions: Mary Baldwin College and Woodrow Wilson's ancestral home in Staunton; Cyrus McCormick's Farm near Steeles Tavern; Southern Virginia University in Buena Vista; Virginia Military Institute and Washington & Lee University in Lexington; and Natural Bridge State Park are all located in the Great Valley west of the Parkway. Crabtree Falls, one of the highest waterfalls in the Blue Ridge, is located a few miles east of the Parkway off SR 56. Wintergreen, a ski resort, is located east of the Parkway near its northern entrance and can be reached by SR 664. The Nature Foundation at Wintergreen **www.twrrf.org** operates out of the Wintergreen resort.

Above: Fig. 6F-1. View to the west from an overlook near Humpback Rocks close to the northern end of the Blue Ridge Parkway. Antietam quartzites hold up the ridges in the foreground. The Great Valley of Virginia is in the middle distance, and ridges of Silurian sandstone are visible in the far distance beyond the Great Valley.

History

Following the creation of the Shenandoah and Great Smoky Mtns. national parks, the idea of connecting them with a road through the mountains gained momentum. The Franklin Roosevelt administration used the Civilian Conservation Corps (CCC) to undertake this job. Much of the Parkway was constructed during the 1930s, but it was not completed until 1983 when the last section around Grandfather Mtn., N.C. was finished. Like the national parks at either end, the Parkway is a remarkable place to study geology and ecology. In southern Virginia the Parkway is used as a part of the local highway system and the park is very narrow there, but along most of its length the Parkway passes through national forests or along the edges of wilderness areas, making it a natural corridor for wildlife through the mountains.

Points of Interest

Overlooks Along the Blue Ridge Parkway from North to South

Most of this section of the Parkway passes through George Washington National Forest. A narrow strip of land beside the Parkway is part of the national park. Between Waynesboro and Roanoke, Va. the Parkway runs along the western edge of the Blue Ridge Geological Province. Overlooks provide superb viewpoints of the Valley and Ridge Province to the west and toward the core of the Blue Ridge to the east. South of Roanoke the Parkway crosses the Upland Plateau and runs a long distance near the eastern edge of the Blue Ridge with views toward the Piedmont. Near its southern end the Parkway crosses the interior of the Blue Ridge Highlands—the widest part of the Blue Ridge, as it passes into the Great Smoky Mtns. National Park.

Reminder

Plants and rock samples should not be collected along the Parkway. However, rock collectors may collect samples not far from the Parkway. The Catoctin greenstone may be collected at a pullout along US 250, 0.3 miles east of the northern entrance to the Parkway. Samples of the Chilhowee Group rocks may be found along roads leading to the Parkway from the west. Good exposures of the Chilhowee are also present in the James River Water Gap (Ch. 6G). Rocks of the Blue Ridge basement complex may be found on roads leading to the Parkway from the east.

Afton Overlook on I-64: Blue Ridge Parkway milepost 1 (mp 1) This overlook is located on the east side of I-64 a short distance east of the Parkway. If you leave the Parkway to reach this viewpoint, you must drive several miles farther east on I-64 to locate a place where you can reach the west bound lane that leads to the top of Afton Mtn. and the Parkway. The view to the east from this overlook is across the Rockfish Valley. Rocks in the valley and in the mountains farther east are Precambrian in age, part of the Blue Ridge basement complex known as the Shenandoah Massif. Ridges in the far distance are located near the eastern border of the Blue Ridge Province. Black Rock Mtn., 3,446 ft. high, is located to the south. Catoctin greenstones (Fig. 6F-2) are exposed in cliffs and at pullouts from

Fig. 6F-2. Slaty cleavage in the Catoctin greenstones was undercut during construction of I-64, leading to repeated landslides.

A house that was part of an early settlers village is preserved at Humpback Rocks.

the Parkway entrance to the exit onto SR 664. A large exposure is located about 200 ft. south of this overlook. Many rockslides occurred during the construction of I-64 along this part of the highway where roadcuts removed the base of the easterly inclined cleavage. High fences along this section keep loose rocks from reaching the paved surface. Two invasive plants dominate the plant community at this first, unnamed overlook. Kudzu is overrunning trees to the east of the viewpoint and ailanthus trees (tree of heaven) cover slopes west of the Parkway.

Humpback Rocks Visitor Center: (mp 8) Picnic tables and restrooms are available here. A trail to Humpback Rocks provides excellent views of the Valley and Ridge Province to the west and across the core of the Blue Ridge to the east. The rocks in the pullouts are Catoctin greenstones (Fig. 6F-3).

Fig. 6F-3. This outcrop of the Catoctin Fm. at Humpback Rocks contains fragments similar to modern aa aa lava flows (pronounced "ah-ah") in Hawaii and at other volcanic areas. The vaguely discernable blocks developed on the surface (of the flow) while the hotter, underlying lava continued to flow, after the top had cooled and become solid rock.

Fig. 6F-4. Slopes covered by talus composed of Antietam quartzite are seen from overlooks along the northern section of the Parkway.

Fig. 6F-5. Cleavage is responsible for sharp edges of the greenstone at Ravens Roost.

Greenstone Overlook and Trail: (mp 10) The ridges of the Valley and Ridge Province visible in the distance across the Great Valley are composed of Silurian-age sandstones. A major thrust fault on which the Blue Ridge moved to the northwest, the Little North Mtn. thrust fault, is exposed at the surface at the foot of the first of these ridges. Quartzites of the Antietam Fm. (see Fig. 6F-1 and the description on p. 80) hold up the ridge in the foreground and the low sloping surface that descends into the Great Valley. The Antietam is repeated here as a result of a thrust fault. Limestones, dolomites, and shales that once covered the quartzites have been eroded away leaving the resistant surface formed by the quartzites of the Antietam Fm. at the ground surface. These steep slopes are covered by fragments of Antietam quartzite (Fig. 6F-4), which fall from the quartzite outcrops. Catoctin greenstone, which lies stratigraphically below the Chilhowee is present at this overlook and along a marked trail that leads away from the overlook.

Ravens Roost: (mp 10) Southeast-dipping rock cleavage (p. 53) in the Catoctin greenstone (Fig. 6F-5) is responsible for the sharp projections that give this exposure its dramatic appearance. The overlooks along this part of the Parkway are good viewpoints for spotting migrating birds and for seeing wildflowers in the spring and fall. Note the talus slopes developed from outcrops of the Antietam Fm. on the ridge in the foreground. A large block of greenstone slid downslope from the east and now lies on top of outcrops in the middle of this overlook (Fig. 6F-6). The retaining wall at this overlook

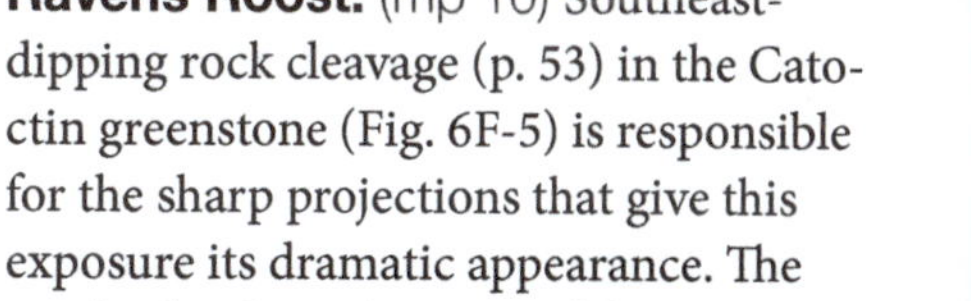

Fig. 6F-6. A block of Catoctin greenstone rests on an outcrop of lavas at Ravens Roost.

Artificial beach at Sherando Lake.

Recreational building near the beach.

is composed of rocks from the Precambrian core of the Blue Ridge. In contrast, the curbstone appears to be from one of the quarries in Paleozoic-age granites of the type found at Mt. Airy and Stone Mtn. State Park (Ch. 8D).

Crossing of SR 664 and Parkway: (near mp 14) Turn west at this crossroad to reach Sherando Lake Recreation Area. Turn east to go to Wintergreen Resort. The Appalachian Trail (AT) which continues south across the Three Ridges Wilderness Area crosses this intersection. A second (gated) road leading into the wilderness area is located near Love, Va., half a mile north of SR 814

Wintergreen Resort: (near mp 14) This popular place to ski in central western Virginia includes a hotel and restaurants, as well as apartments and homes. The Wintergreen Nature Foundation has a small museum featuring archeological specimens found in the area as well as local plants. A hiking guide to the Wintergreen area is available at the Foundation. Directions to the contact between the Blue Ridge basement rocks and the overlying Catoctin Fm. exposed along one of the hiking trails can also be found at the Foundation.

Sherando Lake: (mp 14) A Forest Service campground is located along the shores of the two lakes at Sherando Lake Recreation Area. To reach the lakes take SR 664 west down the mountain about 3 miles on paved roads from the Parkway or take SR 624 and SR 664 from Waynesboro, Va. This facility features swimming, camping, picnicking, hiking, and fishing.

Twenty Minute Cliff: (near mp 20) This pullout provides excellent views to the east down a section of the Tye River Valley (Fig. 6F-7). Shales, siltstones, and sandstones of the Unicoi Fm. (basal unit of the Chilhowee Group) are present in the cliff at this location. Outcrops here exhibit beautifully developed, southeast

Fig. 6F-7. View to the east from Twenty Minute Cliff.

View at milepost 28. The excellent views to the southwest from this point are across the western flank and core of the Blue Ridge.

dipping, slaty cleavage. The cliff gets its name from a phenomenon that occurs during the months of June and July when the sun drops behind the mountains 20 minutes after sunlight hits the rock.

Whetstone Ridge Visitor Center: (near mp 27.1) Located about 2.1 miles south of Tye River Gap. Restrooms and picnic tables are available here.

Tye River Gap and Crabtree Falls: (near mp 30) Exit the Parkway at milepost 30, take SR 56 six miles east to Crabtree Falls. With a drop of 1,200 ft., this is the highest combination of waterfalls and cascades in the eastern United States (Fig. 6F-8). ⚠ **Caution: Do not venture out onto the algae-covered rocks at the cascades;** more than 30 people have slipped and fallen to their deaths here. The falls consist of five major cascades and many smaller drops. Campgrounds and cabins are available near the falls at Montebello, Va. In 1969 when hurricane Camille passed across the region, and again in 1995 during Hurricane Hugo, this region was the site of major landslides caused by exceptionally heavy rains. In both instances, precipitation in excess of 30 in. fell during a single day. This type of rare event causes floods of a magnitude that might be expected once every few thousand years. Floods the size of those in 1969 and 1995 were catastrophic in the areas that received the most rain. During the storms the soil and loose rocks on the mountainsides became saturated with water and slid off of the underlying solid rock, leaving slopes stripped of their vegetation and soil down to the granodioritic (granitic rock containing more amphibole than

Fig. 6F-8. One of the cascades that comprise Crabtree Falls.

Fig. 6F-9. Aerial view of debris flow chutes and flood deposits from the June 27, 1995 flood at Kinsey Run, near Graves Mill in Madison County, Va. (Photo: L. Scott Eaton, from "The debris flows of Madison County, Va.," 34th Annual Va. Geological Field Conference Guidebook)

granite) bedrock. The scars are still visible at hundreds of places in the Blue Ridge (Fig. 6F-9). The mixture of soil, vegetation, and rock that covered the slopes slipped to the foot of the slope, forming fan-shaped deposits of debris. One of these buried a small community. Streams quickly cut their channels into the unconsolidated material of these debris fans.

Fig. 6F-10. McCormick Farm. The lake on this property is a favorite site for local birders.

Cascades and falls are often described as "spray cliffs." Jack-in-the-pulpit, saxifrage, white turtlehead, joe-pye weed, galax, bluets, grass-of-Parnassus, goldenrod, fern mosses, and violets are common ground covers. Tag alder, sweet pepperbush, climbing hydrangea, witch-hazel, smooth hydrangea, doghobble, laurel, and yellowroot are among the most common shrubs and vines to be found in spray cliff communities here. Red maples, black birches, and eastern hemlocks are often among the trees found in this environment (Ch. 10).

McCormick Farm and St. Mary's Wilderness Area: (mp 30) SR 56 to the west leads to two places that are worth visiting. The farm of Cyrus McCormick is located a few miles west of US 11 (Fig. 6F-10). Part of the farm has been restored. McCormick's harvester farm equipment is on display and the farm is still used by Virginia Polytechnic Institute & State Univ. as an agricultural experimental station. The gristmill is still in use. This exit also leads to St. Mary's Wilderness, one of the most popular hiking spots in the Northern Blue Ridge. Note that there are no tourist facilities in the wilderness area. To reach the wilderness area go west from the Parkway on SR 56, turn right (north) at the town of Vesuvius, and drive about 1.75 miles on SR 608 where it veers sharply to the right and passes under a railroad overpass. After about a quarter of a mile turn right onto Coal Rd. and follow it until you come to a gravel Forest Service Road on the left. This road ends in a small parking lot at the trailhead. The trail through St. Mary's passes through the Chilhowee Group. Flooding has filled the valley floor with pieces of the Antietam Fm. quartzite.

Fig. 6F-11. View from the Parkway to the southwest over the western flank of the Blue Ridge. The ridge in the foreground is composed of Chilhowee Group rocks. A ridge in the distance, Purgatory Mtn., is in the Valley and Ridge Province.

Irish Creek Overlook: (mp 36) Chilhowee Group sandstones, shales, and conglomerates hold up the ridge in the foreground. Outcrops of the Antietam Fm. are visible along the edge of the Blue Ridge. The ridges visible in the distance on the west side of the Great Valley are composed of erosion-resistant Silurian-age sandstones. Notice that the Great Valley is much narrower here than it is farther north.

Overlook at the intersection of the Parkway and US 60: (mp 48) The series of sharp peaks in the middle ground, held up by the Antietam Fm. quartzite, are located along the western flank of the Blue Ridge (Fig. 6-11). The Unicoi Fm. is exposed across the Parkway from the overlook. In the middle distance, the western edge of the Blue Ridge is much farther west than it is between this overlook and the James River Gap. It has been thrust out over the rocks of the Great Valley and almost closes the Great Valley in the far distance (about 25 miles to the southwest).

Indian Gap: (near mp 48) A short hike on a well worn trail leading off the parking area takes you to an exceptional exposure of huge, rounded boulders (Fig. 6F-12). This granitic rock, part of the Blue Ridge basement complex exposed here is unusual in that it contains minor amounts of pyroxene minerals combined with large quartz and feldspar minerals. The rounded boulders are a product of spheroidal weathering, a process in which

Fig. 6F-12. Boulders formed by weathering of bedrock at the crest of the Blue Ridge at Indian Gap (left). Blocks of granitic rock at Indian Gap (right).

Sidetrip: Travel along US 60 from the Parkway to Buena Vista

Exit the Parkway near milepost 48, heading west on US 60 toward Buena Vista, Va. (Set your odometer to zero.)

Fig. 6F-13. Huge megaripples in the Unicoi Fm. exposed along US 60 east of Buena Vista.

Stop 1. **Ripples in Unicoi Fm. Exposure:** (0.3 miles) Very large ripples are present on a steeply dipping exposure of the Unicoi Fm. (Fig. 6F-13).

Stop 2. **Unicoi/Blue Ridge Basement Contact:** (1.4 miles) The contact between the Unicoi and underlying Blue Ridge basement complex is visible here.

Stop 3. **Entrance to Reid Hollow Road:** (1.5 miles) Good exposures of the basement complex occur along this road. Some of the hollows along this road are excellent places to look for plants and birds.

Stop 4. **Exposures of Unicoi Conglomerates:** (1.9 miles) Good exposures of intraformational deformation in the Unicoi and Harpers Fms. occur along the next mile and a half along US 60, but there are few places to park.

Stop 5. **Entrance to a Forest Service Maintenance Site:** (3.4 miles) The Antietam Fm. is exposed across the road. A Cambrian-age rock unit, the Waynesboro Fm. that was deposited in a tidal flat, lies beneath Buena Vista. Alluvium and floodplain deposits cover the Waynesboro Fm. in the lower part of the town. The Maury River, constrained by levees and floodwalls, flows through the town. It joins the James River a few miles to the southwest where the James turns abruptly and flows across the Blue Ridge.

Stop 6. **Forest Service Office** (Approx. 4 miles) The office is located on the right side of the road. You can purchase local maps and obtain information about George Washington National Forest here.

Fig. 6F-14. Western flank of the Blue Ridge at Buena Vista. Note the triangular-shaped mountains, called flatirons, which are held up by Antietam (Erwin) quartzites.

Classic Flatirons

If you are approaching the Parkway from Lexington or driving along I-81 (or US 11), you will see classic examples of a topographic feature referred to as flatirons along the mountain front (for a general description see p. 37). The name is applied to inverted v-shaped landforms (Fig. 6F-14) that develop where streams cut across inclined layers of erosion resistant rock. At this locality, the resistant layer is the Antietam (Erwin) Fm., a layer of sandstone and quartzite that is inclined to the northwest (Fig. 6F-15). These flatirons are part of the western limb of the Buena Vista anticline (Fig. 6F-2), which continues to the southwest into Arnold Valley where the anticline plunges and disappears beneath a thrust sheet that carried rocks of the Blue Ridge basement complex over the anticline.

There are excellent views of the mountain front from Glen Maury Park in Buena Vista. The park, which has good camping facilities, is located at the end of West 10th St., west of US 501 in Buena Vista. A large shelter in the park is a fine place to see the flatirons and the dramatic change in landforms that takes place at the contact between the Blue Ridge and the Valley and Ridge Provinces.

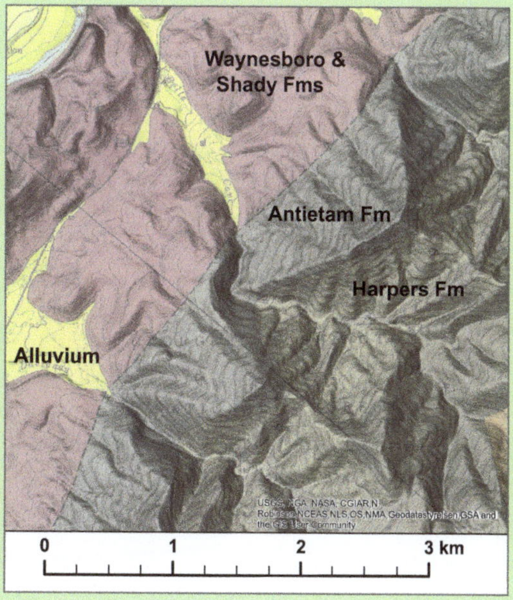

Fig. 6F-15. Geologic map of several of the flatirons formed along the Blue Ridge near Buena Vista. The Antietam Fm. (quartzites and sandstones) is the resistant unit creating the flatiron shape. The Waynesboro Fm. is a shale unit. The Harpers Fm. is composed of sandstones and shales. (USGS map, courtesy of David Harbor).

chemical weathering takes place along fractures and bedding planes. The corners where fractures intersect are subject to more weathering and so gradually decay, leaving behind rounded masses. Some slight gravity-induced movements have opened spaces among the boulders.

House Mountain Overlook: (mp 49.2) Excellent views to the west include House Mtn., a prominent landmark for this area. Silurian sandstones that hold up the ridges to the north and south of House Mtn. also cap this mountain. On the east side of the Parkway a large landslide induced by roadcuts, has left the bedrock exposed (Fig. 6F-16). Note the very thin layer of soil, basically a mat of organic material, at the top of the cut. Chemical weathering along fractures in the Precambrian granitic rocks extending deep into the cut contributed to the instability and eventual slide that was triggered by cutting the toe of the slope during Parkway construction.

Between Parkway mileposts 51 and 52, the AT crosses the Parkway. An unmarked parking area is present on the east side of the Parkway.

Bluff Mtn. Overlook: (mp 52.6) This is a good site to look for wildflowers in the spring.

Rice Mtn. Overlook: (mp 53.4) Excellent exposures of weathered shale of the lower Chilhowee Group are about 100 yards north of the overlook.

White Oak Flats: (mp 55) Chestnut, white, and red oaks, maples, and sycamore are present.

Between White Oak Flats & SR 130: (mp 55-61.5) This section and the pullouts along Otter Creek provide an excellent opportunity to examine plants that prosper along streams in the Blue Ridge.

Otter Creek Facilities: (mp 60.7) A seasonal restaurant and NPS run campground are located here.

Intersection of Parkway & SR 130: (mp 61.5) (Stop 11 on p 161, Fig. 6G-4) Wildwood Campground is located about 1.2 miles east of this intersection on SR 130. The closest motels are located in Lynchburg. A restaurant can be found in Big Island, Va. To reach Big Island continue south on the Parkway across the James River and then take SR 501 east about one mile. SR 130 leads to Lynchburg to the east and to Glasgow to the west. Good exposures of the lower Chilhowee Group siltstone, shales, and phyllites are present along the Parkway near this intersection.

Fig. 6F-16. Landslide along the Parkway was caused by the undercutting of the steep slope. Bedrock is well exposed. Do not climb on this surface.

Lower Otter Creek Overlook and Otter Lake Overlook: (mp 62.7) (Stops 12 and 13, Fig. 6G-4) The rocks exposed at these pullouts are Precambrian augen gneisses (Fig. 6F-17). They obtain their name from the eye-shaped (augen) feldspar minerals that are present in the gneiss. Much of this unit is deeply weathered. Individual feldspar crystals are drawn out producing a streaky appearance. Large and less weathered crystals are present in outcrops at the southern end of the Otter Creek Overlook near the entrance to Otter Lake Loop Trail. This zone of intense deformation continues many miles to the northeast where it is known as the Rockfish Valley fault zone. The Stops along Otter Creek are located near the leading edge of a major thrust fault that grows out of the Rockfish Valley fault zone and places Precambrian basement rocks against the lower Chilhowee Group rock units. This fault is connected with a network of other thrust faults that have been traced to the west into Arnold Valley. This fault continues around Arnold Valley and the mountains to the northeast (Fig. 6G-3) and continues to the southwest along the western edge of the Blue Ridge. The fault contact is located close to the place where the Parkway passes over SR 130. Intensely sheared and recrystallized mylonite is present in the fault zone. This faulting is thought to have taken place during the Alleghanian Orogeny when Appalachian mountain building reached its culmination.

Fig. 6F-17. Photo of a cut and polished slab of the highly sheared augen gneiss seen at Lower Otter Creek Overlook.

James River Visitor Center: (mp 64) (Stop 14, Fig. 6G-4) Picnic grounds by the James River, restrooms, and a seasonal visitor center are located here. A short trail that begins across the Parkway provides excellent opportunities to look for flowers, trees, birds, and rocks. A walkway beneath the Parkway bridge (Fig. 6F-18) leads to a restored river canal lock, one of many that carried canal traffic along the James River and into the Maury River to the west. This canal operated a short time after the Civil War, but was soon put out of business by a railway linking Richmond with towns west of the Blue Ridge.

Another trail across the Parkway crosses excellent exposures of the same fault zone present at Otter Creek Overlook. Many of the trees along this path have identification markers and the trail provides good places to look for birds along the river. ❖

Fig. 6F-18. Walkway under the Parkway Bridge that crosses over the James River.

Chapter 6G

James River Traverse Across the Blue Ridge

Location and Access: The route of the geologic traverse described in the text travels from west to east through the water gap formed where the James River cuts across the Blue Ridge in Virginia. This traverse begins near Glasgow at the intersection of US 501 and SR 130. From the Parkway take the exit onto SR 130, a few miles north of the James River and drive west to the intersection of SR 130 and US 501 where the traverse begins. To reach the beginning of the traverse from I-81, exit I-81 at Lexington, follow US 60 to Buena Vista; continue on US 501 to SR 130, near Glasgow, where the traverse begins. An alternative route from I-81 would be to exit at Natural Bridge and follow SR 130 to Glasgow where it intersects US 501. From Lynchburg take either SR 130 or US 501 to the west. The Blue Ridge Parkway crosses both of these routes in the water gap.

Accommodations: A NPS campground is located on the Parkway at Otter Creek (about one mile north of the intersection of the Parkway and SR 130). The visitor center here is open from April to October. Food and a private campground are located on SR 130, one mile east of the Parkway entrance on SR 130. Motels and restaurants are located in Buena Vista, Lexington, and Natural Bridge near the western end of this traverse. There are also campgrounds at Glen Maury Park in Buena Vista and in Arnold Valley near Natural Bridge Station. Lynchburg is located on the James River at the Blue Ridge's eastern edge.

Nearby Attractions: Natural Bridge State Park and The Caverns at Natural Bridge are about 10 miles west of the James River Gap; Washington & Lee Univ., Virginia Military Institute, and Southern Virginia Univ. are 12-18 miles west of Glasgow.

The James River water gap is an outstanding place to examine the geologic structure of the western flank of the Blue Ridge (Fig. 6G-1). US 501, SR 130, and the C&O Railroad follow the James River across the Blue Ridge between Glasgow and Lynchburg, Va. From Buchanan to Glasgow, Va., the James River flows northeastward along the eastern edge of the Great Valley of Virginia. At Glasgow, the river is joined by the Maury River and abruptly turns to the southeast and flows across the Blue Ridge to Lynchburg, Va., where it turns and flows

Above: Outcrops of the Antietam Fm. in the James River at Balcony Falls, near Glasgow, Va.

Fig. 6G-1. Aerial view toward the west at the entrance to the James River Gap. Glasgow is located near the top of the photo.

northeastward along the eastern edge of the Blue Ridge (Fig. 6G-2). How the river evolved in such as way as to flow directly across the highly resistant rocks of the Blue Ridge remains a subject of debate among geologists (see discussion in Ch. 3). One might expect that the river is following a fault zone, a system of fractures, or some weakness in the bedrock but that is not the case in the James River Gap. Highly resistant rocks, quartzites and sandstones, can be traced across the river and are visible in the channel. We are left with two theories: either the river cut across the Blue Ridge by headward erosion from the Piedmont; or it was superimposed and cut downward into the rocks of the Blue Ridge from a much higher elev. where it flowed from northwest to southeast.

Rock exposures on the James River are also present where it crosses the Blue Ridge's eastern flank, but these are not as readily accessed as those on the western limb.

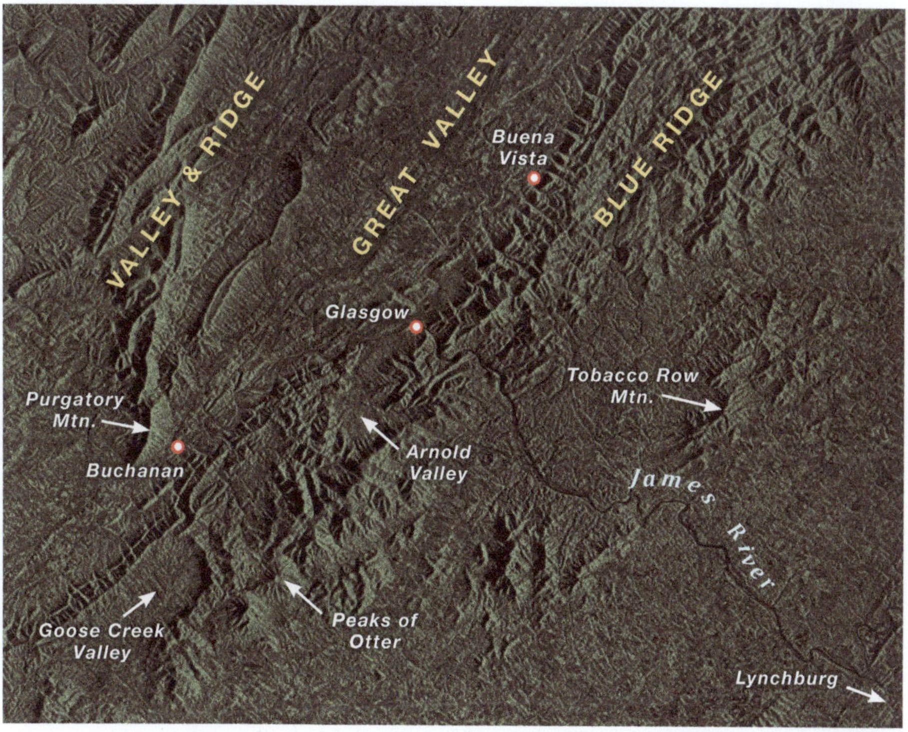

Fig. 6G-2. Side-looking radar image of the James River Gap region. (USGS map, with additions)

Ecological Communities in the James River Gorge

The James River has cut a deep gorge across the western flank of the Blue Ridge between Balcony Falls and Big Island, Va. A number of ecologists have studied plants along the river and in the James River Face Wilderness. Gwynn Ramsey, et al., (1993) worked in the area for fifteen years, collecting and identifying 119 plant families, 468 genera, and 963 taxa. This area of highly diverse flora contains large collections of grasses and sedges (including alfalfa, water chestnut, cotton-grass, sawgrass, nutsedge), members of the pea family, kudzu, soybean, peanut, and many flowering plants (asters, daisies, sunflowers, broom, and members of the rose family), also a great variety of shrubs and trees.

Ramsey, et al., grouped these plants into six ecological communities. Many of these communities are present along the routes taken in this traverse.

Loamy Floodplains

Cliffs and steep slopes are prominent along the southwestern edge of the James River in the gorge. Active rail lines, roads by the tracks, and what remains of an old canal system are present on the narrow floodplain on the northeastern side of the river. Natural floodplains are not well developed anywhere within the gorge. The narrow, undeveloped floodplains, typically a few tens of feet wide, are occupied by ferns, horsetails, club-mosses, and a variety of sedges and grasses. Despite the narrow width, these flat areas with their rich loamy soil are good places to see wildflowers.

Disturbed Areas

Invasive plants are prominent in the eco-communities along the railroad, abandoned canal tow-paths, and trails. Wetland habitats have developed in areas of stagnant water such as sluices and partially inundated canal bottoms.

Kudzu

This alluvial fan formed on the narrow floodplain at the end of the AT bridge across the James River.

Loamy floodplains are good places to see wildflowers such as the Dutchman's breeches (shown).

Small portion of a floodplain near a bridge where the AT crosses over the James River.

Exposed Rock Outcrops & Barrens

A few plants are well adapted to the harsh conditions created by these dry environments that are subject to strong winds, thin soils, and silica-rich bedrock. In large areas of the James River Face Wilderness, barrens underlain by shale and siltstone are similar to those found in the western side of the Valley and Ridge Province. Hickory Stand Mtn. is the best example of these communities where ferns, notably spleenwort are prominent.

Dry, Well Drained Wooded Slopes

A distinctive oak-pine-heath community is present on many of the eastern and south facing slopes that are very dry and covered by thin, acidic soils. Virginia pine, chestnut oak, and blackberry prosper here. This area is near the northern limit of the range of Carolina hemlocks.

Upland Seasonal Depression Ponds

Shallow seasonal ponds located on ridges in the Blue Ridge provide water for game animals. Two of these are located nearby: one on Peavine Mtn. (southwestern corner of the Buena Vista 1:24,000 quadrangle) and another to the northeast of the James River, Sheppe Pond (northeastern corner of the Snowden 1:24,000 quad). Because these ponds are rare in the Blue

Dry stream bed on a steep slope near the James River.

Ridge, they have been sites of intensive study. These two are surrounded by oak-pine woods on acidic, well-drained soils, and are sites for rare, localized "relic" plant species.

Early spring blossoms of the wild cherry tree.

Springs and "Drippy" Rock Outcrops

Water that has infiltrated the soil at higher elevations often finds it way to the surface by moving along bedding planes, fractures, cleavage planes, and faults in the bedrock. These are often found in cliffs, especially those located at low elevations near streams. In winter months ice forms at the surface. Ferns, grasses, sedges, mosses, and many wildflowers prosper on these outcrops.

Drippy rock outcrop.

The combative common snapping turtle lives in and around shallow streams and ponds. (Photo: Bob Cherry, NPS)

One of the largest streams in the James River Face Wilderness flows through Sulphur Spring Hollow. Located in Arnold Valley, the large boulders here are evidence that it becomes a raging torrent during periods of heavy rainfall.

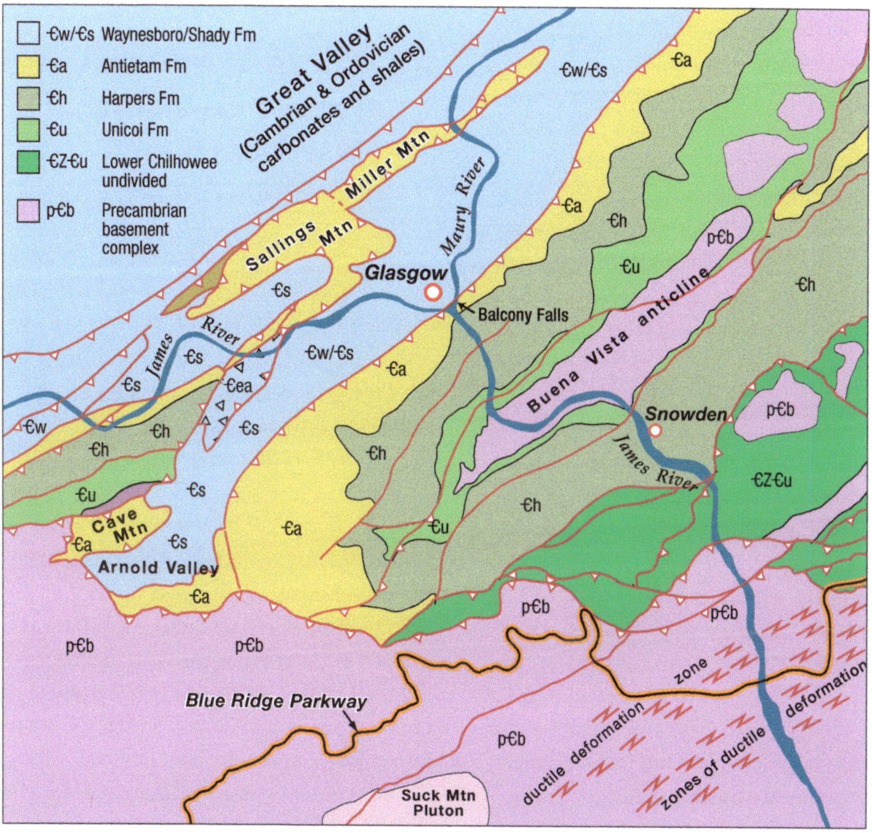

Fig. 6G-3. Geologic sketch map of the western flank of the Blue Ridge at the James River Water Gap. (Modified after Spencer, 1994, and Carter, 2012) Note: a more detailed geologic map of the areaalong the Parkway has been published by the USGS (Carter, 2012).

Geologic Setting

As at Harpers Ferry, the Blue Ridge at the James River has the overall structure of a large anticlinorium. Paleozoic sedimentary rocks floor the Great Valley of Virginia to the west and metamorphic rocks of equivalent age lie in the Piedmont to the east (Fig. 5-1). Several large folds and associated faults involving the sedimentary cover are present along the western limb of the anticlinorium. These folds involve the Blue Ridge basement rocks as well as the sedimentary cover. The southeastern limbs of these folds are cut by thrust faults. One of these thrust faults rises out of a zone of intense deformation in the basement complex, the Rockfish Valley fault zone. Rocks in this zone were deformed under incredibly high temperature and great pressure. They were intensely sheared and yielded to the deformation as ductile material, becoming something like hot asphalt. The younger and more brittle faults that later rose from this ductile zone cut through and carried the sedimentary cover rocks, as well as a sheet of the Blue Ridge basement rocks to the northwest. This sheet moved at least six miles (many geologists think the distance is much greater) to the northwest and passes over the southwestern end of the Buena Vista anticline (Fig. 6G-3). A long slice of this thrust sheet makes up Miller and Sallings mountains (Fig. 6G-3). This fault continues to the southwest along the western edge of the Blue Ridge and is part of a system of faults that extends along the northwestern edge of the Blue Ridge to Alabama where much younger rocks of the Coastal Plain cover it (Figs. 1-2, 1-4). The following traverse identifies places of geologic interest through the James River gap.

The Precambrian rocks were metamorphosed about 1 billion years ago during the Grenville Orogeny (Table 6G-1). The metamorphism took place at great depth, perhaps as much as 12 miles deep in the earth. As mountain building continued these metamorphosed rocks were uplifted. Over the many millions of years following that uplift, the mountains were eroded until the peaks disappeared and the land was near sea level. Streams carrying sand and gravel flowed across this landscape. Much of the sediment we now see in the Unicoi Fm. was deposited in these streams. The continental margin was unstable and subsided as sediment accumulated on the seafloor. These largely marine sediments composed of clay and sand settled, forming the Harpers Fm. Finally the Iapetus Ocean (Ch. 2) advanced inland and beaches formed along the shoreline of this sea that ultimately invaded the interior of North America as far west as Wisconsin.

Points of Interest

Geologic Traverse of the Western Flank of the Blue Ridge

Two distinct traverses across the western flank of the Blue Ridge are included here. The first one is along US 501 and SR 130 with additional information for those who want to know what lies farther east. The second traverse is a side trip into Arnold Valley (marked "AV" on the map).

Stop 1. **Start of traverse in the Great Valley:** (Mileage set to 0)
The first Stop (Fig. 6G-4) is located at the intersection of US 501 and SR 130 on the eastern edge of Glasgow, Va. The broad flat area to the northwest is the floodplain of the Maury River, which joins the James River at the mountain front a short distance downstream. Water covered this floodplain and parts of the town of Glasgow in 1969 and 1986. The river abandoned its old channel and shifted toward the highway during the 1969 flood. It continues to erode the sediments that had been deposited in this area behind a dam, subsequently removed, that stood at the entrance to the water gap (Fig. 6G-5). The mountains to the west are parts of the Sallings Mtn. klippe (Fig. 6G-3). Remember that a klippe is an isolated fragment of a thrust sheet. A nearly horizontal thrust fault lies near

Geologic Period	Description
Cenozoic (0-66 Ma)	Stream gravel; rocks moving downslope (talus, colluvium)
Paleozoic (246-546 Ma)	The upper part of the Chilhowee Group is Cambrian-age (p. 78)
Precambrian (over 546 Ma)	The lower part of the Chilhowee Group is Precambrian-age
	Greenstone dikes and flows
	Blue Ridge basement complex (between 1 and 1.3 Ga)

Table 6G-1. Rocks and rock units in the James River Gap.

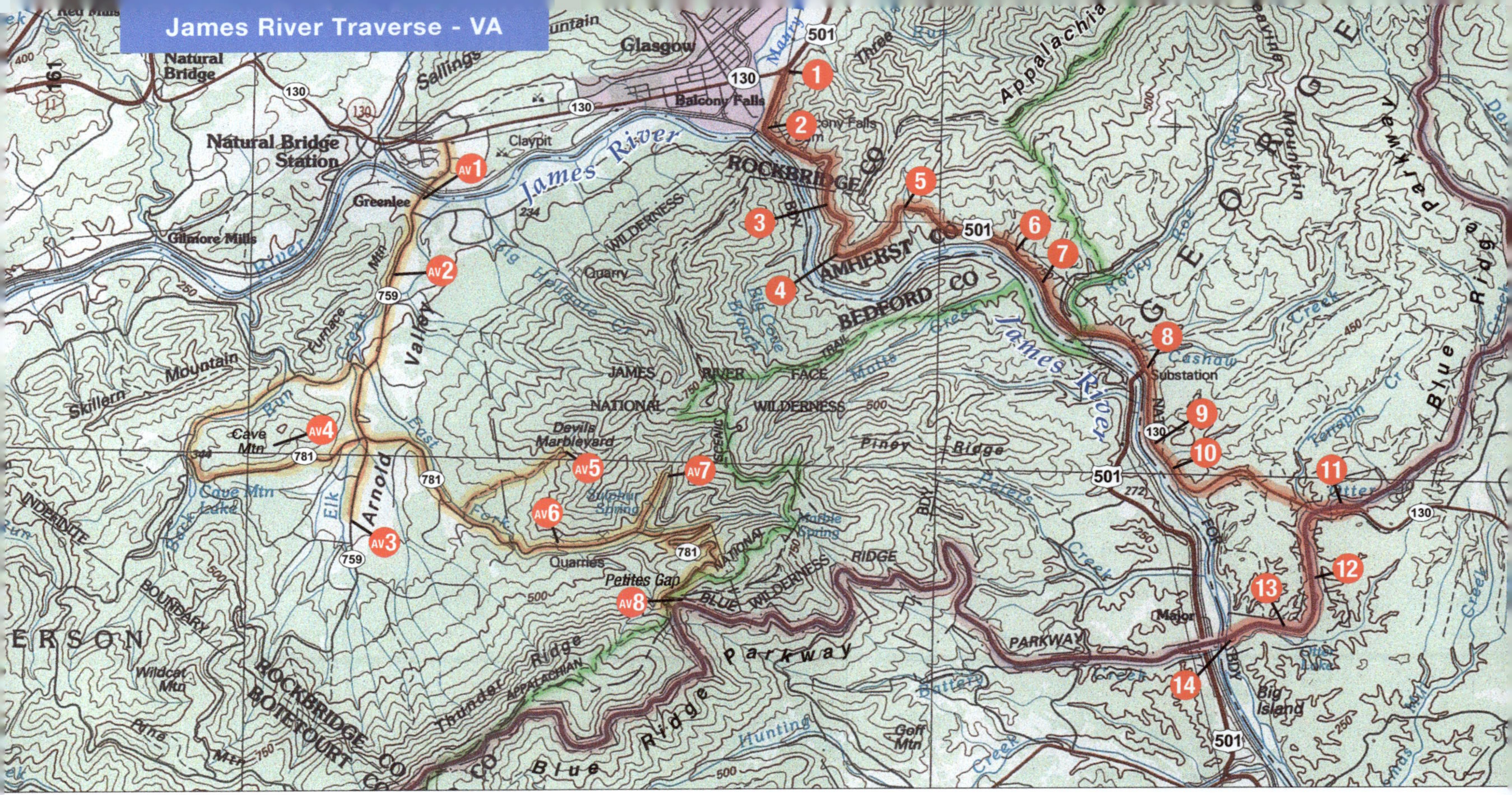

Fig. 6G-4. Route map with Stops along the two traverses described in this area, marked with red dots. Stops 11-14 are detailed on pgs. 152-3. Dots with "AV" before the numbers refer to the Arnold Valley Traverse. The Appalachian Trail is marked in green. (USGS map, with additions)

Fig. 6G-5. Entrance to James River gap at Balcony Falls. Detail of the Balcony Falls area where the Maury and James rivers join (left). The photo above offers a regional view. Carbonate rocks, which are present under the floodplain in the foreground, were deposited on top of the quartzites which are prominent along the mountain front. The carbonate rocks are no longer exposed here, but they are exposed at the surface farther south along the mountain front off the Balcony Falls Trail shown on the USGS 7.5-minute Snowden Quadrangle map. Tobacco Row Mtn. is faintly visible in the distance.

ground level at the base of this mountain. The rocks that compose these mountains, Cambrian-age sandstones and quartzite (Antietam Fm.), were transported at least several miles from the southeast to their present position. It is probable that the Buena Vista anticline was not present at that time, that the folding that formed the anticline occurred after the rocks on Sallings Mtn. had already been moved.

Stop 2. **Balcony Falls, confluence of the James and Maury rivers, the Antietam Fm:** (Mileage = 0.5) A pull-out on US 501 located at the entrance to the James River gap provides a good place to see the Antietam Fm. Watch for traffic at this bend in the road. Safer outcrops are present in Arnold Valley (p. 168). The Antietam is composed of sandstones that formed on beaches along the shore of the Iapetus Ocean that was invading the continental margin. Cross-bedding (Fig. 6G-6) and worm borings left by scolithus linearus that moved vertically down into the soft beach sand are well preserved in the sandstones. The sand grains are cemented together by silica which makes this an exceptionally hard rock that is resistant to weathering and erosion. Directly across the river at this point the mountain slope shows the inclination (dip) of the layer. The carbonate rocks that once lay on top of the Antietam Fm. have been eroded down to ground level.

The contact between the Antietam Fm. and the underlying Harpers Fm. can be recognized by the change from the clean white sandstone of the Antietam to the darker sands and shales of the Harpers Fm. They contain clay and other minerals

Fig. 6G-6. Exposure of the Antietam quartzite with vertical worm borings and cross-bedding that may have formed within a sand dune (left). Shale fragments in the Harpers Fm. (right).

that give the rocks a darker color. Because the Harpers Fm. is over a thousand feet thick and is folded, it is exposed for almost two miles along the highway.

Fig. 6G-7. Rapids form where resistant quartzites of the Unicoi Fm. cross the James River.

Stop 3. **The Harpers Fm. and view of the James River Face:** (Mileage = 2.0) This viewpoint, located in the Harpers Fm., provides good views of the river, the James River Face Wilderness across the river, and the entrance to the water gap. The Sallings Mtn. klippe is visible upstream in the distance. Rapids and riffles in the river reveal the place where the Unicoi Fm. crosses the James (Fig. 6G-7). Anticlines and synclines are exposed in the cliff across the road. A layer of reddish sandstone cemented by hematite, an iron oxide mineral, is exposed at road level. During early exploration of this region, many prospect pits were dug into this layer, but the low quantity of iron and the large percentage of silica discouraged development of large mines in this unit.

Fig. 6G-8. Coarse-grained sandstone and pebble conglomerate tightly cemented by silica makes the Unicoi Fm. very resistant to erosion and explains why it stands out in the river as seen in Fig. 6G-7.

Stop 4. **The Unicoi Fm:** (Mileage = 2.5) The Unicoi Fm., the rock unit that lies immediately below the Harpers Fm., is exposed at this Stop. Granules and sand-sized grains of quartz and pebble

Fig. 6G-9. Aerial view of ridges along the northwestern flank of the Blue Ridge north of the James River gap. The Great Valley of Virginia is near the top of the photo. The ridge in the mid-distance is held up by the Antietam (Erwin) quartzite (Fig. 6G-6). The ridge in the foreground is held up by quartzites of the Unicoi Fm. The road at the bottom is US 501.

conglomerates that were deposited in shallow water make up this unit (Fig. 6G-8). Large ripple marks and cross-bedding found in this layer indicate that it was deposited in a stream. These quartz-rich layers are tightly cemented by silica, making the Unicoi strong and very resistant to weathering and erosion. They form the high ridge across the river to the south. They also extend to the northeast, forming the Rocky Row Ridge (Fig. 6G-9).

Stop 5. **The Great Basement/Cover Unconformity:** (Mileage = 3.5) The contact between the base of the Unicoi and the underlying Precambrian metamorphic complex is exposed on the north side of the road at this pullout. At this outcrop, we see deeply weathered Precambrian gneisses that may have been ancient soils beneath the Unicoi. Fresh, solid gneiss is exposed a short distance east of the contact, and former gravels containing cobbles of gneiss are present a short distance to the west in the stream valley that comes down to road level here.

Stop 6. **The Snowden Fault:** (Mileage = 5.0) This small quarry marks the location of a steeply-inclined fault. Precambrian gneisses crop out of the west (upstream) side, and strongly folded sandstones and shale of the Harpers Fm. are present on the east side. Except for a layer of conglomerate at the contact, the Unicoi Fm. is missing here. The sense of movement is that of a normal fault (Fig. 6G-10), but the folds in the Harpers clearly show that the crust in this area was being compressed. Some of the folds are of a type known as isoclinal (with both limbs of the fold are nearly parallel). Two interpretations of this structure have been advanced. One is that a normal fault formed after the Harpers Fm. was deposited, and was followed later by compression (probably during the late Paleozoic Alleghanian Orogeny) that caused the folds to form in the Harpers Fm. The other interpretation is that the normal fault formed during the Mesozoic Era as Gondwana broke up. Africa pulled away from North America at this time and the Atlantic Ocean began to form.

Stop 7. **Snowden Member of the Harpers Fm:** (Mileage = 5.1) The Snowden is very similar to the Antietam in texture, color, and fossil content (scolithus linearus). It is likely a beach deposit formed when the sea receded during deposition of the Harpers Fm. The abandoned Snowden post office is at 5.3 miles.

Appalachian Trail crosses the James River: (Mileage= 5.8) The AT crosses the river on an abandoned railroad bridge that ends in a parking lot located by SR 130. From this crossing, the AT continues northward to Rocky Row Ridge and the highest peaks along the Blue Ridge in this area. To the south the AT goes into the James River Face Wilderness. About 11 miles across the wilderness area the AT comes back close to the Blue Ridge Parkway at Petites Gap (Fig. 6G-4) where FS 781 crosses the Parkway.

a

b

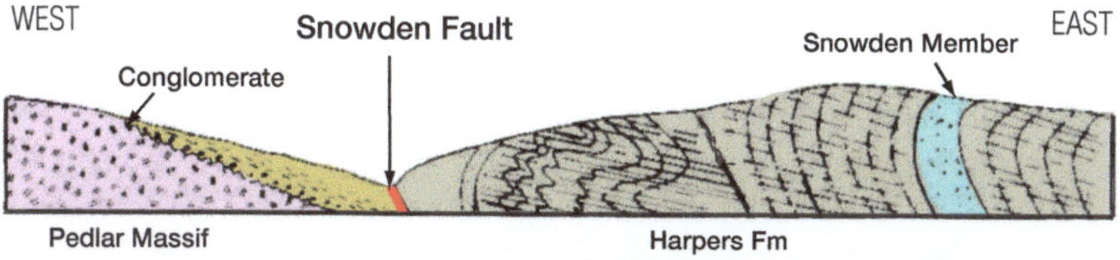

c

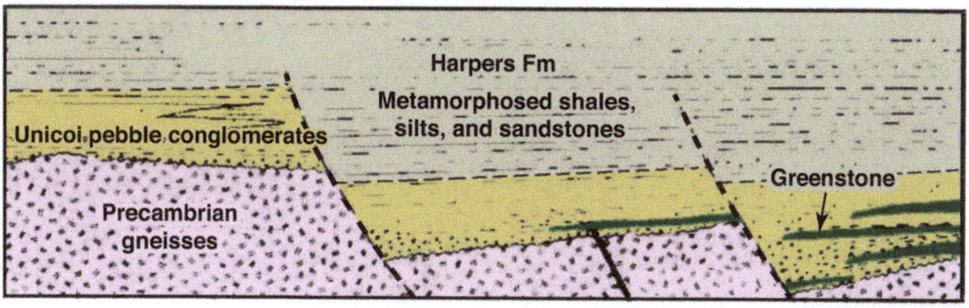

Fig. 6G-10. a) Photo of the folds located close to the Snowden fault on the hanging wall side of the fault. These indicate that the rocks were strongly compressed as they moved to the northwest. b) Sketch of the outcrop at Stop 6 showing the strong deformation (folds and cleavage) in the Harpers Fm. formed as the rocks were moved to the west (right to left). c) Reconstruction of the cross-section as it might have appeared before the final compression and shortening during the Alleghanian Orogeny. (From Spencer, Bell, and Kozak, 1989)

a)

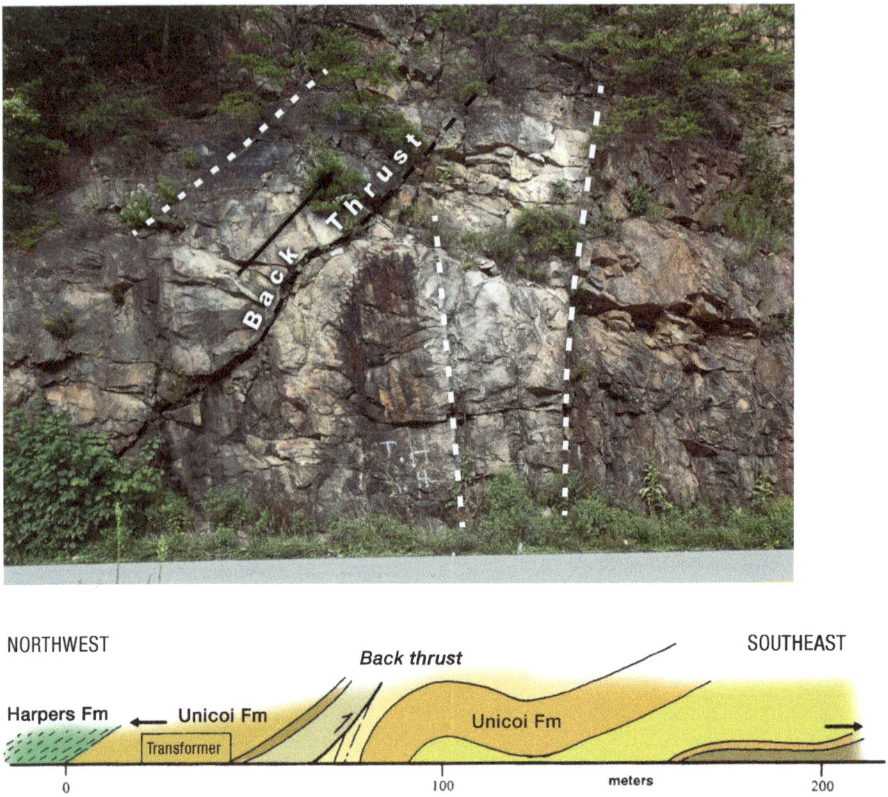

Fig. 6G-11. a) Photo showing the central part of the illustrated cross-section below. b) Cross-sectional sketch illustrating the complex structures found where the contact between the Unicoi and Harpers Fms. occurs. The Harpers Fm. (at the left end) is inclined gently to the northwest. Rock cleavage which is inclined to the southeast is so well developed that it makes it hard to see the bedding. Within the Unicoi, which is a much stronger and more brittle rock than the Harpers shale, a northwest inclined fault cuts across the layers of the Unicoi. The fault and one of the bedding contacts are shown. The fault is called a back thrust because the direction of movement is opposite (in this case from northwest to southeast) from the regional direction of movement (in the Appalachians the regional direction is from southeast to northwest). (From Spencer, Bell, and Kozak, 1989)

Stop 8. **Backthrust in the Unicoi Fm:** (Mileage 6.6) Park in a pullout just north of the bridge where US 501 crosses the river and SR 130 continues to the east. The coarse-grained granule conglomerates exposed at this bridge are part of the Unicoi Fm. The contact between the Harpers and Unicoi Fms. is located about 100 ft. upstream from the bridge where US 501 crosses the James River. Cross the bridge and look back at the structure exposed in the cliff along SR 130 (Fig. 6G-11). The back thrust within the Unicoi Fm. is one of the unusual features at this outcrop.

Stop 9. **Altered basaltic dikes cut Precambrian gneisses:** (Mileage 7.3) Park on the north side of the road in a pullout. About 200 ft. east of this Stop a

large dike cuts across the Unicoi Fm. About 500 ft. east of the Stop the Precambrian basement rocks are thrust onto the Unicoi. During summer months this outcrop may be covered by a thick growth of vines, some of which are poison ivy.

Stop 10. **Blue Ridge Basement Complex:** (Mileage = 7.6) Park a short distance from a small creek labeled "05/1084." Excellent exposures of Precambrian gneiss occur along SR 130 between this creek and the entrance to the Bedford Snowden Hydro Project. The gneiss is a coarse-grained granodiorite containing blue quartz, a sign of high pressure, and hypersthene (a member of the pyroxene group), quartz and feldspar (Fig. 6G-12).

Several basaltic dikes and a lava flow at the base of the Unicoi Fm. crop out along the highway about 350-500 ft. east of the creek. The dikes have not been dated. The lava flow is the same age (Precambrian) as the lower part of the Unicoi Fm.

Fig. 6G-12. Precambrian gneisses (top two photos) and a slightly altered basaltic dike (bottom photo) exposed at Stop 10. An iron stain on a fracture across the basaltic dike gives it an unusual orange color.

Entrance to the Blue Ridge Parkway from SR 130: (Mileage = 9.5) You may continue the traverse east along SR 130 (described below), or you may continue south along the Parkway (p. 174). Otter Creek Campground is located about one mile to the north on the Parkway. This creek is an excellent place to look for birds and wildflowers. It drains a large area to the north and continues southwest to the James River.

Continued Traverse Across the Core of the Blue Ridge Province

Good exposures of the basement complex at Stop 8 are representative on many of the rocks in the western part of the core of the Blue Ridge. The zone in this area varies in width, and is over a mile wide at Big Island. Radiometric dating of rocks in this zone indicates that it formed either late in the Acadian Orogeny or early in the Alleghanian Orogeny. Both US 501 and SR 130 continue east across the central part of the Blue Ridge core. Weathering and erosion have transformed most of the rock outcrops along SR 130 into thick soils known as saprolite (Fig. 8C-19).

Tobacco Row Mtn. lies northeast of the town of Big Island. This is one of the most prominent mountains within the core of the Blue Ridge Geological Province. Exposures of rocks similar to those on Tobacco Row Mtn. occur along US 501. These rocks are exposed nine miles east of the exit from the Parkway onto US 501. The gneisses here are coarse-grained and contain quartz, feldspar, biotite mica, and hornblende.

Arnold Valley, the James River Face & Thunder Ridge Wilderness Areas

To reach Arnold Valley from the intersection of US 501 and SR 130 near Glasgow, drive southwest on SR 130 through the town of Glasgow and continue south to SR 759. Turn left on SR 759 and cross the James River. Glasgow has been flooded a number of times. The most devastating of these came in 1969 during Hurricane Camille. A dam located at the entrance to the pass contributed to flooding caused by the unusually heavy rain that came with that storm. Most of the lower parts of Glasgow were underwater at that time. Subsequent removal of the dam has caused the river to cut deeply into the sediment that had accumulated behind the dam. Work is under way to return the river to it earlier course. This can be seen at the bridge over the Maury at the intersection of US 501 and SR 130.

As you drive southwest on SR 130 toward Natural Bridge the ridge of Antietam (Erwin) quartzite that underlies the Sallings Mtn. klippe lies to your right. The western limb of the Buena Vista anticline (Fig. 6G-3) is to your left. As you approach the town of Natural Bridge Station a quarry in the Antietam (Erwin) Fm. is located along the mountain front. (Note: the famous Natural Bridge of Virginia is located about 6 miles farther west along SR 130.) To reach Arnold Valley from Natural Bridge, take SR 130 east from Natural Bridge to its intersection with SR 759 and proceed across the James River into Arnold Valley.

An industrial park is located at the northern border of Natural Bridge Station.

Students sit on what was once a Cambrian-age seafloor. This inclined surface is located at the quarry in Arnold Valley at AV Stop 6. It can also be seen in Fig. 6G-17.

Take a left turn onto SR 759 at the industrial park. Cross the river and you will be in Arnold Valley. The first road on your left after crossing the river (SR 782) runs parallel to the river, passes the Wilderness Canoe Campground, and ends at a park-like area known as the Locher Tract. This excellent area for wildflowers and birding adjacent to the James River Face Wilderness is open to the public. The Balcony Falls Trail starts in this area and leads into the James River Face Wilderness.

Wilderness Areas: The James River Face, located southeast of Glasgow and on the south side of the James River, was the first wilderness area designated in the eastern U.S. Good views of the James River Face may be obtained from US 501 and SR 130. The Appalachian Trail passes through the wilderness area near Stop 7. The AT and many local trails pass through this area and provide easy access to a great variety of viewpoints and habitats in these areas. Thunder Ridge Wilderness is adjacent to the James River Face to the south. SR 781 (the Petites Gap Road) separates the two areas. The AT crosses the upper edge of Thunder Ridge, but steep slopes inhibit access to much of the area. If you are interested in exploring these areas you should obtain detailed maps such as the Forest Service map of the Snowden Quadrangle. For those unfamiliar with wilderness areas, hunting, fishing, camping, horseback riding, and hiking are permitted, but motorized vehicles are not.

South of the James River, the large Arnold Valley lies at the southwestern end of the Buena Vista anticline (Figs. 6G-3, -4). A thrust fault almost completely surrounds Arnold Valley, part of which is floored by limestones and dolomites typical of the Great Valley. Over many millions of years, erosion has cut through and removed part of the thrust sheet that once covered Arnold Valley and the Buena Vista anticline. We can now see what lies under the thrust sheet.

Points of Interest

Geological Traverse of Arnold Valley and the James River Face Wilderness

This traverse is a side trip into Arnold Valley on the western flank of the Blue Ridge (see red dots with "AV" on p. 161).

Arnold Valley (AV) Stop 1. **View from the James River Crossing:** As you cross the James River, note that the crest of the Buena Vista anticline drops to the southwest, reflecting the plunge of the structure as it passes into Arnold Valley. The Antietam quartzite forms the crest of the anticline. The steep slopes of Thunder Ridge lie in the distance. Blue Ridge basement rocks that have been thrust over the younger Cambrian-age rock in the valley lies to the south of the valley. There is a good view across Arnold Valley from the Thunder Ridge Overlook (Fig. 6H-1).

AV Stop 2. **Glenwood Furnace:** About one mile south of the James River (on SR 759) you will see the ruins of the Glenwood Furnace (Fig. 6G-13). This

Fig. 6G-13. Glenwood Furnace

furnace was built in 1849 it was used to produce pig iron during the Civil War open until 1887. The iron ore used in this furnace crops out about 50 yards east of the furnace. The iron was derived from a breccia (Fig. 6G-14) that developed along the large thrust fault on which the Blue Ridge moved to the northwest. This thrust lies very close to the ground surface in many places in Arnold Valley. It also crops out in a creek near the entrance to the Wilderness Canoe Campground. (Permission is needed to visit this site.) Two interpretations of the origin of this breccia have been advanced. In one, the breccia formed as a result of the breakup of the rock at the base of the fault. In the other the breccia formed as a result of the collapse of the rocks near the base of the thrust into sinkholes present in the underlying dolomite.

Approximately two miles south of the James River on SR 759 a junction with SR 781 occurs at Glenwood Church. A right turn (to the west) at this junction leads to a Forest Service campground at Cave Mtn. Lake. A left turn onto SR 781 passes between the James River Face and Thunder Ridge wilderness areas and eventually reaches the Blue Ridge Parkway at Petites Gap. The AT crosses SR 781 near the Parkway.

AV Stop 3. **View of Arnold Valley:** To get a sweeping view of Arnold Valley and its surroundings proceed on SR 759 past the junction with SR 781 and the Glenwood Church. Continue on the road straight ahead toward Hopper Creek Group Camp. The road crosses a large open field. You will see the large valley surrounded by mountains. It is almost completely encircled by the trace of the thrust fault that carried the Precambrian

Fig. 6G-14. Iron-bearing breccia exposed behind the furnace was one of the sources of ore used at the furnace. This breccia lies in the fault plane of the Blue Ridge fault in Arnold Valley.

basement rocks over the southern end of the Buena Vista anticline. Those basement rocks are exposed near Hopper Creek Camp in the mountains to your right and in the steep slopes of Thunder Ridge to your left. To the northeast, you see the southern end of the Buena Vista anticline and a light-colored patch of rocks that make up the Devil's Marbleyard described below. Return to the church and triple road junction. Turn left onto SR 781/Cave Mtn. Lake Road and proceed.

AV Stop 4. **Cave Mountain:** Much of the floor of Arnold Valley is covered by stream alluvium derived from the surrounding mountains. As you leave the floor of Arnold Valley on SR 781, you cross carbonate rocks that underlie the valley. The road leads to the entrance of Cave Mtn. Lake Campground. If you continue on the road it circles around the top of the mountain, composed of quartzites of the Antietam (Erwin) Fm. The road also lies close to the trace of faults that surrounds the mountaintop. These faults separate the quartzites from the Cambrian-age Shady

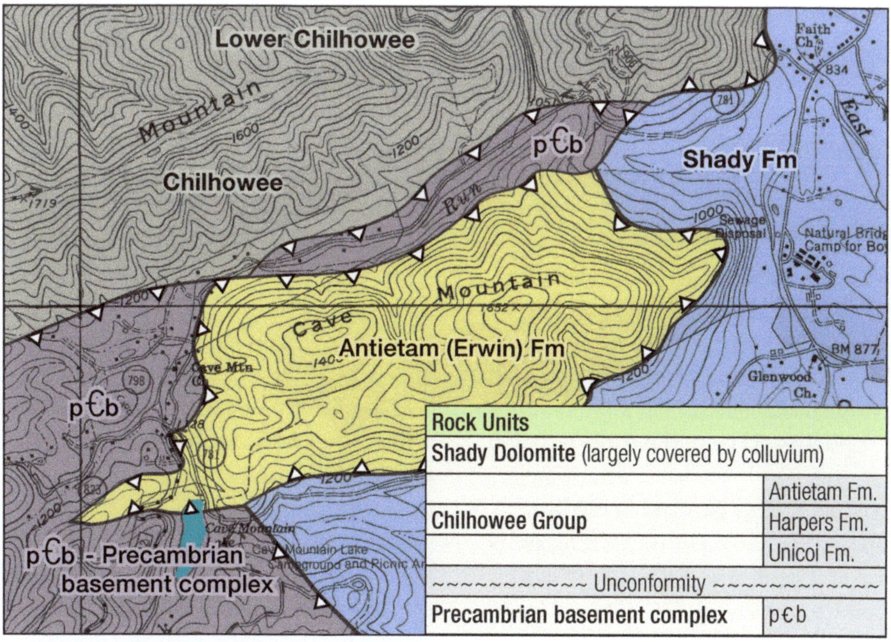

Fig. 6G-15. Sketch map of the Cave Mtn. klippe. The thrust sheet fragment (Antietam Fm.), is now isolated from the rest of the thrust sheet. (Modified after Spencer, 1992, and Carter, et al., 2014)

Fig. 6G-16. The Devil's Marbleyard (left). Visitors sit on large blocks in The Devil's Marbleyard (right).

Fm. on the south side and Precambrian rocks on the north side (Fig. 6G-15). Geologists are still trying to determine if the rocks on Cave Mtn. are remnants of a large thrust sheet once covering this area. In this case it would be a klippe. Alternatively, they could be a slice of rocks that were caught in the fault zone of a large thrust that carried the Precambrian basement rocks onto the Chilhowee Group.

Continue along SR 781 until it reenters SR 759. Turn right and return to the Glenwood Church at the triple junction. The next Stop is located about 1.25 miles east along the Petites Gap Road that leads to the southeast from the triple junction.

AV Stop 5. **The Belfast Trail:** This trail leading to the Devil's Marbleyard is located approx. one mile east of Glenwood Church on SR 781. The trail leads into the James River Face Wilderness and connects with several other trails, including the AT, that cross the wilderness area. Slightly over a mile along this moderately steep trail, you will encounter a large area composed of blocks of Cambrian-age quartzite, the Antietam Fm. (Fig. 6G-16). The rock layers here are inclined at a steep angle near the southwest plunging nose of the Buena Vista anticline. Streams have cut through the layers of quartzite making the slope unstable and facilitating movement of the large blocks down the steep mountainside. The fractures that form the edges of these blocks are visible in the layers of quartzite near the top of the slide area. Many of the blocks are marginally stable. ⚠ **Caution: This site is not safe for small children or people who are unsteady on their feet.** Others will enjoy climbing around on the big blocks and seeing the views into Arnold Valley. If you continue farther into the wilderness area be sure to take a good map and watch out for rattlesnakes. To continue on this

Fig. 6G-17a. View of the large quarry in Arnold Valley. The quartzite was used as a source of road metal (crushed stone). The quarry face is a bedding plane in the Antietam (Erwin) Fm.

Fig. 6G-17b. Closeup of a layer in the Antietam (Erwin) Fm. showing vertical "straw-like" traces of a worm, scolithus linearus, that bored into the sand half a billion years ago.

traverse, return to the Petites Gap Road and go about one mile to the east.

AV Stop 6. **Cambrian-age Seafloor:** An abandoned quarry that was the source of rock used on the roads in this area is located north of SR 781 (Fig. 6G-17a). Limited parking is available along this road. As the soil and layers of quartzite were removed from the slopes here, a huge area of Cambrian-age seafloor was exposed. This ancient seafloor is now steeply inclined (see photo on p. 169) as a result of the folding that produced the Buena Vista anticline. Walking on this slope can be dangerous. Ripple marks formed in the shallow waters covering the sandy seafloor about 540 million years ago extend across the rock. You will also see the ends of the worm borings produced by scolithus linearus on this surface (Fig. 6G-17b). These worms are among the oldest and best-exposed fossils in the Blue Ridge.

AV Stop 7. **Entrance to the Sulfur Spring Trail:** The trailhead is located on the north side of SR 781. This trail leads into the James River Face Wilderness, and joins the Belfast, Appalachian and other trails in the wilderness area. You can enter the trail system near Petites Gap at the Parkway. Walks of about 7-8 miles, mostly downhill from Petites Gap lead you to either Sulfur Spring on SR 781, the AT on SR 130, or the Locher Tract in Arnold Valley. Take a topographic map (of the Snowden Quadrangle) if you plan to go into this area.

AV Stop 8. **Petites Gap:** FS 35 (a continuation of SR 781) joins the Blue Ridge Parkway at this location. An entrance to the AT leads through the James River Face Wilderness to the northeast and along the southern edge of the Thunder Ridge Wilderness to the west. This western section of the trail is a relatively easy walk. ❖

Chapter 6H

Blue Ridge Parkway: James River to Roanoke

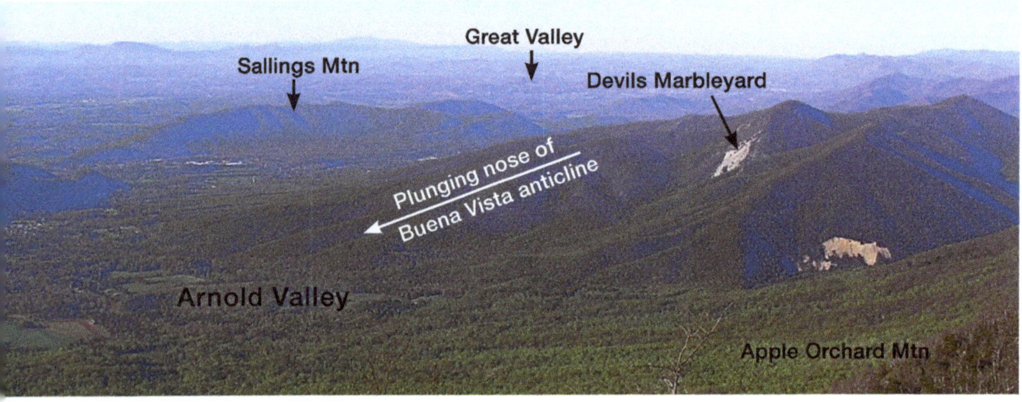

Above: Fig. 6H-1. View of Arnold Valley from the overlook at Thunder Ridge showing the southwest-plunging Buena Vista anticline. The light areas to the right are the quarry and the Devil's Marbleyard.

Location and Access: This section of the Parkway in Virginia passes through the southern most part of the Northern Blue Ridge. The ecosystems and geology are similar to the sections farther north, but the valley of the Roanoke River is much wider than those near the James or Potomac water gaps. This section may be entered by continuing south on the Parkway from the north; by turning onto the Parkway where US 501 or SR 130 cross the Parkway near the James River; or by entering from SR 43 which connects Bedford and Buchanan (this last route is unsuitable for trailers or buses). US 221 and US 460 cross the Parkway near Roanoke. In the Roanoke area, SR 24 and US 220 have entrances onto the Parkway.

South of the James River the Parkway rises from 600 ft. to 3,500 ft. over a distance of about 10 miles. Overlooks along this section provide views to the southeast across the core of the Blue Ridge. Rocks of the Grenville-age Blue Ridge basement complex are exposed along this section. This complex extends close to the horizon near Lynchburg, Va.

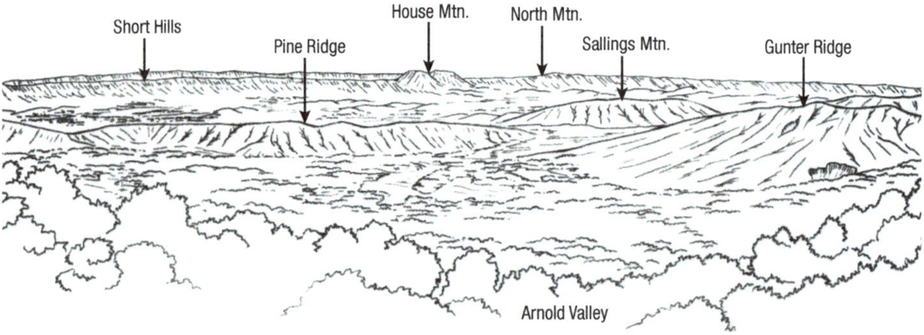

Fig. 6H-2. Sketch of the view across Arnold Valley from the Parkway overlook on Thunder Ridge.

Fig. 6H-3. Distant view of Tobacco Row Mtn. from Terrapin Mtn. Overlook on the Parkway. Tobacco Row Mtn. is a large Precambrian pluton in the Blue Ridge basement core. It has been etched out by erosion from the surrounding metamorphic rocks that are also Precambrian-age.

Terrapin Mtn. Overlook: (mp 72.6) A zone of intense deformation known as the Rockfish Valley zone of ductile deformation lies between the western flank of the Blue Ridge shown in Fig. 6H-3, and the large Precambrian-age pluton, Tobacco Row Mtn., seen in the distance. An example of rock in the deformation zone is shown in Fig. 6F-17.

Good views across the core of the Blue Ridge Province are possible along the Parkway from mileposts 67 to 70. Tobacco Row Mtn., part of the core complex is visible in the mid-distance (Fig. 6H-3). The rocks exposed at the Terrapin Mtn. Overlook are granitic gneiss that are also part of the core complex. Those exposed at the overlook have been dated at about 1.15 billion yrs. old. Part of the layered gneisses are composed of single small crystals of pyroxene and hornblende that lie between layers composed of quartz, feldspar, garnets, and biotite (Carter, Southworth, and Aleinikoff, 2014).

Thunder Ridge Overlook: (mp 74.6) A short detour off the Parkway at milepost 75 leads to a view across Arnold Valley in the foreground, the Great Valley of Virginia, and the first ridges of the Valley and Ridge to the west. This exceptional spot provides a spectacular view of the complex structure that lies along this part of the western flank of the Blue Ridge. Arnold Valley is nearly a structural window (Fig. 6G-3, p. 53). The rocks exposed at this overlook are garnet-bearing gneisses of the Blue Ridge basement complex. They form the steep slope below this overlook and extend to the edge of Arnold Valley below. They also form the mountains that surround the Valley to the south and west. All of these basement rocks have been carried to the northwest as part of the Blue Ridge thrust sheet, a huge mass of mountains that extends to the southern end of the Blue Ridge. The fault that lies at the base of this sheet is present at the foot of the mountains that surround the eastern, southern, and western sides of Arnold Valley. This fault wraps around the mountains to the north (including Sallings Mtn., Fig. 6G-3) and continues along the western edge of the Blue Ridge to the south. The open part of the potential window, called a reentrant, lies along its northeastern edge. Along this edge the Buena Vista anticline (Fig. 6G-3) plunges beneath the valley. The Devil's Marbleyard and a nearby quarry shown in Fig. 6H-1 are located close to the nose of this anticline. From this viewpoint, the floor of Arnold Valley appears almost flat. Thick accumulations of stream gravels cover the flat area to the

Fig. 6H-4. Lichens exposed on rock outcrops along the Parkway at Apple Orchard Mtn.

Fig. 6H-5. Mayapples cover the forest floor at Apple Orchard Mtn. in the spring.

west of Arnold Valley, and Cambrian-age carbonate rocks—the same ones that are present in the Great Valley—lie beneath these gravel deposits. These carbonate rocks (the Shady Fm.) were deposited in marine waters on top of the rocks of the Chilhowee Group that make the Buena Vista anticline prominent in the landscape. The presence of these easily eroded carbonate rocks was a major factor in the development of Arnold Valley.

Most of the rock exposures along the Parkway from the James River by Apple Orchard Mtn. to the Peaks of Otter are Grenville-age (about 1.1 Ga) gneisses and granites. Although the thickness of the soil varies, weathering of the calcalkaline rocks has produced a deep and fertile soil in many places. This provided a good place for the development of an oak-history forest with a luxuriant herb layer. A northern red oak forest occupies the higher ground, and gnarled yellow birches grow on the steep rock-covered north-facing slopes. Cove forests and seepage wetlands are present in hollows and along the headwaters of streams here.

Most of the overlooks to the southwest along the Parkway provide views of the Valley and Ridge. Note that the first

Bent oak tree limbs on Apple Orchard Mtn. show the effects of strong winds.

ridges are progressively closer to the Blue Ridge front. They were 30 miles away at Waynesboro, but at Buchanan they are only a mile away. This is because the Blue Ridge has moved to the northwest, covering thick sections of the sedimentary rocks that lie beneath the Valley and Ridge. Those rocks are exposed in Arnold Valley and in Goose Creek Valley farther south. Seismic lines across the Blue Ridge indicate that they lie beneath much, perhaps all, of the present Blue Ridge Province.

Apple Orchard Mtn. Overlook: (mp 76.6) This area, the highest place along the Parkway in Virginia (4,225 ft.), was the site of an old radar installation

Fig. 6H-6. View of Sharp Top Mtn. across the lake from the Peaks of Otter Lodge.

used for many years as part of the "early warning system" built following WWII. Tobacco Row Mtn. can be seen in the distance to the east. This view is across the core of the Blue Ridge Province. An outstanding view of the Valley and Ridge is located about a hundred yards north on the Parkway. Parking is not possible at this location. Gneiss and granitic rock exposures located about 100 yards north of the station entrance are covered by excellent samples of several types of lichens (Fig. 6H-4). In the spring, Mayapples cover the forest floor at the top of Apple Orchard Mtn. (Fig. 6H-5). The stunted trees near the ridge crest reveal the persistent wind direction and the excellent views of the Valley and Ridge Province. A pullout and parking area for Sunset Field, about .25 miles south of milepost 78, provide access to a beautiful waterfall and one of the best places in the Blue Ridge to see warblers. The AT, several other trails, and a gravel road that leads down into Arnold Valley have trailheads at this parking lot.

Peaks of Otter: (mp 85.9) This popular tourist destination (Fig. 6H-6) is located on the Blue Ridge Parkway at the intersection with SR 43 which connects Buchanan with Bedford. The Peaks of Otter Lodge includes a restaurant that is open year round. A campground, picnic tables, and restrooms are located nearby. A small museum and shop are located at the visitor center 0.75 miles farther south on the Parkway. Fuel, accommodations, and

View of the Peaks of Otter lake and lodge from the summit of Sharp Top Mtn.

restaurants are also available east of the Parkway at Bedford and to the west in Buchanan. Hiking trails are available and buses go to the top of Sharp Top Mtn. from summer to early fall. This locality is of archeological importance as one of the earliest American Indian sites. A memorial to World War II soldiers who died on D-Day is located at Bedford, an attractive small town 10 miles east of the Parkway.

Two high peaks, Flat Top Mtn. (4,004 ft.) and Sharp Top Mtn. (3,870 ft.), and the less prominent Harkening Hill (3,364 ft.) comprise the Peaks of Otter. A high, bowl-shaped depression, known as the Mons area, lies at 2,500 ft. between these peaks. Before humans altered the area, the valley at the base of these peaks was a swamp. Water from springs along the base of the high peaks fed into the swamp that became a watering hole for animals.

History of the Peaks of Otter

Projectile points and arrowheads found in the area around the lake made it clear that American Indians used the site thousands of years before European settlers arrived. In the 1960s, the lake was drained and excavation was started to deepen and enlarge the lake. American Indian artifacts, including hearths, were discovered. One of the hearths contained charcoal from fires along with smoked stones and a Morrow Mtn. projectile point (Fig. 6H-7). The charcoal yielded a radiometric (Carbon-14) age of 3,430 BC. Based on the records from other archeological sites the projectile points found in the excavation are types known as Halifax, Morrow Mtn., Guilford, and Stanly. Most of these belong to the Middle Archaic Period (6,000-3,000 BC). Because no signs of villages or permanent camps were found, it is likely that the sites were temporarily set up for hunting. This was probably the primary use of the site for hundreds, probably thousands, of years. The American Indians who set up these camps probably spent most of the year at lower elevations in the Great Valley to the west or in the lower lands east of the high ridge that forms the western edge of the Blue Ridge in this area.

The first European settlers from Scotland arrived in the area circa 1700 and built their homes in the valley along an American Indian trail that crossed the ridge between Buchanan and Bedford, Va. The trail became a turnpike in 1772, as increasing numbers of settlers used the route to transport freight, including pig iron from the Valley to the east, where it was fashioned into weapons during the Revolutionary War.

By the early 1800s, the Peaks were attracting tourists. One home took in visitors. Called "an ordinary," it was an

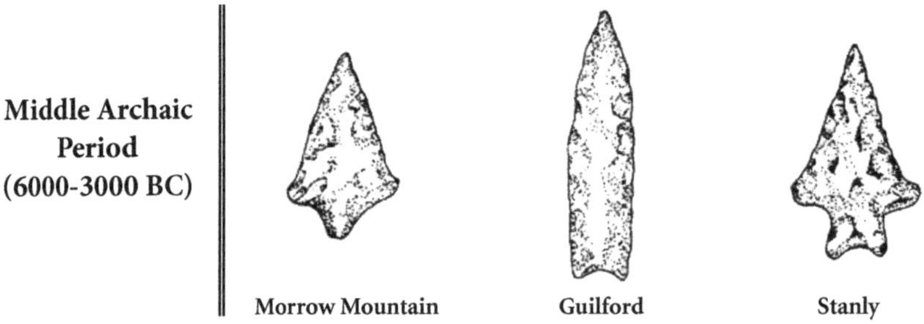

Fig. 6H-7. Drawings of projectile points from the lake at Peaks of Otter. (Griffin and Reeves, 1967. (NPS)

early version of today's bed and breakfast. The swampy area was later dammed and turned into a small lake. By 1857 a hotel that could accommodate 50 people was in operation and attracted many people who came to the peaks on horseback or by wagon to enjoy the mountain air and beautiful views.

Fig. 6H-8. Peaks of Otter Salamander (Photo: J.D. Willson)

A mountain community called Mons grew at the site after the Civil War. During the Depression people from Mons began to seek employment in larger communities. Most of them sold their land to the U.S. Forest Service and by 1935 construction of the Blue Ridge Parkway began here. The planners for the Parkway recognized the scenic importance of the Mons site and made provisions for its future development as an attraction.

Ecological Setting
of Peaks of Otter

Old-growth hardwood forests surround the three high rocky peaks. Today seven nature trails provide access to forests and the recreation area around the lake. From spring through fall, the trails here provide excellent opportunities to see wildflowers. It is a breeding ground for many neotropical birds (from the ecoregion that includes both North and South America) and home to a number of salamander species, notably the Peaks of Otter salamander (Fig. 6H-8). Deer, foxes, black bears, raccoons, opossums, rabbits, and squirrels inhabit the woods. A large lake, cascading waterfalls, and some open meadows provide a variety of habitats for many small animals.

Geologic Setting
of Peaks of Otter

Intrusive granitic (hypersthene granodiorite) and diorite rocks, part of the Blue Ridge basement complex, form the peaks. Why these peaks stand so much higher than the surrounding areas that are underlain by similar rocks has yet to be investigated. You can walk around outcrops of these rocks near the crest of the peaks. During tourist season a bus ride to the top of Sharp Top Mtn. makes these outcrops easily accessible (Fig. 6H-9). When the air is clear you can obtain superb views from here across the Valley and Ridge and the core of the Blue Ridge.

Fig. 6H-9. Granitic rocks exposed at the top of Sharp Top Mtn.

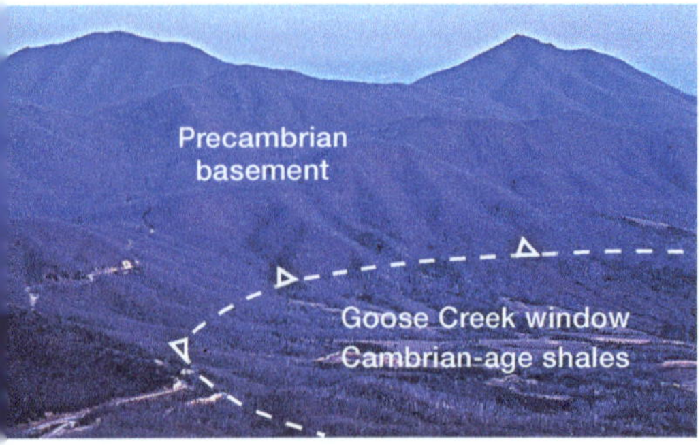

Fig. 6H-10. View toward the Peaks of Otter from the Parkway. The valley in the foreground, part of the Goose Creek window, is underlain by Cambrian-age shales of the Waynesboro Fm. The thrust fault between the Precambrian rocks at the Peaks of Otter and the Waynesboro Fm. is located at the edge of the floor of the valley.

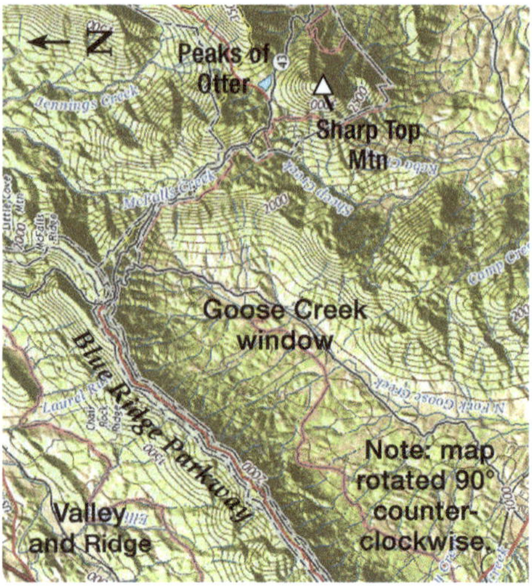

Fig. 6H-11. Goose Creek window (USGS)

Fig. 6H-12. Aerial view of the ridge on the west side of Goose Creek Valley.

Goose Creek Valley: (near mp 90.5) This pullout provides an excellent view to the south of the Goose Creek window (Fig. 6H-10). The rocks in the bottom of this valley are Cambrian shales of the Waynesboro (Rome) Fm. (Fig. 6H-11). A thrust fault separates them from the overlying rocks of the Blue Ridge basement complex, which make up the mountains east of Goose Creek Valley. To the west a long narrow ridge continues farther south toward Roanoke. The Parkway is located along this ridge, which is held up by the sandstones and shales of the Chilhowee Group (Fig. 6H-12).

Roadcuts along the Parkway expose many outcrops of the Harpers Fm. (the middle member of the Chilhowee Group). The Blue Ridge thrust fault is present all along the base of the mountains that border Goose Creek Valley. This fault lies beneath the ridge on which the Parkway is located and comes to the surface on both the east and west side of this ridge (Figs. 6H-13, -14). Continue along the Parkway past the exit to Buchanan on SR 43 at Bearwallow Gap.

Purgatory Mountain: (mp 92) This pullout provides superb views of the regional landscape to the west and the Goose Creek Valley window to the east (Figs. 6H-10, -11, -12). The Great Valley of Virginia is nearly pinched off at Buchanan. Purgatory Mtn., the first ridge of the Valley and Ridge, is less than a mile west of the Blue Ridge flank at this point. Purgatory Mtn. is a southwest-plunging anticline held up by Silurian sandstones and quartzites (Figs. 6H-15, -16, -17).

Two major thrust faults (the Blue Ridge and the Staunton-Pulaski faults) come to the surface at Buchanan (Fig. 6H-18). The Blue Ridge thrust is exposed in the large quarry visible at Buchanan. It continues southward along the base of the Blue Ridge and carries the Blue Ridge thrust sheet over the rocks seen in the valley—just as it does at Arnold Valley. The Staunton-Pulaski fault comes to the ground surface a short distance west of the quarry and can be projected into

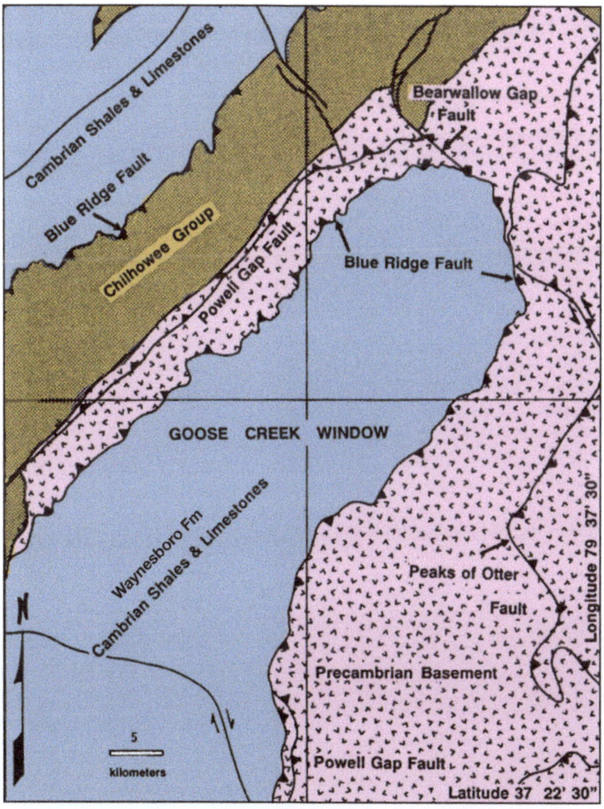

Fig. 6H-13. Geologic sketch map of the Goose Creek Valley window. (Simplified after Henika, 1981)

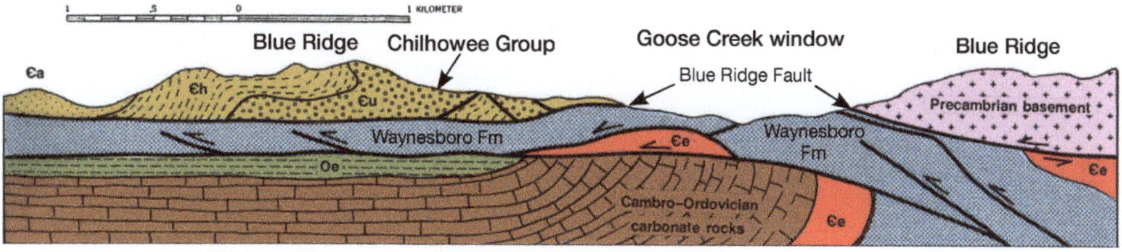

Fig. 6H-14. Cross-section of the southwestern end (lower left corner of map in Fig. 6H-13) of the Goose Creek Valley. (After Henika, 1981)

the air where it continues over Purgatory Mtn. then comes back to ground level on the west side of the mountain. It is likely that the Purgatory Mtn. anticline formed after the thrust fault was in place and caused the thrust to take an anticlinal shape. Cambrian-age shales and dolomites of the Waynesboro (Rome) and Elbrook fms. underlie the large area of low topography south and west of Purgatory Mtn. These formations are part of the Staunton (Pulaski) thrust sheet, one of the largest thrust sheets in the Appalachians. The leading edge of that thrust sheet rises to the surface in the distance, east of the ridge that appears near the horizon. The lateral movement of this thrust sheet is estimated to have covered from 10 to perhaps 100 miles, moving from southeast to northwest.

Roanoke River Water Gap: The Parkway descends nearly 1,500 ft. (from approx. 2,500 ft. to 1,000 ft. in elevation) as it drops into the water gap cut by the Roanoke River across the western flank of the Blue Ridge. The Roanoke River Water Gap is much wider (3.5 miles at its narrowest point) than those of the Potomac and James rivers. Recent geologic mapping indicates that this gap, which separates the northern section of the Blue Ridge from the Blue Ridge Upland Plateau, is a product of erosion along a zone of steeply inclined, northwest-trending faults (Henika, 2010). The fault zones provided lines of weakness along which the river could easily erode its course. It is

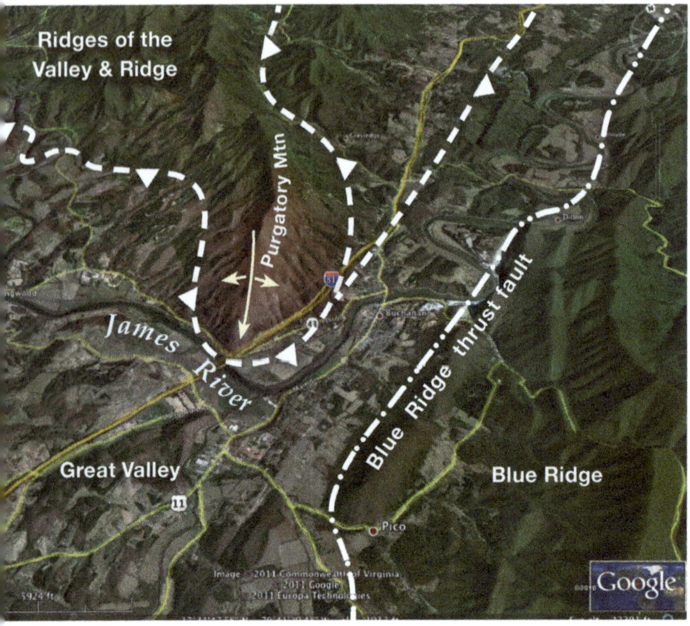

Fig. 6H-15. View of the Purgatory Mtn. anticline, the James River, and the constriction of the Great Valley of Virginia at Buchanan, Va. (Modified from Google Earth, Image © 2011 Commonwealth of Virgina, © 2011 Google, © 2011 Europa Technologies)

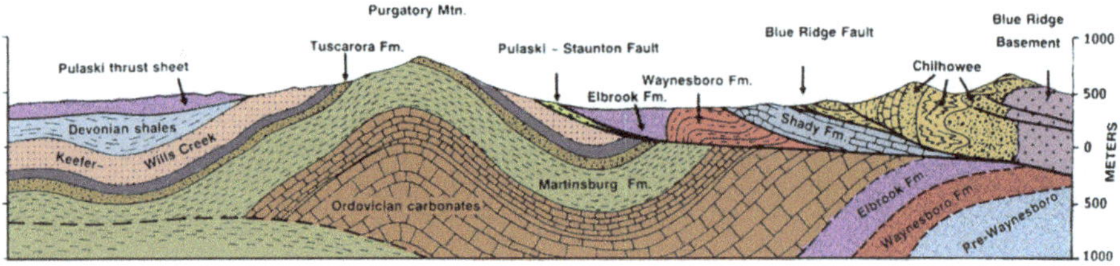

Fig. 6H-16. Cross-section across the Blue Ridge and Purgatory Mtn. at Buchanan, Va. (Spencer, 1989)

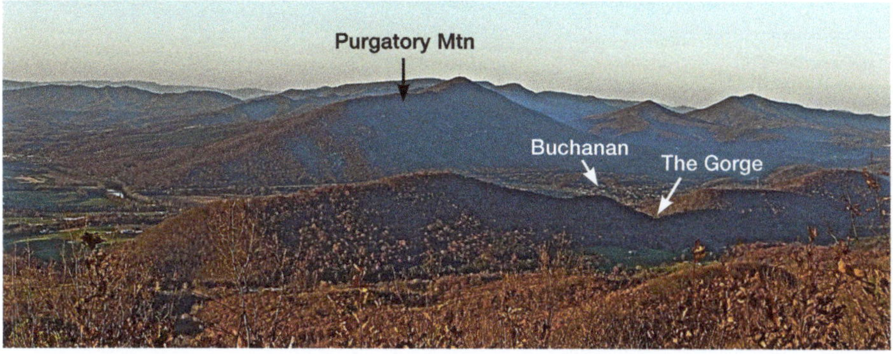

Fig. 6H-17. Parkway view near milepost 92. The Great Valley is only a mile wide here. Just beyond the Valley, the Purgatory Mtn. anticline plunges beneath rocks carried on the Pulaski thrust sheet.

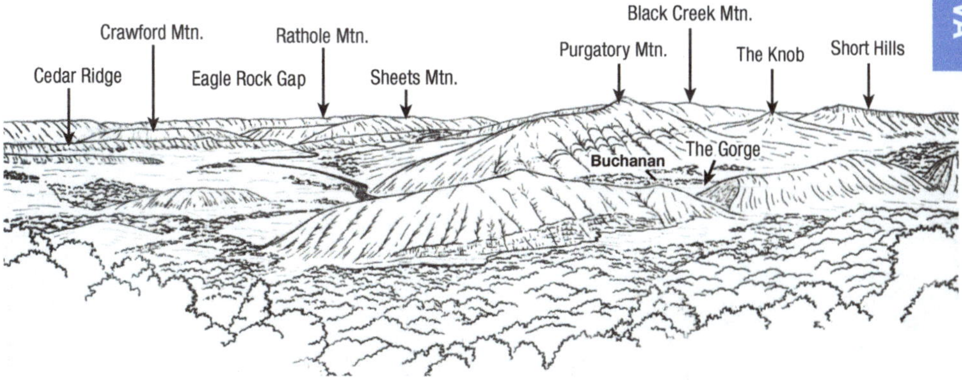

Fig. 6H-18. Sketch of the view of Purgatory Mtn. from the Parkway overlook near milepost 92.

also likely that a section of the Blue Ridge flank dropped along these faults, contributing to the width of the gap.

Smith Mountain Lake: This reservoir, the largest in Virginia, is located along the eastern edge of the Blue Ridge Province. The Roanoke River, the southernmost river that drains portions of the Valley and Ridge into the Atlantic Ocean, provides water for the reservoir. To reach the reservoir from the Blue Ridge Parkway, take the exit near milepost 112.2 onto SR 24. Drive east for about 22 miles, then follow SR 122 south to Moneta, take SR 608 east to White House, and finally take SR 626 south for two miles to the western end of the reservoir where Smith Mountain Lake State Park is located. The lake covers about 32 sq. miles and has a shoreline over 500 miles long. Housing developments surround much of this body of water. Tourist facilities in the park include hiking trails, camping, swimming, and picnicking. Commercial facilities are also present in many of the developments near the lake.

The dam for the reservoir is constructed in a valley cut into Cambrian-age phyllites and schist composed of quartz, muscovite, and chlorite known as the Candler Fm. This unit holds up Smith Mtn. and is located along the eastern edge of the Blue Ridge Province. Its geologic setting is similar to that of the Philpott Reservoir described in Ch. 8C. ❖

Fig. 7-1: Digital elevation map of the Southern Blue Ridge and parts of the Appalachian Plateau, Valley and Ridge and Piedmont. A white dashed line shows the boundaries of the Blue Ridge Mountains. (The National Atlas, USGS, with modifications)

Chapter 7

The Southern Blue Ridge

Major Divisions
of the Southern Blue Ridge

Remarkable changes take place in the landscape and structure of the Blue Ridge where the Roanoke River cuts across the Blue Ridge Province. Unlike the James and Potomac rivers, which flow through narrow water gaps, the Roanoke River flows through a broad valley several miles wide, and separates distinctly different northern and southern portions of the Blue Ridge. Geologists have used the significant changes that take place in the landscape and bedrock geology near the Roanoke River as the boundary between the Southern and Central Appalachians.

The Northern Blue Ridge begins as a well-defined anticlinal structure with a distinct topographically high western limb that forms a prominent ridge from Pennsylvania to the Roanoke River (Fig. 5-1). The eastern limb of this anticline is much lower in elevation than the western limb. Along most of this margin, thrust faults carry rocks of the Blue Ridge up onto Cambrian-age limestones and shales of the Valley and Ridge. On the eastern side, a Triassic basin (the Culpeper Basin) is present in the north. Farther south, Cambrian-age metasedimentary rocks cut by thrust faults separate the Blue Ridge from the Piedmont.

The western side of the Southern Blue Ridge is similar to that farther north. Rocks of the Chilhowee Group are thrust up onto carbonate rocks of the Valley and Ridge, but the anticlinal character of the Northern Blue Ridge is not evident. Instead the central and eastern portions of the Blue Ridge are located on a high level plateau referred to here as the Upland Plateau (Figs. 7-2, 7-3) that gradually changes farther south into a large region of high peaks known as the Blue Ridge Highlands. The Highlands extend into Georgia and Alabama before dropping to lower elevations. The Blue Ridge Province ends where the rocks of the province become covered by the Mesozoic sediments of the Coastal Plain (Fig. 1-4).

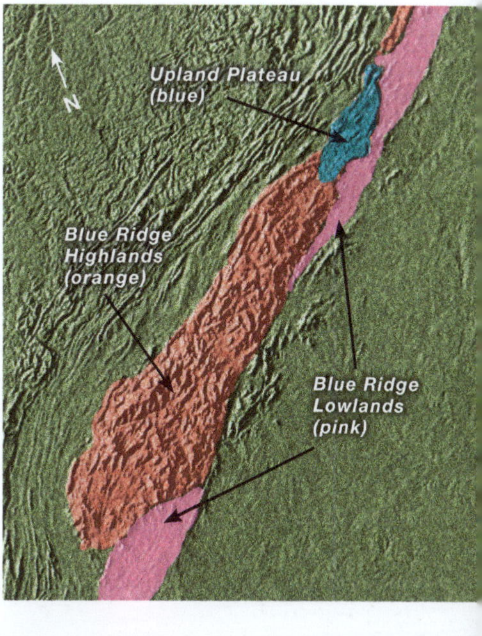

Fig. 7-2. Major divisions of the Southern Blue Ridge. (Modified after The National Atlas, USGS)

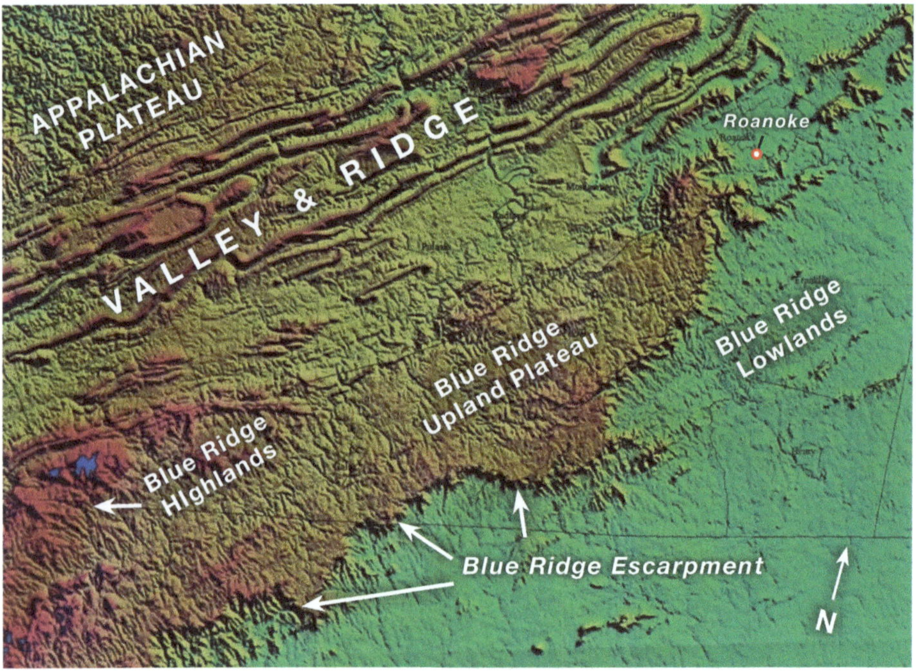

Fig. 7-3. Blue Ridge Upland Plateau area. (The National Atlas, USGS, with additions)

Fig. 7-4. Photo taken across the Upland Plateau.

A major thrust fault named for the town of Brevard, N.C. defines the eastern boundary of the Southern Blue Ridge. It lies east of the Upland Plateau, and the Highlands region (Fig. 7-1). Two groups of rocks are present in the core here. One is similar to the Grenville rocks in the Northern Blue Ridge. The other is essentially the same as the late Precambrian-Cambrian metamorphic rocks found in the Piedmont east of the Brevard fault.

Although the Southern Blue Ridge extends from Roanoke to the coastal plain in Georgia and Alabama, only a portion of it, the section between Roanoke and the Great Smoky Mtns., is covered in this guide.

Upland Plateau

The ridge-like topography seen along the western side of the core of the Blue Ridge north of Roanoke becomes less distinct to the south. It is replaced by the Upland Plateau which is slightly inclined toward the northwest. The elevation of the top of this plateau drops from around 3,000 ft. on the east side to 2,000-2,500 ft. in the west (Fig. 7-5). A few mountains, most notably Buffalo Mtn., rise above the Upland Plateau (Fig. 7-4). These mountains are remnants of the once higher mountains that existed here before erosion reduced them to the present plateau surface. Igneous and metamorphic rocks, similar to those found in the core of the Blue Ridge to the north are exposed across much of the Upland surface.

Fig. 7-5. View of the escarpment from its base. For a broader view of the escarpment and the Upland Plateau see Fig. 8A-5.

Blue Ridge Escarpment

The Blue Ridge Escarpment (Fig. 7-5) defines the eastern edge of the Upland Plateau from Roanoke to the southwest for a distance of nearly 300 miles. The edge of the escarpment is quite irregular (Fig. 7-3). The escarpment rises nearly 1,500 ft. above the level of most of the region to the east. Some geographers and geologists recognize this escarpment as the boundary between the Blue Ridge and the Piedmont. Others use the Brevard fault zone, located farther east to define the boundary (Fig. 7-1). The Brevard fault zone is a relatively straight line, mostly a depression that continues southwest to the coastal plain and northward until it splits into several faults in central Virginia.

Blue Ridge Lowlands

The area located between the base of the escarpment and the Brevard fault zone is referred to here as part of the Blue Ridge Lowlands. These areas of the Blue Ridge

Fig. 7-6. View to the west from Philpott Reservoir across the Blue Ridge Lowlands. The Blue Ridge Escarpment is visible in the distance.

Province are characterized by topography with low relief and below 2,000 ft.

The rocks in the Lowlands located between the escarpment and the Brevard fault are similar to those on the eastern side of the Upland Plateau to the north, but they have been eroded to a much lower elevation as streams that flow to the east have eroded their valleys in a headward (up valley) direction, causing the escarpment to migrate to the northwest. Two localities in the Lowlands zone: Stone Mtn. State Park, (not to be confused with Stone Mtn., Ga.) located close to the escarpment; and Fairy Stone State Park, located near the eastern margin of the Blue Ridge Province, are described in Chs. 8C and 8D.

The Highlands

The Highlands (Fig. 7-3) contain more than 125 peaks that are over 5,000 ft. high including Mt. Mitchell (the highest point in the Appalachian Mtns.), Grandfather Mtn., Mt. Le Conte, and Mt. Rogers (Fig. 7-8). Even the floors of the valleys in the Highlands are higher than the average elevation of the Upland Plateau. Easily recognized changes in ecosystems accompany these shifts in elevation. The high mountain environments resemble those of northeastern Canada more than the nearby Valley and Ridge or Piedmont. A

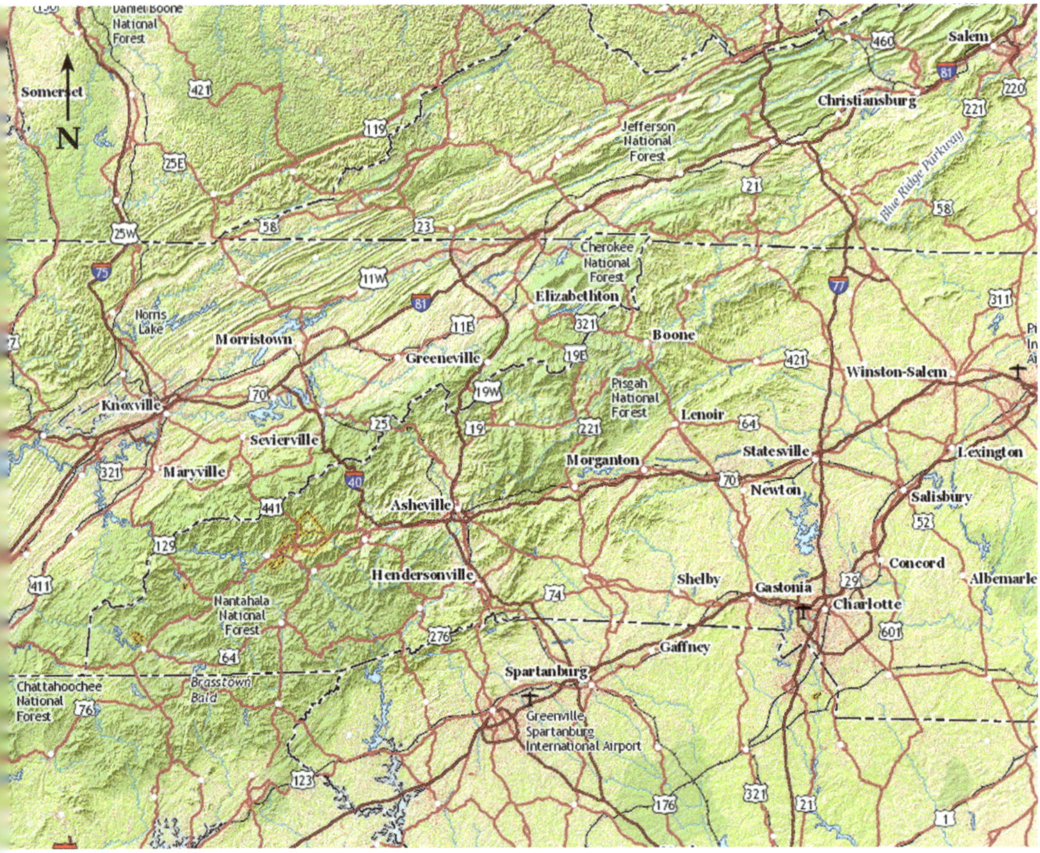

Fig. 7-7. Map of the Blue Ridge Highlands. (The National Map, USGS)

Fig. 7-8. View across the Highlands southwest of the Blue Ridge Parkway between Asheville, N.C. and Great Smoky Mtns. National Park.

few of the high peaks, such as Mt. Rogers and Mt. Jefferson lie in a transition zone between the Plateau and the Highlands. Why the Highlands are so much higher than the Upland Plateau is unclear. In part, this may be because the rocks exposed there are more resistant to erosion than those farther east. This seems to be the case at Mt. Rogers where rhyolites, that are highly resistant to chemical weathering, form the core of the mountains. Farther south tens of thousands of feet of tightly cemented, insoluble quartz sandstones are present in the Great Smoky Mtns. and at Grandfather Mtn. Metamorphism has produced erosion-resistant metamorphic rocks throughout much of the Highlands region. In addition, from Grandfather Mtn. to the south several thick rock masses (terranes described in Ch. 2) that originated off the shores of Laurentia are stacked as huge thrust sheets on top of the Laurentian basement rocks that are exposed at the surface on the Upland Plateau. The Highlands may have been high ever since these durable thrust sheets were emplaced.

Rocks that contain unusually high amounts of magnesium, calcium, and iron underlie many of the mountains that rise above 4,600 ft. in the Highlands. These elements are derived from minerals of the amphibole group (especially a variety known as hornblende). When this mineral undergoes residual weathering the resulting soil contains high amounts of nutrients and has a high pH that buffers acidic soils. Some ecologists refer to these as the "Amphibolite Mtns." Mt. Jefferson (Ch. 8E), Bluff Mtn., Three Top Mtn., Phoenix Mtn. and Paddy Mtn. are all examples of amphibolite mountains in North Carolina.

Geologic Setting

of the Southern Blue Ridge

It is clear that uplift in the Blue Ridge continued for many millions of years after mountain building took place in the later part of the Paleozoic Era. Today streams flowing on bedrock and deep gorges marked by numerous waterfalls and rapids are proof that the streams are still actively cutting their channels lower in the region as they are along the eastern side of the Upland Plateau and in the Northern Blue Ridge. Recently, Cenozoic Era sediments close in age to those exposed in the Coastal Plain have been found in fault-bounded down-dropped blocks located in the lower parts of the Blue Ridge in northern Georgia and Alabama. These prove that crustal movements continued in the Southern Blue Ridge during the Cenozoic Era (Prowell, 2006 and Clark, Knapp, 2001). This is an important story about the region that is only now beginning to unfold.

Many of the geological characteristics of the Northern Blue Ridge continue to the south. The Chilhowee Group of sedimentary rocks derived from Laurentia (North America), lie along the northwestern edge of the Blue Ridge. Grenville-age (1-1.1 Ga) igneous and metamorphic rocks are present on the western side of the Southern Blue Ridge. They, along with the sediments that once buried them thousands of feet below the surface along the edge of North America, have been folded and thrust to the northwest (Fig. 7-9). Many of these faults break the surface along the edges of the Blue Ridge as well as in the Valley and Ridge. Some extend much farther to the west and lie buried beneath the Appalachian Plateau. Thrust faults are far more numerous in the Southern Appalachians than they are farther north and appear to have resulted from more crustal shortening here than took place to the north.

As in the Northern Blue Ridge, numerous faults involving large displacements are present along the eastern margin of the Southern Blue Ridge. The Brevard fault, which defines the eastern edge of the Southern Blue Ridge, shows up as a distinct line in the topography (Fig. 7-1). Defining the sense of movement along this fault puzzled geologists for many

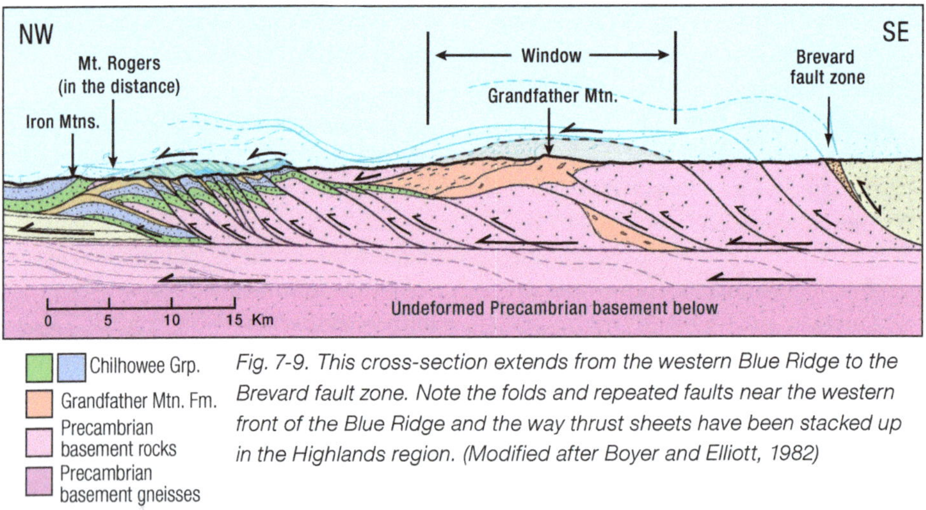

Fig. 7-9. This cross-section extends from the western Blue Ridge to the Brevard fault zone. Note the folds and repeated faults near the western front of the Blue Ridge and the way thrust sheets have been stacked up in the Highlands region. (Modified after Boyer and Elliott, 1982)

Rock Units along the Western Side of the Southern Blue Ridge Note: Although the Ocoee Supergroup, Grandfather Mtn. Fm., and Mt. Rogers Fm. are similar in age, they are separated by faults. Thickness estimates are given for some units.		
Smoky Mtns.	**Grandfather Mtn.**	**Mt. Rogers**
Chilhowee Group Ocoee Supergroup (50,000 ft. +/- thick)	Chilhowee Group Grandfather Mtn. Fm. (Between 3,000 & 9,000 ft. thick)	Chilhowee Group Konnarock Fm. Mt. Rogers Fm. (9,000 ft. +/- thick)
~~ Unconformity (a post-Grenville land surface) ~~~~~~~~~~~~~~~~~		
Blue Ridge basement complex (Most are Grenville-age igneous and metamorphic rocks)		

Table 7-1. The various rock layers found on the western side of the Southern Blue Ridge.

years. Some structural features along the Brevard fault indicate that the eastern side was thrust from southeast to northwest. Other evidence indicates that a pronounced movement took place parallel to the fault with the Piedmont side of the fault moving to the southwest relative to the Blue Ridge. It now seems likely that both are right—that the movement in the fault plane involved a diagonal motion in which the thrust sheet moved up and to the left (Fig. 7-10). Movements may also have taken place several times, probably in different directions.

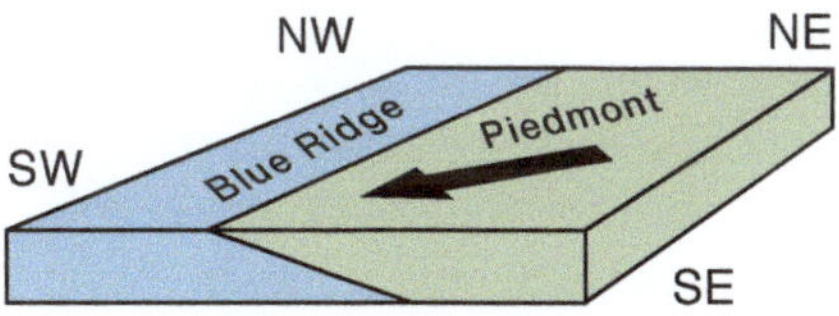

Fig. 7-10. The arrow indicates the sense of motion on the Brevard zone.

The Western Side
of the Southern Blue Ridge

The western side of the Southern Blue Ridge includes continental crust formed during the Grenville Orogeny over a billion years ago. These rocks form much of the Precambrian-age basement of eastern North America today. The western side also includes volcanic rocks formed when the Iapetus Ocean began to form as Rodinia began to break apart. The sediments that later became the Chilhowee Group (p. 78) were deposited along the new continental margin as the Iapetus Ocean (Ch. 2) invaded Laurentia about 500 to 600 million years ago. They are still present and almost continuously exposed along the western edge of the Southern Blue Ridge.

Three other thick sequences of stratified sedimentary and volcanic rocks lie between the Precambrian Grenville basement and the base of the Chilhowee Group in the Southern Blue Ridge. These three groups, described in more detail later, are the Ocoee Supergroup, the Grandfather Mtn. Fm., and the Mt. Rogers Group (Table 7-1). They are estimated to be as much as 50,000 ft. thick in the Great Smoky Mtns. and between 3,000-9,000 ft. thick at Mt. Rogers and the Grandfather Mtn. area. These units contain the record of what was happening in the Southern Blue Ridge for nearly half a billion years during the Neoproterozoic Era (Fig. 2-9).

Fig. 7-11. Outcrop of the Thunderhead Fm. (top). It is one of the most prominent units in the Ocoee Supergroup. The close-up (directly above) shows the coarse-grained texture of this sandstone.

These rock sequences are not connected so they are described as three distinct packets of sedimentary and volcanic rocks. The first, located in the Great Smoky Mtns., is known as the **Ocoee Supergroup** (Fig. 7-11). A similar sequence of sandstones, called the **Grandfather Mtn. Fm.** is present around Grandfather Mtn. The third, known as the **Mt. Rogers Group**, is present near Mt. Rogers. The units present at Mt. Rogers include rocks that are similar to the Catoctin Fm. in the Northern Blue Ridge. Both contain thick sections of volcanics, and are located in similar stratigraphic positions. Despite these similarities, these two volcanic centers are not the same age; they are not physically connected, nor are they present farther south in the Highlands region.

Blue Ridge Basement Complex

The Blue Ridge basement complex is located on the western side of the Blue Ridge and contains a number of subdivisions. Two of these will be discussed here. **Cranberry gneiss** (Fig. 7-12) (a member of the Elk Park Group) is the name applied to a variety of metamorphosed granitic rocks that are parts of the Precambrian-age (specifically, Mesoproterozoic-age, Fig. 2-9) continental basement of Laurentia (North America). Some of these rocks may have originated as layered sedimentary rocks but many appear to have originated as igneous intrusions. They were deformed and metamorphosed during the Grenville Orogeny over a billion years ago. At that time the texture of the original rocks changed from equigranular and porphyritic textures (p. 315) to banded and sheared gneisses or augen gneisses containing large feldspar crystals (Fig. 7-12). The presence of fragments of the basement in the sedimentary rocks of

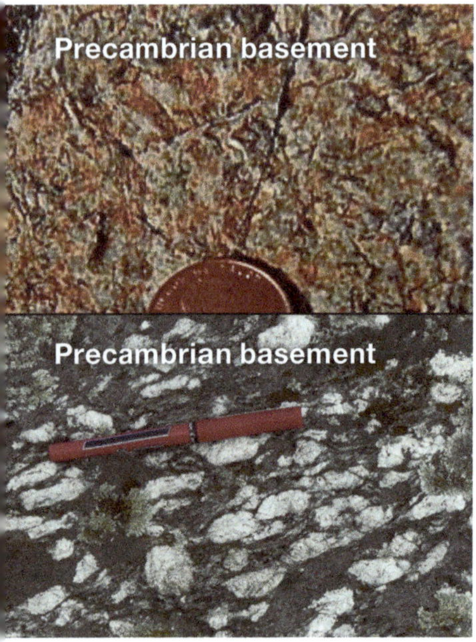

Fig. 7-12: Closeups of Precambrian basement gneisses. Both have been mapped as Cranberry gneiss.

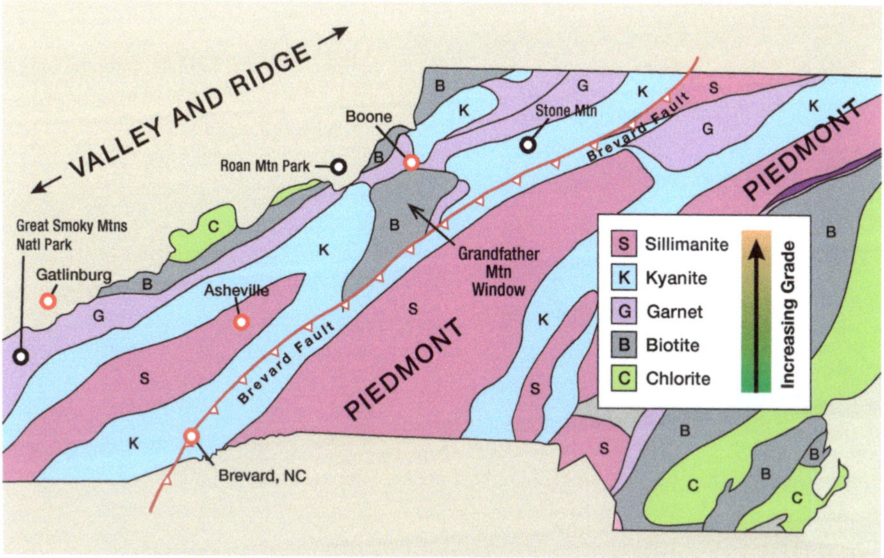

Fig. 7-13. Schematic map of the Southern Blue Ridge. (After the Geologic Map of N. Carolina, 1985)

the Mt. Rogers Fm. indicates that the basement was exposed at the ground surface when the Mt. Rogers Fm. was deposited.

A second group of Precambrian rocks, the **Crossnore Complex**, is present in the Upland basement. This is a highly diverse group of rocks including intrusions and volcanic rocks, many of which were later altered to amphibolites and schistose greenstones. They formed late in the Precambrian Era. Fragments of these rocks are also present in the Mt. Rogers Fm. and in other metasedimentary rocks of late Precambrian and early Paleozoic-age (the Ashe Fm.), proving that they were exposed at the surface when these formations were being deposited (Fig. 7-13).

Exotic Terranes

There is a large area between Boone, N.C. and Great Smoky Mtns. National Park that is underlain by rocks that are not directly related to those surrounding them. In addition, they are separated from the surrounding areas by thrust faults. These terranes are called "exotic" because they were moved from a distant location. The specific terrane described is referred to as the Mars Hill terrane (Fig. 2-12).

In studying the terranes of the Southern Blue Ridge, described in Ch. 2, geologists identified several that are not related to either the Laurentian (North American) mainland or to the rocks exposed on the eastern side of the Blue Ridge. These rocks came from much farther east. One of these is the **Mars Hill terrane** (Ch. 8H). It contains rocks such as eclogite that are rarely found at the Earth's surface (Ch. 9). They appear to have been derived from Earth's mantle and are accompanied by Grenville-age gneisses found elsewhere in the Appalachians. They are discussed in a geological guide (Merschat, et al., 2012). The Mars Hill terrane was added to the eastern margin of Laurentia during the orogenies that took place during the Paleozoic Era and may well have come to the surface during crustal compression that forced the contents of a subduction zone upward

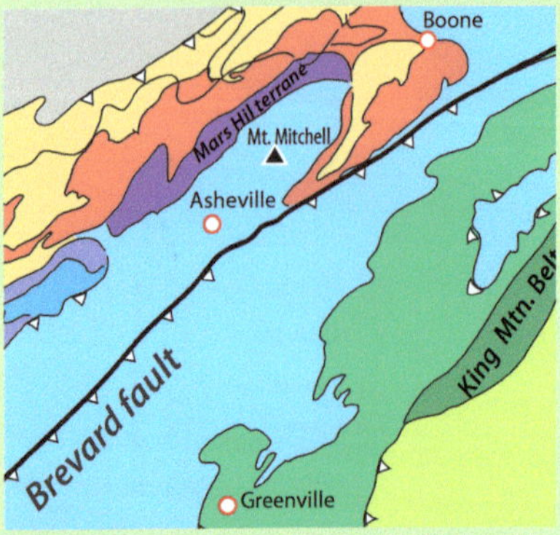

Section of exotic terrane map (Fig. 2-12).

Interpreting Geologic History in Metamorphosed Terranes

Geologists face much greater difficulty interpreting the history of the central and eastern parts of the Southern Blue Ridge than they do with its western section. Along the western flank, the rock units (especially the Chilhowee Group) have been slightly metamorphosed, but the primary features used to identify the environment in which the sediments were deposited have been preserved. One may easily distinguish sediments deposited on a beach from those laid down in much deeper water. It is also possible to recognize unconformities, times when sedimentation stopped and the area became a land surface rather than being submerged under marine waters.

In the central and eastern parts of the Southern Blue Ridge, many of the rock units occur as strips or lens-shaped packets of rocks bounded by faults. Exposures of these rocks are scattered over large areas covered by forests or grasslands. All of the rocks have been altered by metamorphism and most have been subjected to more than one period of mountain building. Those that were deformed at high temperatures, close to the melting point of the rocks, often exhibit complex structures such as those exposed along I-26 (Ch. 8K). This makes resolving the details of their origin much more difficult. Establishing the geologic history of these rocks is even more complicated because conditions of metamorphism such as temperature and pressure varied from one region to another, making it difficult to correlate the rocks from one area with those from other areas. Even their ages may be confused as a result of recurring episodes of metamorphism. The modern methods of radiometric dating can only determine the ages for the most recent metamorphism. Despite all of these problems, geologists are slowly putting together models that can be tested against new discoveries. In the true fashion of science, the mysteries of this long lost history are slowly being unraveled. The effort continues, and improved explanations emerge.

and onto the continental margin. This type of crustal inversion is known from other parts of the world and farther north in the Appalachians, but deep crustal inversions extending into the mantle are not commonly found in the Blue Ridge.

Boundary Between the Eastern and Western Sides of the Southern Blue Ridge

Major thrust faults usually occur along the boundary between the eastern and western sides of the Southern Blue Ridge. North of Grandfather Mtn., the Fries and Gossan-Lead thrust faults are at the edge of the thrust sheet that underlies the eastern Blue Ridge. The Fries fault was named for exposures located near the town of Fries, northwest of Galax, Va. (Fig. 8B-8). Across the Upland Plateau this entire thrust sheet is part of the Tugaloo Terrane (Fig. 2-12). Several terranes, bounded by thrust faults, are present in the central part of the Blue Ridge Highlands south of Grandfather Mtn., N.C. (Fig. 7-1).

The Eastern Side
of the Southern Blue Ridge

Northwestward-directed thrusting is evident on both sides of the Blue Ridge. The late Precambrian and early Paleozoic sedimentary rocks in this part of the Blue Ridge as well as the Paleozoic rocks in the Valley and Ridge were moved great distances to the northwest. However, the two sides have very different geological histories. On the western side, the sedimentary rocks of the Chilhowee Group were deposited along a passive continental margin, one similar to that found along the margin of North America and Africa today where there is no volcanic activity, few earthquakes, and no mountain building taking place. These sediments were deposited on continental slopes and shelves that subsided as Rodinia split apart and Laurentia and Gondwana separated. The rocks now found on the eastern Blue Ridge initially formed farther to the east. They have been cut into slices by thrust faults and deformed by multiple periods of deformation associated with Paleozoic mountain building. Geologists are still debating the geologic evolution of this eastern section of the Blue Ridge and the Piedmont.

Metasedimentary Rocks of the Southern Blue Ridge

Some students of the eastern Blue Ridge think many of the metasediments found here were originally deposited on rocks formed at least half a billion years earlier, subjected to the Grenville mountain building, and later eroded to a land surface much lower in elevation. We still refer to these Grenville and pre-Grenville-age rocks as the Blue Ridge basement complex. The younger sedimentary rocks were laid down on top of this basement in rift valleys formed as Rodinia broke apart (Fig. 2-11) These younger metasediments in the eastern Blue Ridge are thick and parts of them are largely composed of amphibolite, a metamorphic rock formed by the alteration of basaltic rocks or volcanic ash. For this reason, some geologists think that these rocks may have originated in volcanic island arcs—island chains similar to those now found in the western Pacific. These islands would have been located much farther offshore away from the rift basins. If this is true, the rocks formed in the island arcs must have later been shoved great distances to the west—up and onto the basement rocks of Laurentia (North America). This process is referred to as accretion or docking. This accretion took place when the ocean in which the island arcs existed closed during two major episodes of mountain building known as the Taconic (about 450 Ma) and the Acadian (about 320-360 Ma) orogenies (Fig. 2-9). It occurred many millions of years before the culminating deformation of the Southern Appalachians, which happened late in the Paleozoic Era when Gondwana collided with Laurentia. At that time all of the rocks we now see in the Blue Ridge and the Valley and Ridge were pushed to the northwest, creating the super continent, Pangaea (Ch. 2). After more than 200 million years of erosion, we now see the end product of these processes, including rocks that were formed more than 500 million yrs. ago when Rodinia broke apart.

When geologists began studying the post-Grenville rocks on the eastern side of the Blue Ridge they described them and assigned local names to the rock units they examined. The complex job of correlating these names continues today.

Fig. 7-14. Ashe Fm. at Meadows of Dan, Va.

Fig. 7-15. Alligator Back Fm. in Doughton Park.

Fig. 7-16. Outcrop of the Tallulah Falls Fm.

Lynchburg Fm.: In the vicinity of Lynchburg, Va., a thick section of metasediments and volcanics were named for that city (Fig. 5-1). Metamorphosed sandstones and some conglomerates lie at the bottom of this group of rocks. The texture and features, such as cross-bedding that formed at that time indicate that the sediments were deposited as huge alluvial fans, similar to those now found in Death Valley (p. 261). These fans were and still are clearly forming in deep basins that were created when the crust in the western United States rifted apart. Their similarity suggests that many of the rocks in the Lynchburg Group were also deposited in large rift basins formed when Rodinia split apart.

Ashe Fm.: In Ashe County, N.C., rocks that were altered to amphibolite, schist, and gneiss were likely originally deposited on the much older Grenville basement. Originally named for Ashe County, they were later correlated with the lower part of the Lynchburg Group, but nevertheless still retain their original name (Fig. 7-14).

Alligator Back Fm.: A sequence composed of amphibolites, greenstones (altered basalt), finely laminated gneiss, schist, and phyllite with some marble, is stratigraphically above the units of the Ashe Fm. These were named for excellent outcrops at Alligator Back Mtn. in Doughton Park (Fig. 7-15). The contact between the Ashe and Alligator Back Fms. is present at Philpott Reservoir. Stream channels along this contact show that there must have been a lapse in deposition, a time when the older Ashe Fm. was instead being eroded across an ancient land surface. In this same area, it is possible to correlate the metasediments of the Alligator Back with gneisses (especially the Moneta Gneiss) that occurs north of the Philpott Reservoir and with the upper parts of the Lynchburg Group.

Tallulah Falls Group: Farther south, in North Carolina, many of the post-Grenville rocks were altered by high-grade metamorphism. They became hot enough to melt, and formed mixtures of gneiss and igneous rock, called **migmatites** (Fig. 8L-2). These units are grouped together and named the Tallulah Falls Group after

Eastern American toad

The eastern coyote, a "coywolf" hybrid with mostly coyote DNA, started moving into the region in the 1970s after larger predators like mountain lions and wolves were eradicated. They primarily feed on small mammals, fruits,and berries. (Photo: ForestWander†)

a well-known waterfall located in Georgia. How these units correlate with the units farther north is still under investigation. All of these accreted rocks (the Lynchburg, Alligator Back, Ashe, and Tallulah Falls fms., Fig. 7-16) are part of what was described earlier as the Tugaloo terrane (Fig. 2-12) (see Labotka & Hatcher, 2006). All are now metamorphic rocks, schists, gneisses, and amphibolites containing minerals that are keys indicating the levels of temperature and pressure to which they were subjected. In general, the highest grades of metamorphism (p. 319) occur southwest of the Grandfather Mtn. area where kyanite and sillimanite indicate increasingly high levels of metamorphism have occurred. Two main phases of metamorphism associated with the Taconic and Acadian orogenies are present.

Ecological Setting
of the Southern Blue Ridge

The southern section of the Blue Ridge from Roanoke, Va. to Birmingham, Al. extends over 500 miles. The region is easily subdivided into zones by elevation, but the climatic changes over this distance, much of which is along a north-south line, also play an major role in determining what plants are present. The following brief introduction to the ecological characteristics of the three major divisions: Upland Plateau, Lowlands, and Highlands illustrates the great diversity of eco-communities present in this region.

Upland Plateau

(Fig. 7-4) This plateau surface slopes from about 2,000 ft. on the southeastern edge to 1,500 along the northwestern edge. Streams along the eastern edge of the escarpment drain into the Atlantic while those on the plateau surface sloping toward the west are part of the New River drainage system. Spruce and fir trees are rare except for Christmas tree farms in parts of the uplands. Most of this region is used for farming and the grazing of cattle. No old-growth forests remain. Vegetation across the plateau is much the same as that found in the same elevation range in the Highlands.

Upland Plateau

Fig. 7-17. View to the west from the Blue Ridge Parkway showing the large area known as the Lowlands (in the foreground). The mountains in the distance are the Tobacco Row Mtns.

Lowlands

(Fig. 7-17) A long strip of lowland topography characterized by hilly countryside and streams flowing down low slopes lies south of the Roanoke River between the edge of the Blue Ridge Escarpment and the Brevard/Bowen's Creek fault zone, which is used here as the eastern boundary of the Blue Ridge Geological Province (Fig. 7-1). The steep slopes at the foot of the escarpment, at roughly 1,500 ft., give way to lower topography between 600-1,000 ft. to the east. Two sites within this zone, Philpott Reservoir and Stone Mtn., are described in more detail in Ch. 8.

Much of this region is covered by a thick and fertile saprolitic soil (Fig. 8C-19) that formed from the underlying bedrock. Almost all of the region is covered by a patchwork of second-growth trees, pastures, and fields. The old-growth forests that contain a mixture of oaks and large chestnuts were either cut or destroyed by the chestnut blight (caused by a fungus) long ago. They have been replaced by oak-hickory-pine forests that contain a number of trees rarely seen in the higher parts of the Blue Ridge. These include sweet gum, persimmon, pin oaks and Virginia, and shortleaf pine. The forests also contain oaks, maples and sycamores along streambanks as they do at higher elevations.

Highlands

(Fig. 7-18) Low temperatures, higher intensities of solar radiation, the presence of large areas of exposed rock, and an increase in toxic substances in the air all have a strong influence on the environment in the high mountains of the Blue Ridge. The physical conditions and plant communities are similar to those of eastern Canada. Remnants of plant communities that were widespread during the Pleistocene are still present. At that time patches of ice remained at these heights throughout the year. Environmental conditions and the plant communities present in the highland valleys are similar to those found on the Upland Plateau. The forests in the Highlands include the natural ecological communities described in Ch. 4. More detailed descriptions are available on the Natural Heritage Program website.

Fig. 7-18. The Highlands south of Asheville, N.C.

Three communities: spruce-fir forests, shrub and grass balds, and high outcrop barrens are rarely found in the Blue Ridge outside the Highlands.

Spruce-fir Forests: These evergreen trees prosper at high elevations in the Highlands. Although they may be scattered and make up part of the canopy at lower elevation, they are the dominant canopy at the top of mountains above 6,000 ft. The understory at these high elevations is mainly composed of evergreen shrubs, especially rhododendrons. At lower elevations between 5,000-6,000 ft. a greater variety of plants make up the understory. These may include a variety of deciduous shrubs and rhododendrons.

Southern Appalachian Shrub and Grass Balds: Balds constitute a group of rare communities found only on high elevation summits and upper slopes in the Southern Blue Ridge. Rocky, cold, windswept habitats probably contribute heavily to the creation and persistence of shrub balds. At least some may have originated after catastrophic logging or fires that removed the forests many years ago. One of three types of vegetation is often present: evergreen shrub-land dominated by catawba rhododendron, shrub-land dominated by American mountain-ash and minniebush, or deciduous shrub-land dominated by blackberry. Rare rocky summit plants like those seen on balds (Fig. 7-7) are readily accessible on Grandfather Mtn. Some are visible on the rocks near the swinging bridge.

High Outcrop Barrens are wind-blasted rock outcrops subject to severe winter temperatures and ice. Soil, if present, is usually a thin veneer of organic matter, gravel, or silt. Communities in this group include shrub and herbaceous vegetation. The lower edge of these barrens ranges from about 3,000 ft. in the north to over 4,000 ft. in the south. Vegetation is usually a patchwork of shrub thickets such as mountain ash, red chokeberry, huckleberry, and mountain laurel, herbaceous mats composed of such herbs as Rand's goldenrod, mountain sandwort, and hairgrass, and lichens. Threats to the plant communities include trampling and destruction of fragile vegetation mats and the introduction of invasive weeds such as flat-stemmed bluegrass and sheep-sorrel. ❖

(The National Map, USGS, with additions)

Chapter 8

Places of Special Interest: Southern Blue Ridge

from the Upland Plateau to the Highlands

The southern section of the Blue Ridge begins near Roanoke, Va. and extends all the way to central Georgia where the undeformed sedimentary rocks of the coastal plain cover the highly deformed and metamorphosed rocks of the Appalachian Mtns. that include the Blue Ridge. The deformed belt we know as the Appalachians continues beneath the coastal plain and reappears in western Arkansas where it is known as the Ouachita Mtns. This range extends a short distance into Oklahoma before the coastal plain sediments overlap and bury it once again. No equivalent of the Blue Ridge is present in the Ouachitas. The areas described in this guide only extend as far south as the Great Smoky Mtns. The story of the Blue Ridge continues into the Carolinas, Georgia, and Alabama.

The prominent topographic ridge-like expression of the Blue Ridge disappears for several miles where the Roanoke River flows across the Blue Ridge as it passes from the Valley and Ridge into the Piedmont. It rises a short distance south of Roanoke and takes on the form of a high plateau, slightly tilted to the northwest. Our journey continues across this plateau that is drained by the New River. The tributaries of this river start along the eastern edge of the Upland Plateau and in the Highlands farther south. The edge of the Plateau is a prominent escarpment that separates the Upland Plateau from a subdued region referred to here as part of the Blue Ridge Lowlands. Over geologic time, erosion along the edge of the plateau has gradually moved the escarpment to the northwest, creating the lower surface that is described in sections devoted to Fairy Stone and Stone Mtn. state parks. The character of the Upland Plateau is examined along the Blue Ridge Parkway between Roanoke and Doughton Park. Higher mountains rise along the southwestern part of the Upland Plateau. We will review them at Mt. Jefferson, Mt. Rogers, the region around Grandfather Mtn., and at Mt. Mitchell. The guide concludes by looking at the southern section of the Parkway between Mt. Mitchell and the final section in and around the Great Smoky Mtns. National Park. ❖

Places of Special Interest: PAGE

8A. The Upland Plateau 202
8B. BR Pkwy - Roanoke to Doughton Park 208
8C. Philpott Rsvr & Fairy Stone State Park 219
8D. Stone Mtn State Park 230
8E. Mt Jefferson State Natural Area 236
8F. Mt Rogers National Recreation Area 238
8G. Grandfather Mtn Area & Linville Falls 258
8H. Mars Hill Terrane & Roan Mtn. State Pk... 268
8I. Spruce Pine Mining District 272
8J. Mt. Mitchell State Pk & Craggy Gardens... 273
8K. Precambrian Basement Rocks on I-26..... 278
8L. BR Parkway - Asheville to Oconaluftee..... 279
8M. Great Smoky Mtns National Park 280

Chapter 8A

The Upland Plateau: New River Basin & Blue Ridge Escarpment

Location and Access: Many of the places of interest to naturalists in the Upland Plateau and Highlands sections of the Blue Ridge are located off of the Blue Ridge Parkway in Virginia and North Carolina. They are listed here from north to south. Long sections of the Parkway south of Roanoke, Va. are used as local highways, hence it is accessible at all local highway crossings and from many state highways extending east of US 221 in Virginia and east from US 19 in North Carolina. Interstate I-77 crosses the Parkway a short distance north of the Va.-N.C. border and Doughton Park. The New River State Park is located near Scottville, N.C. on US 221-N.

Accommodations: Motels or cabins, restaurants and fuel are available at many places described in this section. From north to south: Roanoke, Fancy Gap, Mt. Airy, Galax, Sparta, and Doughton Park. Excellent campgrounds are present along the Parkway at Rocky Knob and Doughton Park.

Geologic Setting
of The New River Basin

The New River (Fig. 8A-1) differs from many other streams draining the Blue Ridge. To the north the Potomac, James, and Roanoke rivers all rise in the Valley and Ridge Province and flow eastwards across the Blue Ridge and Piedmont into the Atlantic Ocean. In contrast, the headwaters of the New River lie solely within the Blue Ridge, and it flows northwest into the Ohio River and eventually the Gulf of Mexico.

The New River also differs in that it flows in a meandering pattern as it crosses the Upland Plateau. This pattern (Fig. 8A-1) is generally characteristic of streams that flow on very low slopes (Fig. 8A-2), usually at low elevations. The southern part of the Mississippi River between Memphis and New Orleans is a classic example.

Above: The New River near Independence, Va.

Like most other meandering streams, the New River flows on low slopes along much of its course as it crosses the Upland Plateau. The headwaters of the New River rise near Boone, N.C. close to the Blue Ridge Escarpment along the eastern edge of the Upland Plateau. Meanders begin nearby and continue as the river flows northward across the Upland. Meandering occurs as the New River crosses the resistant rock units that define the northwestern flank of the Blue Ridge to Radford, Va. where it enters the Valley and Ridge Province. It also meanders as it follows a more westerly course across the Valley and Ridge. After passing through deep gorges of the Appalachian Plateau in West Virginia, the New River joins the Gauley River creating the Kanawha River which flows into the Ohio River about 100 miles downstream.

The meanders on the Mississippi River flow across a broad floodplain cut into the semi-consolidated sediments of the Gulf Coastal Plain. In contrast, the meanders along the New River have cut into consolidated bedrock flanked by cliffs rather than broad floodplains.

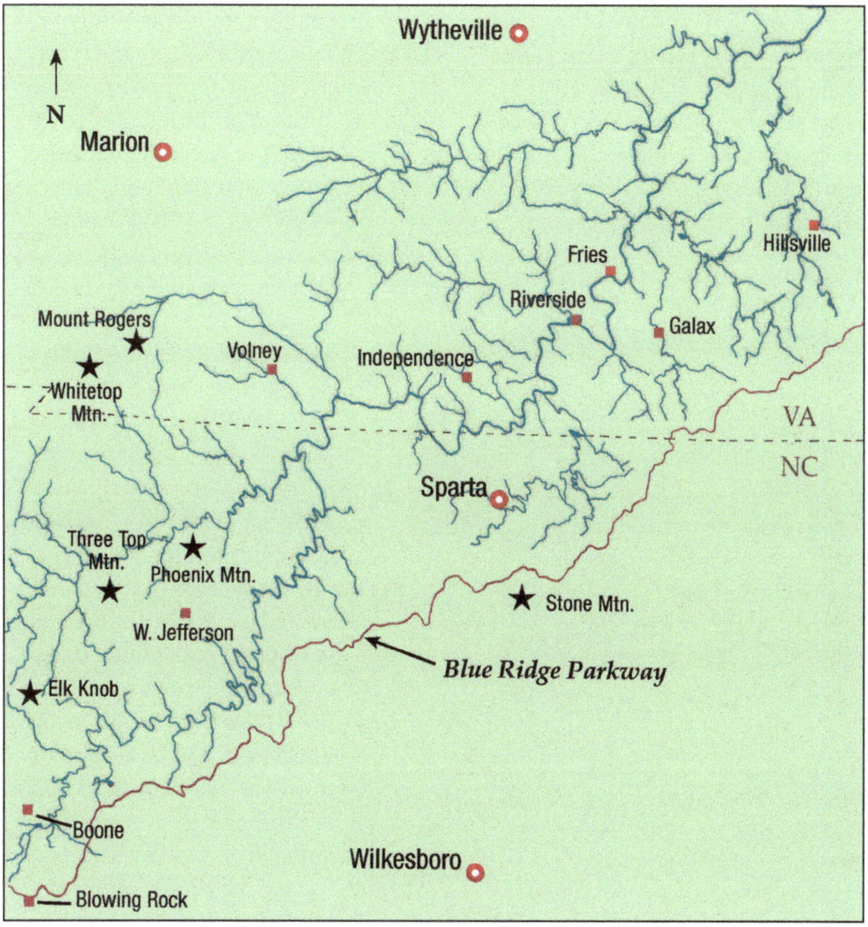

Fig. 8A-1. Sketch of the New River drainage on the Upland Plateau. The Blue Ridge Parkway runs close to the top edge of the Blue Ridge Escarpment.

Fig. 8A-2. View across the Upland Plateau. Buffalo Mtn. stands high above the plateau surface.

Geologists call these "entrenched meanders," and many think that the streams exhibiting this type of meander pattern were once flowing across surfaces of low relief close to sea level. If this is correct the New River is a very old stream that was flowing across a relatively flat or gently sloping region before it was uplifted to its present elevation. Early geomorphologists, William Morris Davis and Douglas Wilson Johnson, thought that nearly flat land surfaces very close to sea level, called peneplains, existed extensively across the Southern Appalachians. Those ideas were later abandoned, but similar ones have reappeared in recent years.

Students of the New River point to evidence that it was once a tributary of a much larger stream, known as the Teays River (Fig. 8A-3). The Teays River had its headwaters in the Blue Ridge or possibly even in the Piedmont. It flowed across the Valley and Ridge and the Appalachian Plateau in West Virginia, then across Ohio, Indiana, and Illinois, until it joined the Missouri River and eventually the Mississippi River in Illinois. This drainage pattern was disrupted when continental ice sheets covered the area north of the Ohio River during the Pleistocene Ice Ages that began about two million years ago. The ice sheet advanced and retreated a number of times, and reached its maximum extent about 17,000 years ago during the last major advance. As the ice stopped, debris carried by the ice accumulated at the edge of the melting ice sheet. The Teays River (and its headwaters, the New River) were blocked by ice. As the ice began to melt large lakes formed

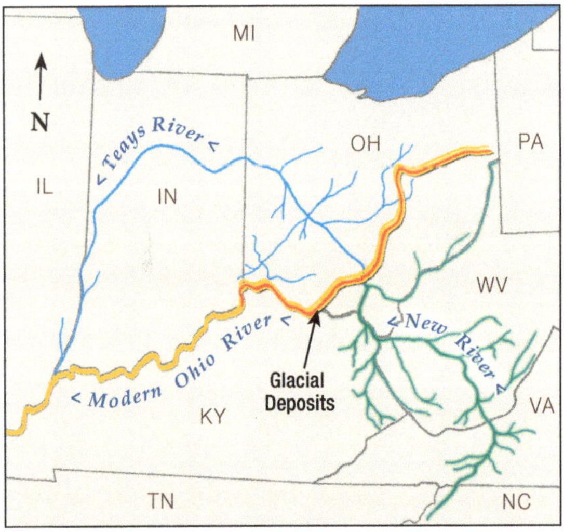

Fig. 8A-3. Reconstructed drainage of the Teays River as it appeared before the last major advance of ice that reached its peak about 17-18,000 years ago. The glacial deposits formed during the last glacial advance (red line) diverted the flow of the New River away from the Teays River and into the modern Ohio River. (After Karl Ver Steag, 1946)

behind the dams of debris left at the edge of the ice sheet. These lakes covered much of the northern Midwest, and remnants of them remain as the modern Great Lakes. Glacial deposits lie around the edges of the Great Lakes and determine the present course of the Ohio River. During the last advance the ice covered the course of the Teays River and sediment from the glaciers filled in and covered the Teays River channel. Nevertheless, it has been possible to reconstruct the Teays drainage system by using cores taken from wells.

The Blue Ridge Escarpment

The headwaters of the New River are located very close to the cliffs and steep slopes, referred to as an escarpment, that define the eastern side of the Upland Plateau. This abrupt change from gentle to steep slopes forms what is called the Blue Ridge Escarpment (Fig. 8A-4). Most escarpments gradually retreat, as masses of soil and rocks move down the steep slopes and erosion by streams gradually cuts into the cliff face. Streams extend by eroding in a headward (up valley) direction. That is now taking place along the Blue Ridge Escarpment (Figs. 8A-5, 3-20). Many escarpments form where movements along faults have caused the land on one side of the fault to rise relative to that on other side. For this reason one might suspect that the Blue Ridge Escarpment has retreated from a major fault zone along which the western side, the Upland, was uplifted relative to the eastern side. The edge of the Blue Ridge Escarpment is not straight as one might expect if it had been formed by a fault, and no fault is found at the foot of the escarpment. The Brevard fault zone located many miles east of the escarpment is the closest major fault, but there is no general agreement among geologists that the Blue Ridge Escarpment was initiated there. It might have come from other fault zones farther to the east.

In any case, the escarpment is slowly migrating to the west as streams flowing down steep slopes cut into it along

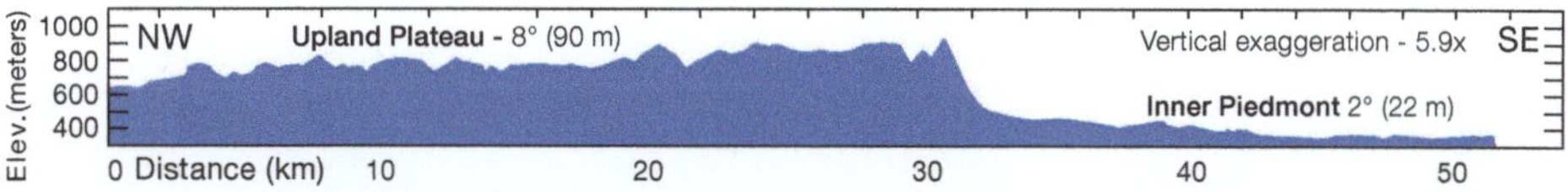

Fig. 8A-4. Profile of the Blue Ridge Escarpment and plateau surface. (After Spotila, et al., 2004)

Fig. 8A-5. View across the Upland Plateau and Blue Ridge Escarpment as seen from Elk Point, N.C.

the eastern edge of the Upland Plateau. Tributaries of the New River, the main stream on the Upland Plateau, flow through channels that are gently sloping to the north. It is clear that over time the streams that flow down the steep slopes of the escarpment toward the east have intercepted the tributaries of the New River and diverted their drainage to the east as a result of headward erosion.

Some students of the evolution of landscape in the Appalachians think a plateau or mountainous surface existed completely across the Northern Blue Ridge and that headward erosion and stream piracy, like that now taking place along the eastern edge of the Upland, gradually reduced the landscape in the Northern Blue Ridge to its present form. Today the ridge along the western edge of the Blue Ridge Province remains as the primary barrier between the Valley and Ridge and the Piedmont. Some think that headward erosion and stream piracy is responsible for the courses of the Potomac, James, and Roanoke rivers. In the future, tributaries of the Roanoke River are likely to capture the eastern part of the New River drainage.

The Blue Ridge Escarpment at Elk Spur The Blue Ridge Parkway is located near the edge of the Blue Ridge Escarpment. An excellent place to view the escarpment is located near the end of Elk Spur road that runs along a ridge that is an extension of the plateau (Fig. 8A-5). To reach Elk Spur, enter the Parkway at Fancy Gap and turn north. Take a right turn onto NC 608; continue 1.5 miles and turn right onto Elk Spur Road. Drive 1.3 miles, turn onto Pop's Peak Rd. (FS 1146) until you reach the TV towers visible from Elk Spur Road. Near the towers you will find excellent views of the escarpment to the north. The property and house at the end of the road belong to Wake Forest University.

Ecological Setting *of the New River Basin*

In 1998, the entire New River was designated an American Heritage River, expanding the earlier recognition of a large part of the river as a National Scenic River (Fig. 8A-6). The outstanding recreational opportunities, natural features, unusual ecology, and geologic history contribute to making this river exceptional. Those portions of the river basin that lie on the Upland Plateau are

Fig. 8A-6. Picnic grounds on a narrow floodplain located on the New River near Independence, Va.

rural in character. Areas of low slope are used mainly for farming while the 50% of the area with steeper slopes is forested or used for cultivation of Christmas trees. The eastern headwaters of the basin extend up to the edge of the Blue Ridge Escarpment while streams coming from the southwest have their headwaters in the high mountains located along the zone between the Upland Plateau and the Highlands. Some of the most important ecological sites in the New River Basin are in these mountains. Southern Appalachian bogs and what are called "hanging valleys" are prominent among them.

The areas between land and water called **riparian zones** are characterized by water plants and plants requiring large quantities of water. Bogs, floodplains and riverbanks are riparian habitats. The New River Basin is host to many of these plant communities. Virginia spriaea, a member of the rose family found only in the Southern Appalachian region, is one of the rare plants found in the riparian zone in the New River Basin. Other rare plants that can be found here are the goldenthread, robin runaway, bog rose, marsh marigold, and fen orchid.

Because a diversity of aquatic organisms inhabit the rivers and streams of the New River Basin several rivers here (including the North Fork, South Fork, New River, and Little River) are designated Aquatic Significant Natural Heritage Areas. Among the rare fish to be found are the tongue-tied minnow, the Kanawha darter, Kanawha minnow, and the sharpnose darter. Brook trout, North Carolina's only native trout species, thrives in these cool, oxygen-rich waters at high elevations. Amphibians such as the hellbender and common mudpuppy are also present.

Fig. 8A-7. Bog turtle (Photo: USF&WS†)

Only about 10% of North Carolina's mountain bogs remain in their natural state, most having been drained and developed because the nutrient-rich, flat-lying grounds yielded by draining are highly desirable for agricultural purposes and settlement. Bog turtles appear in abundance in the New River Basin (Fig. 8A-7). They are the smallest turtles in North America (adults average 4 in. long) and have been disappearing due to illegal pet trade collection and habitat destruction. They hide in the mud most of the time, hibernate in winter and feed mostly on small invertebrates.

New River State Park: This scenic park, designated part of the National Wild and Scenic River System in 1976, consists of a 26.5-mile stretch of the New River with several access areas containing campsites, picnic areas, and trails. The park is popular for canoeing on placid waters that wind their way south-to-north through the mountains. The visitor center, located in the center of the park at the US 221 Access Area, features exhibits about the natural and cultural history of the park. Rangers here offer regular programs. See the N.C. state parks website for details: **www.ncparks.gov.** ❖

Chapter 8B

Blue Ridge Parkway – Roanoke to Doughton Park

Location and Access: Many of the places of interest to naturalists in the Upland Plateau and Highlands sections of the Blue Ridge are located along the Parkway in Virginia and North Carolina (Fig. 8B-2). These locations are listed here from north to south. Sections of the Parkway south of Roanoke, Va. are also used as local highways, so access to the Parkway is possible at all local highway crossings and from many state highways that extend east off of US 221 in Virginia and east of US 19 in North Carolina. Interstate Highway I-77 crosses the Parkway a short distance north of the Va.-N.C. border north of Doughton Park.

Accommodations: Motels, restaurants and fuel are available at Roanoke, Va., Fancy Gap, Va., Mt. Airy, N.C., Galax, Va., Sparta, N.C., and Doughton Park. Excellent campgrounds can be found along the Parkway at Rocky Knob and Doughton Park.

Above: Fig. 8B-1. This view across the Upland Plateau was taken at Doughton Park near where the transition between the Upland Plateau and the Highlands takes place.

Roanoke Valley Overlook

The overlook at milepost 130 on the Parkway offers excellent views of the city of Roanoke and the valley of the Roanoke River. The Roanoke Valley marks the break between the Northern and Southern Blue Ridge. The change in topography is most obvious along the western edge of the Blue Ridge, which forms a distinct ridge to the northeast of Roanoke and a high plateau to the southwest. The valley of the Roanoke River once provided an easy route for American Indians and early European settlers traveling between the Piedmont and the Valley and Ridge Province.

The Roanoke River has cut its valley along a zone marked by several steeply inclined northwesterly trending faults. The Precambrian basement rocks have been dropped down in this zone. They rise higher to the northeast toward the Peaks of Otter, and are several thousand feet higher on the plateau to the southwest. The Roanoke Valley Overlook is located

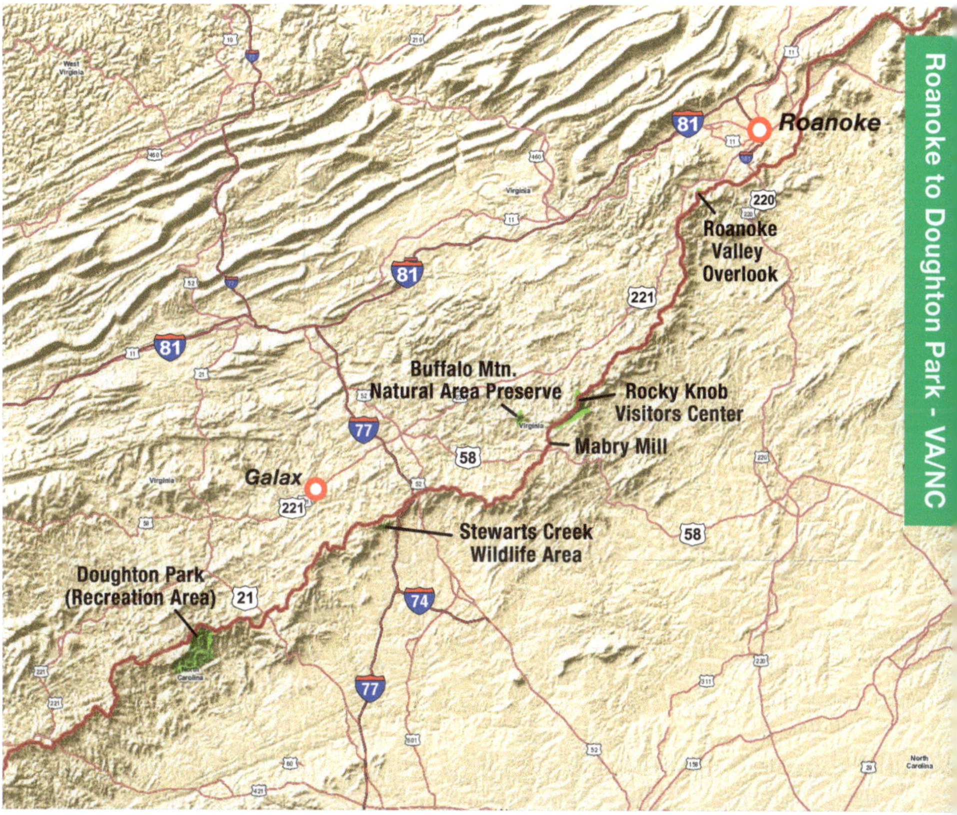

Fig. 8B-2. Map showing points of interest along the Parkway from Roanoke to Doughton Park. (The National Map, USGS, with additions)

Fig. 8B-3. Late Precambrian gneiss with blue quartz exposed at the Roanoke Valley Overlook. The blue color in the quartz indicates that the rock contains needle-shaped crystals of other minerals such as rutile.

on the Upland Plateau and the southern side of the Roanoke River valley (Fig. 8B-2).

A rock exposure at the overlook provides an excellent opportunity to see one of the rocks that is characteristic of the North American (Laurentian) basement. It is one of the gneisses of Precambrian-age (over 1 Ga). It is widely exposed along the western side of the Upland Plateau and into the Blue Ridge Highlands region (Figs. 7-1, 7-2). Examples of blue quartz are present at this outcrop as well. Inclusions of tiny mineral crystals such as rutile in the quartz often cause the blue (Fig. 8B-3) though in some cases the color is caused by refractions of light.

Fig. 8B-4. View of Buffalo Mtn. from across the Upland Plateau.

Buffalo Mountain Natural Area Preserve

Buffalo Mtn. (Fig. 8B-4) is located southwest of Floyd, Va. Take US 221 south about 6 miles to its junction with SR 727. Drive south on SR 727 to the entrance road to Buffalo Mtn. Natural Area Preserve. A parking area is located about one mile up this road. The trail to the summit is steep, rising about 570 ft., and is about a mile long. A campground is located at Rocky Knob near milepost 170 on the Blue Ridge Parkway. Motels are available at Floyd, Va. The preserve is located approx. 30 miles south of Christiansburg and Radford, Va. The preserve is closed at times for prescribed burning and protection of the plant resources. For more information contact the Natural Heritage Program (540-676-5673).

Geology

With an elev. of 3,971 ft., Buffalo Mtn. rises nearly 1,300 ft. above the surface of the Upland Plateau and forms a dramatic feature in the landscape. It is a remnant of the mountains that once rose high above the modern level of the plateau. Geologists call such remnants **monadnocks**. The bedrock at Buffalo Mtn. is composed of amphibolite (rich in amphibole minerals and iron-magnesium silicates), garnet amphibolite, biotite-muscovite gneiss and mica schist—all of which are parts of the Ashe Fm. These late Precambrian rocks were moved into the Blue Ridge in the middle part of the Paleozoic Era during collisions between Laurentia (North America) and the volcanic island arcs that had formed offshore.

Ecological Communities

An unusually large number of rare plant species thrive on Buffalo Mtn. (13 are listed by the Virginia Natural Heritage Program: **www.vmnh.net**). These rare species are nurtured by a great variety of microclimates, variations in soil depth and chemical composition. The soil is particularly rich in magnesium derived from weathering of the amphibole-rich bedrock here that shows up in seepage wetlands. Mountain sandwort, Pains frostweed, and mountain rattlesnake root are three of the alpine species left over from the Ice Age climate that prevailed here.

Openings in the forest cover, called glades, found on the southern and southeastern flanks of the mountain contain eastern red cedar and native grasses.

Fig. 8B-5a. Mealy bug that resembles the Buffalo Mtn. mealy bug. (Photo: Mokkie†)

Grassy, prairie-like glade communities including poverty oat grass occur near the summit of Buffalo Mtn. The large Buffalo Mtn. mealybug (Fig. 8B-5a), an insect found only on this mountain, feeds on these grasses. Late in the summer these glades have many wildflowers including purple blazing star and stiff goldenrod. Woodland seeps, places where water containing high levels of iron and magnesium comes to the surface from underground, occur near the base of the southern side of the mountain. These give rise to a large population of rare plants including the large-leaved grass-of-Parnassus and the bog bluegrass (Fig. 8B-5b). These herbs are normally found in swamps or bogs at high altitude (even in the arctic where the soil is acidic, with a low pH).

Fig. 8B-5b. Northern grass of parnassus flowers. (Photo: Alan Cressler)

Much of the region along the Parkway has been affected by human activities. The forests that covered this region when European settlers first arrived were cut and the land cleared. Very few remnants of the original forest remain. Modern forests here are successional. In many places, both north and south of this section the Parkway passes through mountains that are parts of national forests. Farms, pasture lands, and Christmas tree plantations cover large areas of the low sloping fertile soil found in this region. Consequently, the borders of the Parkway are narrow through much of this section and the Parkway serves as an important highway for local use.

The flame azalea is a native wildflower.

Rocky Knob Recreation Area

Rocky Knob Visitor Center: (mp 169) The visitor center is only open on weekends (Fig. 8B-6). In the Rocky Knob Recreation Area you will find camping, cabins, picnic grounds, and several trails.

Amphibole and biotite-mica schists of the Lynchburg Fm. are exposed near the visitor center. These rocks, characteristic of much of the eastern part of the Blue Ridge, were accreted onto the North American continent (Laurentia) during the Paleozoic Era. Milky quartz veins are abundant (Fig. 8B-3). Old mining prospects for iron and copper are located near the northern end of the parking area. Pyrrhotite (an iron sulphide mineral that resembles pyrite) and limonite (an iron oxide commonly called rust) are present at many prospecting sites and can often be seen on outcrops along the Parkway (Figs. 8B-7a-b).

Fig. 8B-7a. Iron-stained rocks (metagraywacke) at Rocky Knob Park.

Fig. 8B-7b. Pebble conglomerate, part of the Ashe Fm., found in the overlook directly across from the visitor center at Rocky Knob.

Fig. 8B-6. The visitor center grounds at Rocky Knob Recreation Area.

Symbol marking the grounds of FloydFest located in Floyd, Va., near Rocky Knob.

Music in the Mountains

Southern Virginia and the northern section of North Carolina are famous for bluegrass and old-time music. Many communities have regularly scheduled events that attract tourists from far and wide. In Virginia you are likely to see signs along state roads pointing out that you are on "The Crooked Road"—the roads connecting places like Abingdon, Bristol, Floyd, Galax, Marion and many others in this region where fiddles, banjos, and other string instruments are an important part of local traditions.

Blue Ridge Music Center: (mp 213) Farther south on the Parkway you will pass the Blue Ridge Music Center, a place created and operated by a partnership between the National Park Service and the Blue Ridge Parkway Foundation. Concerts featuring the banjo and fiddle are performed May through September.

Mabry Mill

Mabry Mill: (mp 176) This restored mill located near the Parkway is a popular tourist attraction. Edwin B. Mabry, great great-grandson of Isaac Mabry, who received a land grant for 183 acres of land in 1782, built this mill in the late 1800s. Initially he installed a water-turned lathe used to make chairs. He built his home nearby and the site became a blacksmith and wheelwright shop. Later a sawmill was added, and finally a gristmill was put into operation. Ed Mabry was clearly an independent, self-sufficient sort of person. The mill has been restored by the Park Service and is used to demonstrate "old-time" skills such as basket weaving, spinning, weaving, and the production of apple-butter. Old-time music is played on Sunday afternoons throughout much of the tourist season.

Mabry Mill

Beaver dam close to the Parkway near Mabry Mill.

Stewarts Creek Wildlife Mgmt. Area

This area is located about seven miles southeast of Galax, just south of the Blue Ridge Parkway. Parking and access to the area is available at two places. The lower parking lot is located at the end of SR 795 and is adjacent to a stream providing easy access for anglers. To reach the upper parking area from the Blue Ridge Parkway turn south on SR 715, then left on SR 975 and follow it until it ends. To reach the lower parking area exit the Parkway at Pipers Gap and travel south on SR 620 until you reach SR 895, turn right and continue to SR 795 where you turn right again. Continue to the lower parking area. Trailblazer signs indicate the correct route to both parking areas. Camping is permitted in the management area. The closest motels are at Fancy Gap, Va., Galax, Va., and Mt. Airy, N.C.

The area is well known for hunting, fishing, hiking, and camping. Several trails and remnants of logging roads provide good foot access to the property. It is also an excellent place to look for birds and wildflowers. The home of Confederate General J.E.B. Stuart is located nearby. About 7 miles north of Mt. Airy, N.C., the 1,087-acre Stewarts Creek Wildlife Management Area is located on the Blue Ridge Escarpment. The upper parking area is on the Upland Plateau at an elevation of 2,955 ft. Parking is also available far down the escarpment at 1,580 ft.

Ecological Setting

The upper reaches of Stewarts Creek flow over outcrops of the Alligator Back Fm. Rhododendron thickets cover the steep slopes and much of the lowland areas.

Bloodroot blossoms

The open lowland areas are wooded with yellow birch, tulip poplar, and several species of magnolias. Older, mature-growth forests at higher elevations are primarily composed of oaks, hickories, and maples. This is an excellent area to look for a great diversity of wildlife. Canada warblers can be found in rhododendron thickets. In the woodlands, look for hooded and black and white warblers, American redstart, and wood thrush. In the more mature forests of upland areas, listen and look for black-throated warblers, blue/black-throated green warblers, scarlet tanagers, and veery. Wild turkey, ruffed grouse, white-tailed deer, and gray squirrel are abundant at Stewarts Creek. The elusive red fox may be seen as well.

Deer, grouse, turkey, and squirrel also thrive at Stewarts Creek. Abundant wild grapes especially draw grouse near the upper portions of the mgmt. area. The numbers of grey squirrels in areas of upland hardwoods varies from year to year depending on mast (nut) production.

Pool habitats and large native brook trout are abundant partly due to the fact that the streams here are managed under a "no harvest" regulation; all fish caught must be released back into the stream.

Both forks of Stewarts Creek begin high in the mountains and then cascade down through steep rocky gorges. Trails along these forks provide a pleasant hiking experience for those who wish to enjoy the scenic streamside beauty and mtn. views.

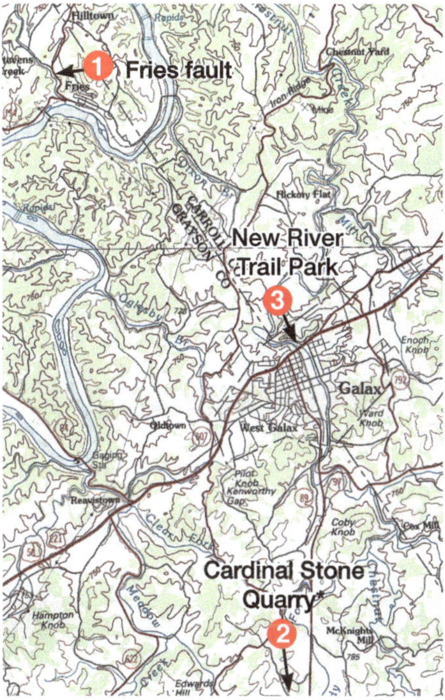

*Fig. 8B-8. Geological points of interest near Galax, Va. *Stop 2 is south of the map. (USGS)*

Galax Area

The town of Galax, Va. widely known for the Appalachian folk music festivals held there in the summer is an important site along the "Crooked Trail." If you plan to be at Galax during the festivals, be aware that accommodations sell out quickly.

Points of Interest

Several localities of geological importance are located at or near Galax (Fig. 8B-8). Many will also find the New River Trail Park an interesting place to hike.

Stop 1: **The Fries Fault:** The Fries fault zone (also known as the Gossan-Lead fault) is the name applied to one of the major fault zones in the Upland Plateau (Fig. 8G-5). This fault zone separates rocks formed in the island arcs and subcontinents that lay off the eastern margin of Laurentia (N. Am.) from rocks that were part of Laurentia after the Grenville Orogeny took place about a billion years ago. Faults with other names but sharing this position are present in the Northern Blue Ridge and south of Grandfather Mtn. throughout the Highlands.

The type locality for the Fries fault lies along SR 94 about 0.3 miles north of the town of Fries, Va. From Galax you can reach the fault zone by taking SR 94 from Galax to Fries, or you can take SR 607 (Fries Rd.) to New River Trail State Park, an abandoned railroad right of way, and turn left onto SR 72 (Fries Road). Drive through Fries to SR 94; turn right onto SR 94 and park on roadside. The Fries fault is exposed along a large curve on SR 94. It contains breccia and mylonite (indicating high shearing in which the rock was recrystallized) (Fig. 8B-9). At this location the Precambrian gneiss has been thrust onto the Unicoi Fm.

Fig. 8B-9. Close-up view of the mylonite in the Fries fault zone.

Fig. 8B-10. Cardinal Stone Quarry (left) and close-up of ultrabasic rock found at the quarry (right).

Stop 2: **Peridotite at the Cardinal Stone Quarry, Galax:** (Fig. 8B-10) Ultrabasic rocks are bodies of rock that contain much higher levels of magnesium and iron than most crustal rocks. Minerals present in these bodies include serpentine, talc, magnetite, and some olivine, all of which are thought to originate deep in the crust or perhaps even further down in Earth's mantle. Many of them originated in subduction zones and were uplifted during the intense deformation that occurred during mountain building. Most of these bodies are small and lie within outcrop belts of the Ashe Fm. One such body is exposed in the Cardinal Stone Quarry located south of Galax. Take SR 89 south from Galax and turn right (west) on SR 613. Entrance into the quarry is limited. However, sample blocks from the quarry lie along the entrance driveway (Fig. 8B-10).

Stop 3: **Ecological sites in Galax:** (Fig. 8B-8) The New River Trail Park, located along Chestnut Creek, passes through Galax. A parking lot is located at the entrance to the trail on the north side of US 58 and US 221 where T. George Vaughan, Jr. Ave. intersects US 58. This "trail park" continues along the creek to Fries where it joins the New River and continues across the Upland Plateau and into the Valley and Ridge Province. The trail is an exceptional place to look for birds and wildlife on the Upland Plateau. Camping, hunting, fishing, horse trails, boating, and hiking are available at many places along the trail park. Check the website for details.

The town of Fries, Va., located along the New River.

Doughton Park (Doughton Recreation Area)

View of exposures of the Alligator Back Fm. along the Parkway at Doughton Park.

This recreation area managed by the NPS is located along the Parkway near milepost 240 in North Carolina, south of Sparta and northwest of Wilkesboro. It encompasses a lodge, campgrounds, picnic tables, a restaurant, and miles of trails.

Geologic Setting

The park is situated close to the edge of the Blue Ridge Escarpment at an elevation of 3,700 ft. Overlooks along this section of the Parkway provide views across the eastern flank of the Blue Ridge. Erosion has cut the escarpment back to the west and reduced the elevation of the section of the Blue Ridge between the escarpment and the Brevard fault zone (Fig. 7-1). From some overlooks along the Parkway you can look down on the balds at Stone Mtn. Park (Ch. 8D). Beyond the Brevard fault zone, you can see the Sauratown Mtns. and a region of low-level relief typical of the Inner Piedmont. Layers of quartzite hold up the highest parts of the Sauratown Mtns.,

Fig. 8B-11. Close-up photographs of the Alligator Back Fm. exposed at Doughton Park. Small isoclinal folds are evident in the photograph at left. The "pin stripe" texture commonly found in the Alligator Back Fm. is prominent in the photo at right. Both were taken at the end of the road passing through the picnic area at Doughton Park.

The Blue Ridge Parkway near Doughton Park.

which rise to over 1,700 ft. in elevation. The quartzite cliffs there are well known for rock climbing. Some geologists have suggested that these quartzites may be equivalents of the Chilhowee rocks found on the western flank of the Blue Ridge. If this is correct, huge sheets containing schists and sedimentary rocks were pushed over the rocks in the Sauratown Mtns. and onto the Blue Ridge. Erosion eventually removed some of the schists revealing the quartzite in the mountains surrounded by older schists. Thus, the Sauratown Mtns. are known as a structural window (p. 53).

Doughton Park is the type locality for the Alligator Back Fm., one of the most widely exposed rock units in the eastern part of the Blue Ridge. These late Precambrian rocks were accreted to the Blue Ridge as part of a huge thrust sheet during Paleozoic mountain building. The distinct fine layers in the schists that make up this formation have prompted geologists to refer to it as a "pin stripe" texture (Fig. 8B-11). Good exposures are present along the Parkway. One of the most convenient exposures occurs at the end of the road that passes through the picnic area. For superb views, especially to the west, take the half-mile long trail from this Stop to the cliff face. Veins of milky quartz are present in many parts of the Alligator Back Fm. ❖

The iconic monarch butterfly (left) is called a "milkweed butterfly" because its caterpillars feed exclusively on milkweed plants. Milkweeds are known for their milky sap. Above is a milkweed pod that has split open to release its seeds in autumn.

Chapter 8C

Philpott Reservoir and Fairy Stone State Park

Location and Access: Virginia's Fairy Stone State Park is located around the large reservoir created by Philpott Dam. The dam is situated at the end of SR 905, a local road off of US 57 northwest of Martinsville. The main entrance is at the end of SR 346, a dead end road off of US 57. For details about reaching other facilities around the lake obtain a map from the park headquarters.

Accommodations: Cabins overlooking the lake and campsites are located at the park headquarters. Campsites, picnic grounds, and boat ramps are located at other sites around the reservoir. Restaurants and grocery stores are located in Ferrum, north of the lake, in Martinsville, and along US 220 east of the lake.

Outstanding Features: Fishing, boating, camping, and many trails are present in this large park. This is one of the few places one can find the staurolite crystals known as "fairy stones."

Nearby Attractions: Martinsville is well known for the Virginia Museum of Natural History and for its auto racetrack. The museum contains interactive geological and ecological displays that make it especially attractive to young people. Ferrum College is located a short distance north of the park. Smith Mtn. Lake, a large reservoir surrounded by many small communities and housing developments, is northeast of the Philpott Reservoir. Its geological and ecological settings are similar to those at Philpott Reservoir.

Philpott Reservoir covers about 3,000 acres of land and is surrounded by 7,000 acres of forest that is managed by the U.S. Army Corps of Engineers (Fig. 8C-1). This is one of the best places in the Blue Ridge to go for lakeshore camping. Campsites and cabins overlooking the lake are available. Boats to rent and loading ramps are available around the lake. Birders will have a good chance to see water birds here. Geologists will find outcrops that reveal the various rock types found in

Above: Fig. 8C-1. View of Philpott Reservoir, taken from the wayside at the dam. Note the Blue Ridge Escarpment in the distance.

Picnic area by Smith River near Stop 1.

the Alligator Back Fm., one of the most widely distributed units on the eastern side of the Blue Ridge. You will also see the topographic expression of one of the major faults used to define the eastern edge of the Blue Ridge Province, and to collect rarely found crystals of staurolite, the famous "fairy stones."

Large areas around the reservoir are open for hunting and fishing. Turkey, deer, bass, crappie, sunfish, walleye, catfish, and carp are abundant in these areas. Contact the Virginia Dept. of Game and Inland Fisheries for more detailed information. Fairy Stone State Park, which covers about 5,000 acres on the west side of Philpott Reservoir, is closed to hunting but fishing is permitted in the streams and in Fairy Stone Lake. The park provides the most complete recreational facilities in the area. It has a large beach, cabins, campgrounds, picnic areas, and boat landings. No restaurants or motels are available in the park. Some supplies can be obtained from local rural stores.

Ecological Setting

The shorelines of the Philpott and Smith Mtn. reservoirs, which are about 1,000 ft. above sea level, extend for many miles, creating an environment strongly influenced by these large bodies of water. White pines, yellow poplars, maples, and oaks are prominent in the upper story of the forest that covers the rolling hill country around the reservoir.

Birds seen at Philpott Reservoir

The Philpott and Smith Mtn. reservoirs are attractive environments for many birds including waterfowl that are rarely seen in the higher parts of the Blue Ridge (Fig. 8C-5). Look for brown-headed nuthatches and yellow-throated warblers in the pine treetops near the dam. Barn swallows nest on the Philpott Dam. Many waterfowl are present on the lake especially during migrations These include horned grebe, bufflehead and a variety of terns. Kingfishers are often present along the shore, as are killdeer and mallard ducks. In addition, most of the birds found on the Plateau and in the Highlands including eagles, hawks, blue jays, Carolina chickadees, tufted titmice, red-eyed vireos, chipping sparrows, cedar waxwings, Carolina wrens, and goldfinches also appear at the reservoir.

View across Philpott Reservoir.

Snow geese

Foresters tern (Photo: Dick Rowe)

Kingfisher (Photo: Dick Rowe)

Sea gull (Photo: Dick Rowe)

Mallard ducks (Photo: Dick Rowe)

Blue heron

Killdeer (Photo: Alan D. Wilson†)

Wood duck (Photo: Frank Vassen†)

Canadian geese

Bufflehead (Photo: Mdf†)

Fig. 8C-5. A great variety of water birds can be seen at Philpott Reservoir.

Japanese silver grass is an invasive ornamental grass that grows by local roadsides.

Fig. 8C-6. Philpott Dam

Away from the lakefront, eastern woodland birds including the mourning dove, ruby-throated hummingbird, downy, red-bellied and pileated woodpeckers, wood-pewee, blue jay, Carolina chickadee, tufted titmouse, nuthatch, gnatcatcher, bluebird, cedar waxwing, red-eyed vireo, indigo bunting, towhee, chipping sparrow and American goldfinch inhabit the woods and fields. See Ch. 11 to identify birds.

Several interpretative trails with plant identification markers are available around the reservoir. One is located along Oak Hickory Trail. Others are located near the visitor center for the dam, along the Goose's Roost Trail at Goose Point (off SR 882), along Salthouse Branch Trail (off SR 773), and along the trails at Jamison Mill (off SR 778). These last two sites must be approached from the east side of the reservoir. Obtain maps and directions from the dam's visitor center. Smith Mtn. Lake, located northeast of Philpott Lake, is the second largest lake in Virginia. Most of its shoreline is privately owned, but public access is available at a 1,248-acre park located on the north shore. Cabins, campgrounds, a beach, docks, hiking trails and a visitor center are available here. During the tourist season the park features night hikes, canoe trips, naturalist led programs for children and arts and crafts programs. The ecological environment, including species of birds and wildlife is similar to those found at Philpott Reservoir. An osprey nesting-platform built at the end of one of the trails at Smith Mtn. Reservoir attracts a lot of attention.

Geologic Setting

The area around the Philpott Reservoir is geologically important for several reasons. The Bowens Creek fault, the eastern boundary of the Blue Ridge Geological Province in this area, is exposed here. Several varieties of the rocks and small-scale structural features found in the Alligator Back Fm. are exposed around the reservoir. The contact between the Alligator Back and underlying Ashe Fms. is exposed nearby, and this is one of the best places to find staurolite crystals in the Southern Appalachians.

Fairy Stone State Park is also located at the southeastern edge of the Blue Ridge Geological Province. Like Smith Mtn. Lake, the Philpott Reservoir lies in the Lowlands region between the Blue Ridge Escarpment and the faults (in this area, the Bowens Creek fault) that are used to define the eastern edge of the Blue Ridge Province. The Lowlands formed as stream erosion caused the escarpment to retreat to the northwest. The Bowens Creek fault, the northeastern continuation of the Brevard fault, crosses the reservoir about 0.75 miles north of the dam (Fig. 8C-6). From the visitor center one gets a beautiful view across the Lowlands to the distant Blue Ridge Escarpment.

The park gets its name from an unusual occurrence of the mineral **staurolite**, a metamorphic mineral formed at high temperature and pressure. The mineral is an iron aluminum silicate that commonly occurs as two intersecting hexagonal crystals grow in the form of a cross (Fig. 8C-7). A legend suggests that these crosses formed from the tears of fairies after hearing of Christ's crucifixion.

The staurolite minerals grew in micaceous schist, but because they are more resistant to weathering than the schist, they weather out and may ultimately be found lying loose on the ground surface. Staurolite is a product of medium-grade metamorphism that took place at temperatures of about 500°C and pressures found inside the earth at depths in the range of 2-9 miles. This gives us some idea of the conditions that existed in the rocks that were ultimately moved during mountain building from a position at great depth on or east of the Laurentian continental margin to their present position in the Blue Ridge. The schist, part of the Alligator Back Fm. (equivalent to part of the Lynchburg Group), was derived from sediments, mixtures of shale, silt, and volcanic ash and lava flows formed along the continental margin during late Precambrian time. Metamorphism and the movement of these rocks to their present position in the Blue Ridge took place during the middle to late part of the Paleozoic Era.

A variety of rock types occur in the Alligator Back Fm. including finely laminated quartz and feldspar-rich gneiss, small amounts of marble, graphitic mica phyllite, schists, and quartzite. In contrast, the Ashe Fm. consists of conglomerates, sandstones, and thick sections of lava and metavolcanic rocks many of which have been metamorphosed to amphibolites and amphibolite gneiss.

Fig. 8C-7. Staurolite crystal collected in Fairy Stone State Park.

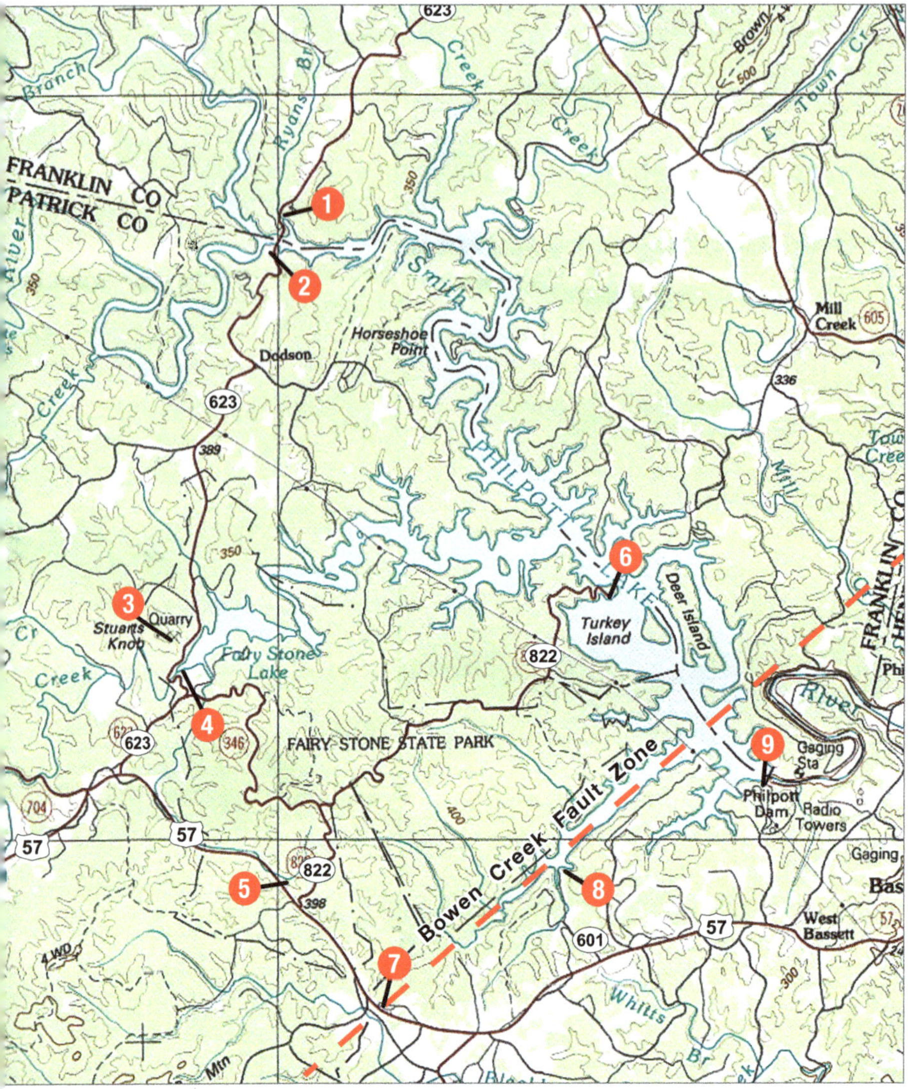

Fig. 8C-8. Stops in the Philpott Reservoir area. (USGS map, with additions)

Points of Interest

Most of the Stops listed here are located along SR 57, SR 623, and SR 601 or along their side roads (Fig. 8C-8). More details about these and other Stops in the Piedmont are described in Henika, et al., 2000. Those interested in additional Stops in this area should refer to Conley and Henika, 1973, in the bibliography.

Stop 1. **Alligator Back Fm:** This exposure is conveniently located at the northern end of the reservoir near the Ryans Branch campsite and boat landing. Stop on the north side of the bridge at the parking area and boat landing. A dark green amphibole schist (part of the Alligator Back metabasalt) (Fig. 8C-9) is located at water level. Dark-gray-to-black metasiltstone and micaceous quartz-

Fig. 8C-9. Amphibole schist at Stop 1.

Fig. 8C-10. Graphitic schists found at Stop 2.

Fig. 8C-11. This red rock is a variant of what is found in the tailings at the Stuart's knob mine. This one is not rich enough in iron to have been used, but bears the reddish color of iron oxide.

ite beds are exposed in roadcuts at the bridge. (See Ch. 9 for rock descriptions.)

Stop 2. **Graphite-schist:** An excellent exposure of quartz-rich graphite schist (part of the Alligator Back Fm.) is visible in cliffs located 0.1 miles south the bridge over Smith River along SR 623 (Fig. 8C-10). The black color and shiny surfaces of the foliation are enhanced by the presence of the rarely seen mineral graphite.

Stop 3. **Abandoned iron mine at Stuart's Knob:** Two mines are located near Fairy Stone Lake. The first one, on Stuart's Knob, can be reached by following a trail from a parking area located between SR 623 and the reservoir. **Tailings** are the leftover materials from mined rock. At the mines near Philpott Reservoir, the tailings often contain a black iron oxide mineral called magnetite. The ore here appears to have been concentrated along 3-6 ft. wide bands of schist located at the contact between the schist and a metamorphosed intrusion of gabbro (a coarse-grained equivalent of basalt). At the ground surface the soil is rich in limonite (an iron oxide formed by the weathering of magnetite) (Fig. 8C-11).

Fig. 8C-12. The beach at Fairy Stone Lake. The iron mine is located to the left of this part of the beach.

Fig. 8C-13. Entrance to the iron mine near the beach at Fairy Stone Lake.

Fig. 8C-14. Tailings from the iron ore near the beach at Fairy Stone Lake.

Stop 4. **Abandoned iron mine at Fairy Stone Lake:** The second mine is located near the shore of Fairy Stone Lake. Take SR 346 to reach Fairy Stone Lake beach. Turn left just past the entrance station where you can see the beach. Walk to the southwest along the beach to its end then continue a short distance along a paved road until it ends. From there take a dirt path until you come to an opening to one of the mines (Figs. 8C-12, -13, -14). It is now gated, but the rocks of a shear zone in the Alligator Back Fm. in which the iron ore became concentrated are exposed. (Note: the map shows SR 623 crossing a creek that feeds into the lake—that bridge has been removed.) The shorelines at both of these Stops afford good viewing of birds and wildflowers.

These mines were operated from the late 1700s until the early 1900s. During most of that time the mining was done by hand with pick and shovel. The ore was used locally because it could not be exported. In 1905 a rail line was installed and iron exporting began, resulting in a short-lived

Fig. 8C-15. Alligator Back Fm. at Stop 5. The white mineral found here is quartz. The darker layers contain more micaceous minerals. The lighter colored layers contain more quartz silt and sand. The orientation of the micaceous minerals produces a strong alignment in the dark layers.

Fig. 8C-16. A block of pegmatite located near Goose Point Park. The light colored shiny mineral is muscovite mica. The black mineral is hornblende.

population explosion lasting only a few years until milled ore could be imported from Germany for about the cost of local unmilled ore. By 1920 mining had ceased, most local timber had been cut, the rail line failed, and the Great Depression set in.

Stop 5. **Structures in the Alligator Back Fm:** This exposure is located on SR 57 about 0.2 miles northwest of the intersection of SR 57 and SR 822 (Goose Point Road). This is an excellent place to see small-scale structural features such as well-developed foliation formed by the alignment of platy minerals such as mica and veins of quartz (Figs. 8C-15, -16). Look for the small-scale folds with nearly vertical axes that are present in the micaceous schist.

Stop 6: **Goose Point Park** has lake access with a boat landing, picnic area and campground with showers and other facilities. Trails here provide good opportunities to see birds and wildflowers.

Continue on SR 57. A trail used to look for staurolite crystals is located by a store located on the east side of SR 57, opposite the intersection of SR 57 and the northern entrance to SR 687. Crystals are sold in this store but may also be purchased or found at Stops 7 and 8.

Stop 7. **Candler Fm. at Bowens Creek Fault Zone:** Exposures of the highly sheared mylonite rocks of the Bowens Creek fault zone are present along SR 57 northwest of the Haynes 57 Market (another store where staurolite crystals may be purchased). The fault zone lies within the Candler phyllite, a Cambrian-age unit that was slightly metamorphosed. The Candler Fm. is a gray colored muscovite-chlorite schist and phyllite (Fig. 8C-17). Erosion along the Bowens Creek fault zone has produced this long northeast-southwest trending branch of the Philpott Reservoir (Fig. 8C-8). The Bowens Creek fault is a thrust fault that is inclined toward the southeast. Along some sections, the fault appears to be subhorizontal. At other places it is steeply inclined—always toward the southeast. The hanging wall (the upper part of the thrust sheet) is always displaced (by distances considered to be measured in at least tens of miles) toward the northwest.

Fig. 8C-17. The Candler Fm. crops out close to the Bowens Creek fault zone at Stop 7.

Stop 8. **Bowens Creek Fault:** Continue south on SR 57, turn left on SR 601 and proceed to the boat ramp. Walk along the shore beyond the beach area. You will find staurolite crystals that have been replaced by the mineral sericite. Some of these crosses are 2-3 in. long. They occur in a type of mica schist known as the Fork Mtn. Fm. These may be found in slabs of rock in the water along the shore. A campground and picnic areas are available at this site.

Stop 9. **U.S. Army Corps of Engineers Museum:** Return to SR 57 and continue east to Philpott Dam Road, then take SR 905 to the end of the road. Here you will find a small museum, marked trails, and a very nice picnic ground overlooking the dam (Fig. 8C-18). There are excellent views of the reservoir and the escarpment at the edge of the Blue Ridge Upland in the distance.

Fig. 8C-18. The U.S. Army Corps of Engineers museum and information center are located at a site overlooking the Philpott Dam.

Stop 10. **Contact between the Alligator Back and Ashe Fms:**
Note: this Stop is located to the east off of the map on (Fig. 8C-8). Take SR 605 (Henry Road) from US 220. The contact between these two formations is located near the post office at Henry, close to the intersection of SR 605 and 606. Outcrops of biotite-muscovite gneiss and amphibole schist located along SR 606 are near the top of the Ashe Fm. (part of the Ashe known as the Moneta Fm.). The lower part of the Alligator Back Fm. is composed of metagraywacke, which is exposed south of the intersection of SR 606 and 605. The basal rock unit in the Alligator Back, a metamorphosed conglomerate, is exposed in Town Creek about 0.5 miles south on SR 606 from Henry Post Office. Access to the creek is a difficult and steep slope.

If you drive from the Philpott Dam to Bassett be sure to notice places where you can see a deeply weathered rock called saprolite (Fig. 8C-19). You can still see the texture and structure of the rock from which the saprolite formed mainly by chemical alteration of the bedrock. This weathering caused the feldspar minerals to become clay. ❖

Fig. 8C-19. Saprolite exposed along SR 57 near Bassett, Va.

Chapter 8D

Stone Mountain State Park

Above: Fig. 8D-1. Granite dome exposed at Stone Mtn., N.C. as seen from the Blue Ridge Parkway overlook.

Location and Access: This park is located south of the Blue Ridge Parkway and west of US 21 in North Carolina. From the Parkway, drive south on US 21 to Roaring Gap. Take SR 1100 (Oklahoma Road) into the park. If you are approaching from the south along US 21, take SR 1002 and the John P. Frank Parkway into the park.

Accommodations: A campground with facilities for trailers is present near the park visitor center. A large number of backpack camping sites are available. Motel accommodations, restaurants, and groceries may be found in Sparta, about 8 miles northwest of the Blue Ridge Parkway, and in Elkins about 20 miles to the southeast along US 21.

Outstanding Features: This park is best known for the beautiful bare-rock bald exposed here. It also contains many trails providing excellent places for nature walks, some of which are along rhododendron-lined streams. Rock climbing, fishing, and horse trails are available.

Ecological Setting

Stone Mountain State Park covers an area of over 15,000 acres with elevations that range from 1,900 to about 3,500 ft. The park's main features are a 600-ft. high granite dome (Fig. 8D-1) and approximately 17 miles of designated trout waters. The primary habitats are mountain streams with rhododendron thickets, oak and mixed hardwood forests with pines, and rocky outcrops. The understory is composed of blueberry, rhododendron and mountain laurel. Small, slow growing pines and cedars fringe the rock outcrops, while mats of lichens, mosses, and small ferns grow on areas of open, exposed granite. To see these types of communities, take the Stone Mtn. Loop Trail. Trail maps are available at the park office (Fig. 8D-2).

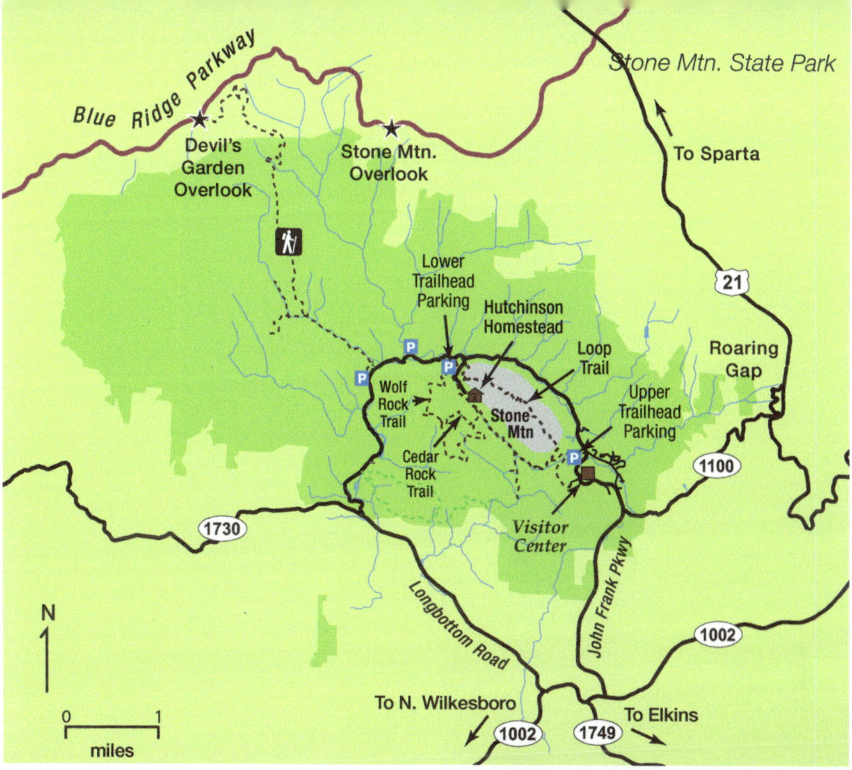

Fig. 8D-2. Stone Mtn. State Park is located south of the Blue Ridge Parkway and west of US 21. A paved road that follows a beautiful stream allows access to the points of greatest interest.

Fishing is permitted and trout are plentiful in the streams throughout the park (Fig. 8D-3). A great variety of birds are present including wild turkey, ruffed grouse, whippoorwill, red-bellied, downy, and northern flicker woodpeckers, as well as water thrush. Belted kingfishers may be seen along the streams. Chestnut, black, and scarlet oaks, red maples, hickory, white pine, and dogwood are common in wooded areas (see Part 3 for identifications).

Fig. 8D-3. Streams in the park are popular for trout fishing.

Succession of Plant Communities on Granite Outcrops

The Stone Mtn. balds are classic examples of the specialized habitats found where granite rock is exposed over relatively large areas (Fig. 8D-4). These habitats have no soil or very shallow soil that has a very low moisture-holding capacity.

Fig. 8D-4. Only specialized plants prosper in the extreme conditions found on bare-rock balds.

Fig. 8D-5a. Pale green lichen outlines one of the shallow depressions that have developed on the granite. Water stands in these depressions for a short time after rain falls. A patch of dark green moss is present close to the center of the depression, which contains fragments of quartz and feldspar freed from the solid granite by weathering.

Fig. 8D-5b. Dwarf Cinquefoil (also known as Canada Cinquefoil) grows on a very thin mat of pine needles atop decaying granite.

Fig. 8D-5c. This small depression along the upper part of the loop trail near the crest of the dome contains an accumulation of pine and cedar needles that will slowly decay into soil, fragments of granite, and patches of grass.

Fig. 8D-5d. Pine trees grow out of the solid granite. The roots extend into cracks in the otherwise uniformly solid rock. This site is located along the upper loop trail about a mile from the trailhead.

Fig. 8D-5e. The thin soil on the bare rock supports mainly moss and grasses.

The thin soil usually contains few nutrients. Water runs off of the outcrop quickly after a rain. Consequently, plants have abundant water for short periods of time followed by long periods without moisture. Extremes of temperature and availability of water create a habitat that resembles a desert. Plants in this stressful environment respond to these conditions in several ways. Some plants grow when conditions are favorable and leave seeds or spores to revitalize during the next time conditions are favorable. Lichens are able to survive even during the most stressful times. Others, such as prickly pear cacti store water in their stems or leaves.

Fig. 8D-6. View of the surface produced by exfoliation of the granite dome. The dark stripes are stains left where water flowed down the surface.

Several plant communities are present on the balds (Figs. 8D-5a-e). These range from those that can survive in the small pools formed on the bare rock granite surfaces after rains to pine and hardwood forests that surround the exposed granite surface where the soil is thicker. The exposed bare rock is often covered by slow growing lichen, and inhabited by lichen-colored grasshoppers, spiders, and beetles. Where a layer of soil begins to form you will find herbs that grow year round accompanying the lichens. The plant communities change progressively to a stage where perennials such as spiderwort, daisy, broom sedge, and jasmine can thrive. Where the soil is thicker these are followed by shrubs and small trees, and finally by the pine-hardwood forest characterized by loblolly pine, small oak trees, sumac, and red cedar. Stone Mtn. Loop, Wolf Rock Trail, and Cedar Rock Trail are some of the best places to see the plant communities that exist on Stone Mtn.

Geologic Setting

Stone Mtn. is located close to the edge of the 1,000-ft. high escarpment that defines the eastern edge of the Blue Ridge Upland. As viewed from the Blue Ridge Parkway (Fig. 8D-1), the steeply inclined southwestern edge of Stone Mtn. stands out clearly because it is almost devoid of vegetation (Fig. 8D-6). Similar bare-rock mountainsides are seen at a number of places in the southern part of the Blue Ridge. The origin of these features is unclear. Some suggest that they form where fires burning on steep slopes destroy the vegetation. In recent years, we have also seen soil stripped from parts of mountainsides where unusually heavy rains saturate the soil that then slides off leaving exposed bedrock surfaces (p. 42). Once the bedrock is exposed in this way it is difficult for new soil to form and much easier for subsequent rains to continue the process of stripping away soil and vegetation. Because pine trees require little nourishment they may put down roots into cracks. Look for them along the cracks high on the Stone Mtn. bald.

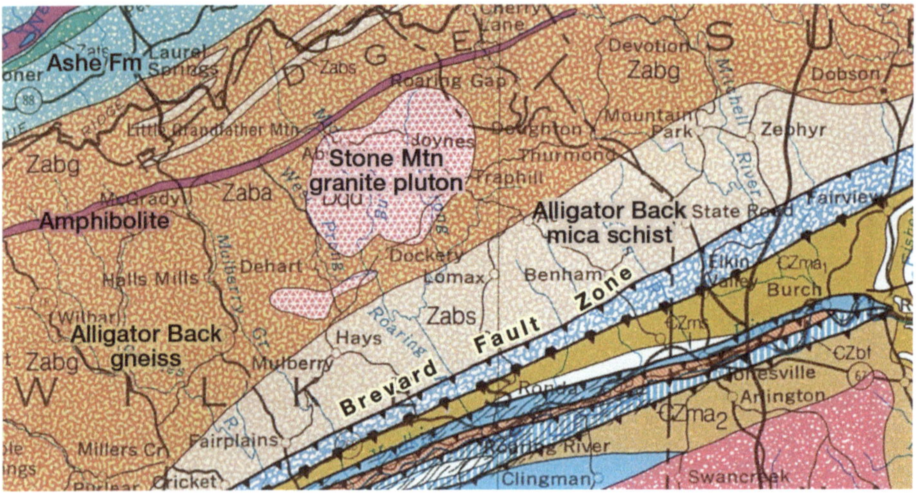

Fig. 8D-7. Geologic map of the Stone Mtn. area. (From the Geologic Map of North Carolina)

The geologic map of the area around Stone Mtn. (Fig. 8D-7) shows that the granite dome is oval in shape. The granite contains feldspar, quartz, biotite, and muscovite micas. (Technically, it varies in composition between quartz diorite and granodiorite.) Similar rocks are found at Mt. Airy, Spruce Pine, and Whiteside Mtn. These intrusions of granite took place during the Devonian-age Acadian Orogeny (about 390 million years ago). At that time molten rock (magma) rose from deep in the Earth's crust and intruded the rock that now surrounds the intrusions. Because the magma was injected at great depth, the overlying rock provided insulation, causing the molten rock to cool very slowly. When magmas cool slowly the minerals in the resulting rock are large. They increase in size because there is more time for the minerals to grow larger. The intrusions accompanied metamorphism of the rocks that surround the intrusions. As a result of the metamorphism these rocks were transformed from sedimentary rocks to a finely laminated gneiss now known as the Alligator Back Fm. (p. 196).

Uplift of all these rocks took place during Paleozoic Appalachian mountain building. After millions of years of erosion the once deeply buried rocks were exposed at the ground surface where streams and mass wasting continue to remove the material at the surface.

Once the rock that had covered the granite was removed, the pressure caused by the weight of that rock was released and fractures parallel to that surface began to form. This process, known as **exfoliation**, results in the formation of scale-like layers near the surface of the rock body (Fig. 8D-8).

Fig. 8D-8. Exfoliation on the surface of the bald.

Points of Interest

For access, see the map of Stone Mtn. State Park (Fig. 8D-2).

Decomposing granite: In some places along the loop road that passes through the park, you can see exposures of the granite in which the feldspar has decomposed to clay (p. 39) The original texture is still visible in some of the exposures even though you can scrape off the surface of the rock with your bare hand. The rock eventually breaks down to form a granular mass (Fig. 8D-9).

Fig. 8D-9. Weathered granite. This is part of a pile of fragments that weathered out of a massive exposure of granite. The white fragments are feldspar; the slightly darker, grayish minerals are quartz.

The Hutchinson Homestead was built by settlers in the mid 1800s.

Views from Hutchinson Homestead: Excellent views of the bald surface can be seen near the Hutchinson Homestead, a preserved pioneer home and outbuildings. At this location you can see the exfoliating on the bare rock surface of Stone Mtn. The parking lot at this site is the access point for those who are hiking the loop trail starting at the lower trailhead. Those who are unable to make the quarter mile walk from the parking area to the homestead may drive up a road that leaves from the eastern end of the parking area to a small (perhaps temporary) parking lot close to the homestead. Be sure to visit the homestead since this is perhaps the best place in the park to see Stone Mtn.

Balds plant communities along trails: The granitic domes have distinctive plant communities. You can reach these by taking the Stone Mtn. Loop Trail, the Wolf Rock Trail, or the Cedar Rock Trail. The primary succession of the thin vegetation mats is an ecological classic. You may take the loop trail, which is 4.5 miles long from the lower parking area (a walk that includes about 500 steps cut into the mountain), or you can enter the upper trailhead. For those who would prefer a shorter walk use the northern part of the upper trail loop. You can see many of the important features on rock exposures located about a mile along this low-to-moderately difficult path. ❖

Chapter 8E

Mt. Jefferson State Natural Area

Location and Access: The entrance to Mt. Jefferson State Natural Area is located along US 221 at West Jefferson, North Carolina. A paved road leads up the mountain to a parking area with access to trails near the top.

Accommodations: Motels and restaurants are located close to the park in the towns of Jefferson and West Jefferson.

Above: Fig. 8E-1. View of Three Top Mtn. from Mt. Jefferson.

The paved road to the top of Mt. Jefferson provides easy access to exceptional viewpoints, two nature trails, and rock exposures (Fig. 8E-1). When the skies are clear the overlooks provide exceptional views of Mt. Rogers, Whitetop Mtn., and the New River Basin to the north as well as the high peaks located to the west and southwest (Fig. 8E-2). This is one of several mountains located near the edge of the Upland Plateau and the beginning of the Blue Ridge Highlands.

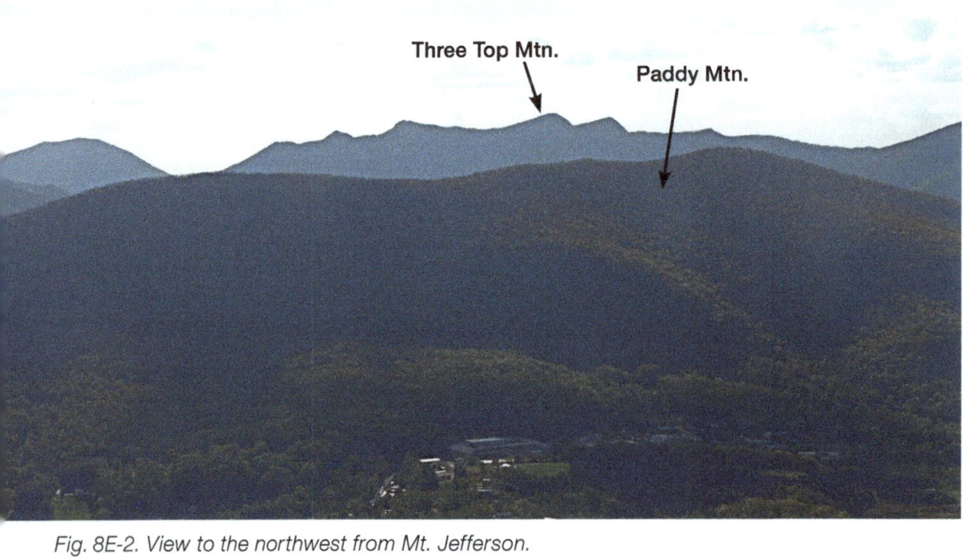

Fig. 8E-2. View to the northwest from Mt. Jefferson.

Fig. 8E-3. Amphibole-rich gneiss on Mt. Jefferson.

Geologic Setting

Rocks exposed along the road and the trails at the mountaintop are amphibolite gneisses (Fig. 8E-3) and schists, part of the Neoproterozoic/Cambrian-age Ashe Fm. The most abundant minerals in these gneisses are hornblende (black) and plagioclase feldspar. Their chemical composition suggests that they may have been formed near an oceanic ridge (Misra & Conte, 1991). Originally they were basalts or sediments consisting of volcanic ash, but the original rocks were later transformed due to metamorphism during mid-Paleozoic mountain building. At that time they were uplifted and moved to the northwest as part of a major thrust sheet that became part of the Laurentian continent, now part of the eastern Blue Ridge. The traces of the Fries and Gossan-Lead thrusts faults crop out between Mt. Jefferson and Mt. Rogers in a location northwest of the map in Fig. 8G-5.

Ecological Setting

A diverse population of trees, shrubs and wildflowers mantle the steep slopes and summit of Mt. Jefferson which stands at 4,665 ft. Trails at the top of the mountain make it possible see the plants that have prospered at this elevation. Acidic soil yields a dense heath shrub layer along the trail near the top of the mountain. Strong winds and frequent ice storms during winter months have produced dwarfed and gnarled trees on the crest and northern slopes of the mountain. At lower elevations open forests occupy areas covered by more basic soil. Near the top of the mountain the cove forest includes oak-chestnut forest with red maples, yellow birch, tulip trees, and basswood that are well developed. Slightly below the crest of the mountain, big-toothed aspens are present. These are rarely found this far south. Dogwoods, azaleas, mountain laurel, and Catawba rhododendron make up the understory, and pink lady slipper, trillium, and lily of the valley are among the most common wildflowers.

Forests on the north facing slopes include red maple, yellow birch, tulip tree, and basswood with an understory of mayapple, hobblebush and a variety of other shrubs and herbs on the forest floor. Strong northerly winds and frequent heavy ice in the winter have stunted many of the trees on the summit and north-facing slopes producing short, gnarled trees. ❖

Chapter 8F

Mt. Rogers National Recreation Area

Location and Access: The Mt. Rogers area located in southwestern Virginia encompasses a large region between the Va.-N.C. state line and I-77. The gateway to the area at the visitor center and headquarters of Mt. Rogers National Recreation Area, is located 6 miles from I-81 on US 16 south of Marion, Va. in a mountain setting close to the AT. This is an excellent place to learn more about the area, get directions, and pick up maps and nature guides. The Iron Mtns. and Grayson Highlands State Park are located within this area. It includes many hiking and horse trails widely distributed across the southern Virginia Blue Ridge. This guide concentrates on the area around Mt. Rogers and includes Grayson Highlands State Park. To reach the state park from I-81, leave I-81 at Marion, Va., and drive southeast on SR 16 to US 58 (Fig. 8F- 3). Alternative routes are on US 58 from Abingdon, or east on SR 600 from Chilhowee, Va. From the Blue Ridge Parkway leave the Parkway at milepost 229 onto US 21 going west, drive through Sparta, N.C. and continue on SR 93 to US 58. Alternatively, leave the Parkway at Deep Gap (north of Boone, N.C.) and drive on US 221 to West Jefferson and then on SR 194 to US 58.

Accommodations: The Mt. Rogers Natl. Rec. Area contains numerous campgrounds, several of which accommodate horses. Grayson Highlands State Park offers an excellent campground and picnicking facilities, but you should bring food with you. The closest motel accommodations are in the Virginia towns of: Marion, Abingdon, Jefferson, West Jefferson, and Chilhowee. Restaurants in the area are available at Troutdale and at the intersection of US 58 and SR 16 at Volney.

Trails: The AT crosses Whitetop Mtn. and passes close to the top of Mt. Rogers. Several roads provide access to many parts of the park including Whitetop Mtn., The Rhododendron Trail and a trail along the summit of Wilburn Ridge. All are recommended. Maps showing trails in the Mt. Rogers area are available at the Mt. Rogers National Recreation Area's Pat Jennings Visitor Center located off SR 16 a few miles south of Marion, Va., and also at Grayson Highlands State Park.

Above: Fig. 8F-1. Meadows and forests at Elk Garden. Outcrops of rhyolite, seen here in the foreground, are present throughout the area.

Fig. 8F-2. View to the south from SR 600 between Stops 6c and 7. Pines and deciduous trees cover the lower slopes here. Northern hardwood forests are present higher in the mid-distance, and spruce-fir forests cover the top of Mt. Rogers.

The Virginia Creeper Trail was originally built as a rail line. It is now used by hikers, cyclists, and horseback riders. It extends over 33 miles from near Whitetop Station to Abingdon, Va. The section across the Blue Ridge from Whitetop Station to Damascus, Va. is especially beautiful as it follows a stream valley through a mature northern hardwood forest.

Outstanding Features: The highest mountains in Virginia are located here. The area is famous as the type locality for the Mt. Rogers rhyolitic lavas and for evidence of a late-Precambrian glaciation. This area also provides an introduction to the ecosystems of the Highlands section of the Blue Ridge (see Chs. 4 and 7).

Nearby Attractions: Hiking and horseback riding are available in the park, and a museum is located in the Grayson Highlands State Park and at the Jennings Visitor Center off SR 16.

Ecological Setting

A great variety of ecosystems are present in the Mt. Rogers Natl. Recreation Area. (See the general discussion of the Highlands ecology on p. 198.) Fraser fir and red spruce forests, birch trees, rhododendron thickets, sphagnum bogs, blackberries (along the forest edge), moss and lichen-covered rocks and trees cover the top of the mountain. At lower elevations, northern hardwood forest habitats prevail (Figs. 8F-1, -2). Several grass and shrub covered balds are present on the two highest peaks in Virginia: Mt. Rogers (5,729 ft.) and Whitetop Mtn. (5,520 ft.) (Stop 10). See more information about the ecology of the Blue Ridge Highlands is in Chs. 4 and 7.

Hayscented fern (above) and Running cedar (also called fan clubmoss) (below). Ferns and mosses are common at mid to high elevations on the shady, and hence damp sides of the high mountains. These plants were found along SR 859 near Whitetop Mtn.

Fig. 8F-3. Regional relief map of the area around Mt. Rogers. The ridges in the upper left are held up by Silurian-age sandstones. I-81 runs through the large valley that is underlain by Cambrian and Ordovician carbonate rocks. The ridges south of the valley are held up mainly by Chilhowee sandstones and quartzites. The Mt. Rogers volcanic zone appears as a large eye-shaped area. Most of the southeastern corner of the map is underlain by Precambrian-age basement rocks. (The National Atlas, USGS, with additions)

Fig. 8F-4a. The area around Mt. Rogers was the site of explosive volcanic eruptions about 760 million years ago. (Photo: USGS)

Geologic Setting
of the Mt. Rogers Area

The Mt. Rogers area is of special geologic interest because it was the site of volcanic activity in the late Precambrian (about 760 million yrs. ago) (Figs. 8F-4a-b). Three volcanic centers have been recognized, with Mt. Rogers capping the largest central area. The two others are at Pond Mtn. to the southwest and Razor Ridge to the northeast. The volcanoes probably formed in a region of low relief. They eventually grew to be thousands of feet high and were composed of rock formed from a viscous silica-rich magma. The lowest part of the volcanic pile is composed of a rhyolite that contains an abundance of large crystals (phenocrysts) of feldspar. A porphyritic rock that contains less silica called the Buzzard Rock member lies above that (Fig. 8F-8d). The highest layer of rock is a rhyolite lava flow and a thick porphyritic welded tuff formed when the ash that erupted was so hot that the particles fused together (Fig. 8F-4b). Generally, volcanic rocks that contain a lot of silica are so viscous that they tend to form domes. In contrast, welded tuffs may spread long distances laterally. Most basaltic lavas are so fluid that they cover vast areas, like those seen in the Catoctin flows in Maryland

Fig. 8F-4b. Welded tuff is a rock made from ash that was so hot when it erupted that the particles fused together.

Fig. 8F-5. The repeating sequence of layers of similar color, composition, and grain size, often described as rhythmic layering is one of the distinguishing features of the Konnarock Fm.

and Virginia from about 200 million years after the ones at Mt. Rogers. A few basaltic lava flows are present at Mt. Rogers, but most of the lavas found there are composed of rhyolite, an igneous rock that has a high silica content like that found in most continental crust. Based on the similarity of composition, melting of granitic continental crust is the most likely source of the rhyolites found at Mt. Rogers.

Geologists think that the volcanic activity took place early in the history of the breakup of the huge continent Rodinia. The continental crust stretched, became thinner and broke as normal faults formed (Fig. 3-2). The temperature at which a rock melts depends on the pressure on it. When there is less pressure then lower temperatures are required to cause melting. Stretching and thinning of the continental crust lowered the weight on the rocks at depth and hence caused melting. The molten crustal materials found their way to the surface along the faults that accompanied the stretching. As stretching continued the continental crust subsided. Initially large lakes formed where the crust pulled apart. Eventually an ocean, the Iapetus, filled the depression formed as the continents moved apart.

The Mt. Rogers area is also of special interest because after the volcanic activity subsided and the volcanoes had been eroded to a low level, the area was covered by water. Sediments carried into this body of water settled leaving a sedimentary rock unit now known as the Konnarock Fm. (Fig. 8F-5). It buried the earlier volcanic rocks. The Konnarock Fm. contains an unusual primary structure known as varves or rhythmic layering. Varves form as a result of seasonal variations in the thickness, size, and often color of sedimentary layers. The thicker and coarser layers form during warm seasons when more sediment is reaching the sea. During winters snow cover on the land and the frozen surface of the water reduces the supply of sediment. Thin, fine-grained, and dark-colored layers form at this time of year.

The Konnarock Fm. contains several features formed in bodies of water located close to glaciers (Stops 2 and 3). When glaciers extend into a lake or sea, the edge of the glaciers break off, creating icebergs that drift offshore carrying rocks that have been frozen into the ice as it moves across the ground surface. When the ice melts, the rocks drop to the bottom of the water body, which may be already covered by soft, fine-grained sediment. These rocks are called **dropstones** (Fig. 8F-6).

Evidence of ice sheets (Fig. 8F-7) during the late Precambrian has been found in a number of localities around the world prompting some geologists to suggest

Fig. 8F-6. A dropstone derived from the Blue Ridge basement complex now resides in the Konnarock Fm.

that the Earth went through an extraordinarily cold period at that time. Because the glacial deposits formed at the same time at many different latitudes, it seems possible that most or perhaps all of the Earth was covered by ice and snow. This idea is known as the "Snowball Earth" hypothesis.

Fig. 8F-7. Modern glaciers in the Arctic Ocean (above and left). As glaciers move into the ocean, chunks of ice fall off into the water. These icebergs then float out into the ocean carrying rocks that were frozen into the bottom of the glacier as it moved over land. As the icebergs melt, these rocks sink to the bottom and become embedded in the sediment as dropstones.

Fig. 8F-8a. Rhyolitic volcanic breccia showing flow structures. The dark blocks are fragments of older rocks that were incorporated into the lava.

Fig. 8F-8b. The dark-colored rock is a rhyolite porphyry. The pink crystals of feldspar formed in the magma when it was deep in the earth. The darker, fine-grained rhyolite cooled and crystallized at shallow depth where cooling was rapid.

Rock Units in the Mt. Rogers Area

(In sequence with the youngest at the top.) Note: The Chilhowee Group and Precambrian basement rocks continue to the northeast and southwest of Mt. Rogers, but the Mt. Rogers volcanic rocks and the Konnarock Fm. are only present in the Mt. Rogers area.

Rock Unit	Subdivisions in the Rock Unit
Cambrian-age carbonate units of the Valley and Ridge Province	-
Chilhowee Group	Erwin Fm. (Antietam Fm.) Hampton Fm. (Harpers Fm.) Unicoi Fm. (Weverton Fm.)
Konnarock Fm.	-
Mt. Rogers Fm. (Fig. 8F-8a-d)	Wilburn Member (rhyolitic welded tuff) Whitetop Member (phenocrysts-poor rhyolite) Buzzard Rock Member (porphyritic rhyolite low in silica) Fees rhyolite Member (porphyritic, silica rich rhyolite)
~~ An unconformity – an ancient land surface ~~~~~~~~~~~~~~~~~~	
Precambrian basement complex (more than a billion years old)	

Table 8F-1. Various rock units found in the Mt. Rogers area.

The Tectonic Setting of Mt. Rogers

The region around Mt. Rogers is located in the western part of the Blue Ridge. Here, as elsewhere along the western flank, folded and thrust-faulted rocks of the Chilhowee Group (Fig. 8F-10) lie between the carbonate rocks of the Valley and Ridge and the Precambrian units found in the core of the Blue Ridge. This belt of Chilhowee Group units is unusually wide here because some of the units in the Chilhowee and other, lower, Cambrian-age carbonates and shales are duplicated as a result of repeated thrust faults exposed near Marion, Va. (Fig. 8F-9). The complexity of this deforma-

Fig. 8F-8c. White porphyritic rhyolite with phenocrysts of feldspar.

Fig. 8F-8d. Large fragments, 5-6 in. across are present in the Buzzards Rock rhyolite exposed near the visitor center.

tion can also be seen around Mt. Rogers (Figs. 8F-10, -11) where thrusts and folds cause the Chilhowee to be repeated in a prominent ridge, Holston Mtn. and in the mountains farther to the northeast (see Stop 6). A belt of Cambrian-age carbonate rocks comes to surface in a structural window (p. 53) known as the Mtn. City window, located southwest of Mt. Rogers.

The distinctive volcanic and glacially related rocks of Mt. Rogers are also involved in thrust faults. The thick pile of rhyolitic lavas that compose most of the Mt. Rogers area have been displaced to the northwest. The lowest part of the Mt. Rogers group lies on an ancient land surface composed of crystalline rocks of the Precambrian basement that had formed during the Grenville Orogeny. Later the lavas as well as the basement were folded and faulted as they were moved to the northwest during Paleozoic mountain building as seen at Stop 17.

Major thrust faults, the Fries and Gossan-Lead faults, separating the eastern and western parts of the Southern Blue Ridge are located a few miles southeast of Mt. Rogers (Fig. 8F-3).

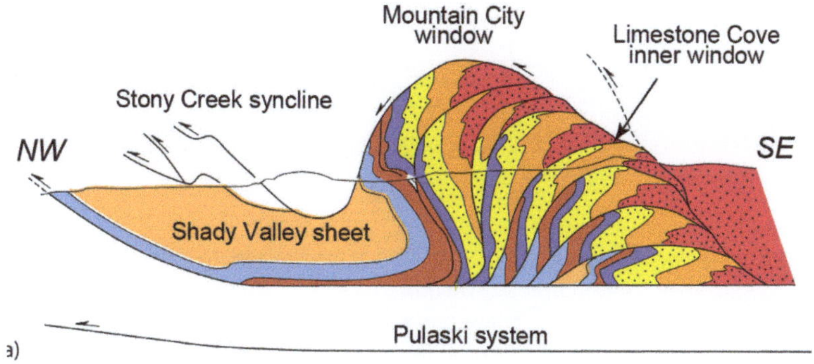

Fig. 8F-9. This schematic cross-section shows how a number of thrust faults have caused duplication of parts of the Chilhowee Group and lower Cambrian carbonate and shale units that are exposed along SR 16 between Marion and Troutdale, Va. (After Diegel, 1986)

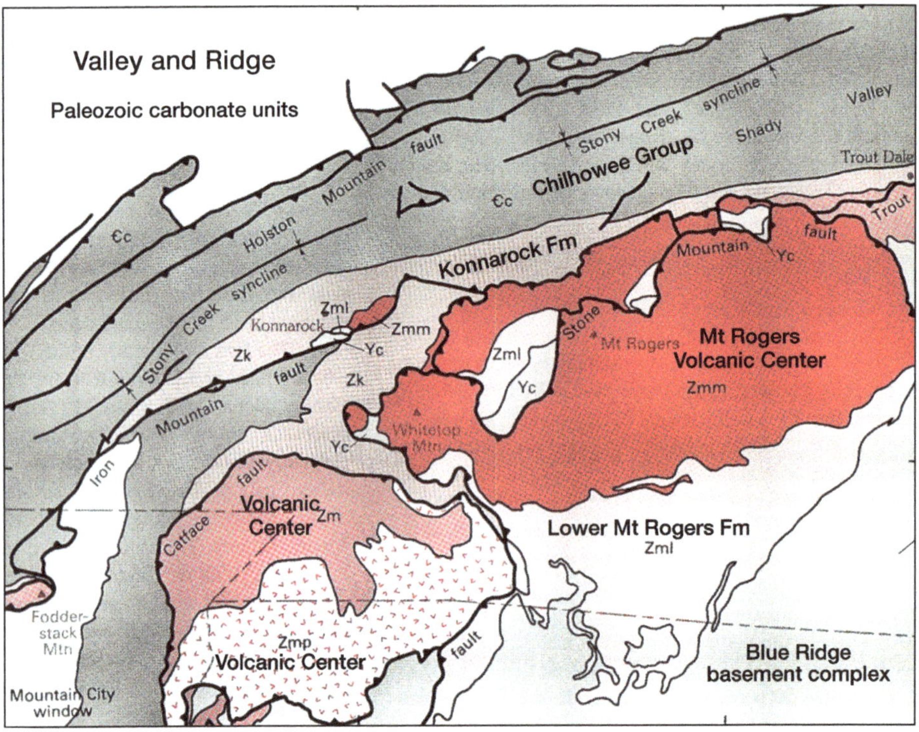

Fig. 8F-10. Geologic map of the Mt. Rogers area. Revisions to this map are being made that will appear on a new map that is being prepared by Merschat and Southworth at the USGS. (From Rankin, 1993)

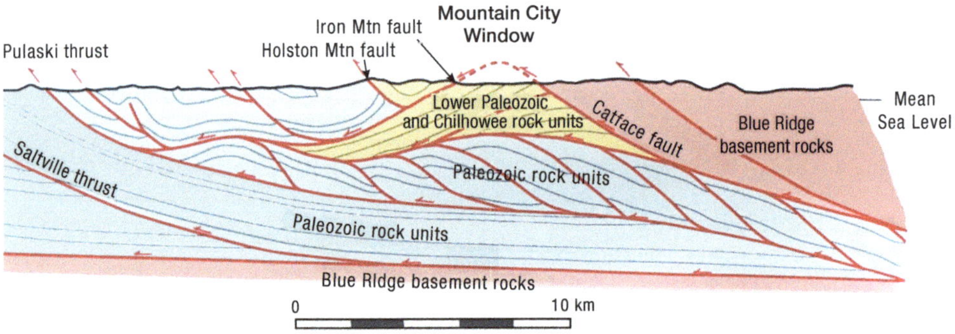

Fig. 8F-11. This cross-section depicts an interpretation of the faults that lie along a line running NW to SE across the Mountain City window. Note that the area west of the Mt. Rogers volcanic rocks, the Mountain City window, contains much younger (Cambrian-age carbonate rocks) that were deposited on top of the Chilhowee Group rocks and far above the rocks of the Mt. Rogers formation. (After Woodward,1985)

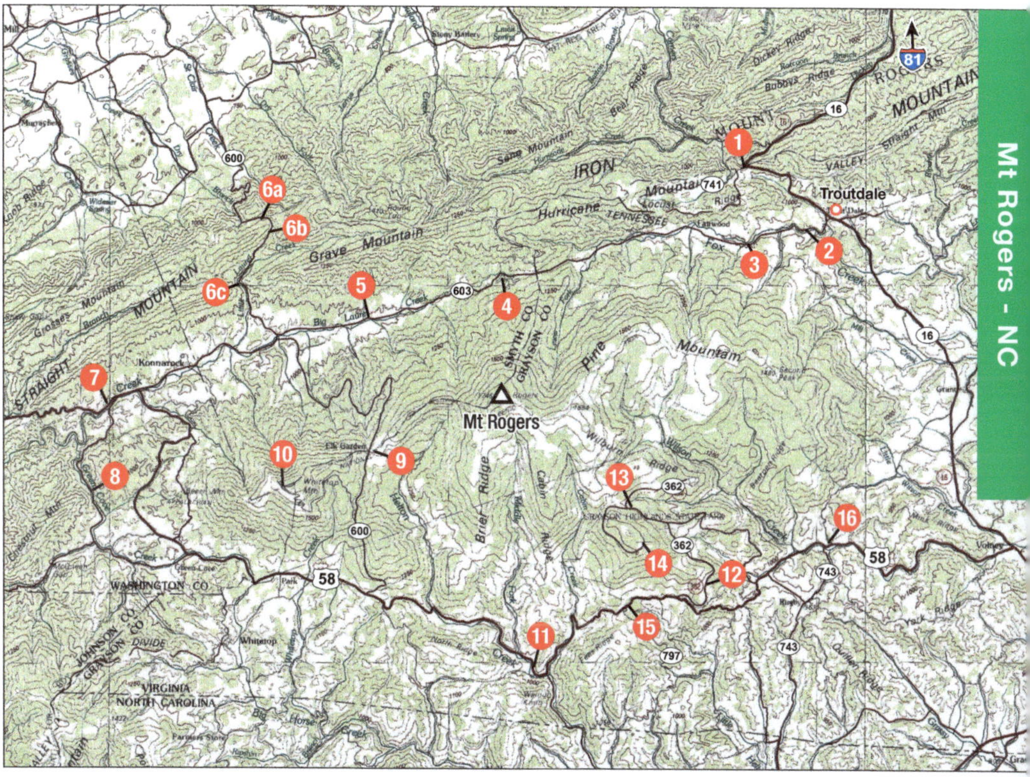

Fig. 8F-12. Points of Interest in the Mt. Rogers area. (USGS map, with additions)

Points of Interest

in and near Mt. Rogers National Recreation Area

This traverse is laid out assuming that the traveler leaves I-81 at Marion and drives south on SR 16 toward Troutdale (Fig. 8F-12). The route begins at the intersection of I-81 and SR 16. One of the largest thrust faults in the Valley and Ridge is exposed along the northbound entry ramp to I-81 but pedestrian traffic is not allowed here. Red shales of the Cambrian-age Rome Fm. are exposed a short distance southeast along SR 16. As you continue southeast toward Sugar Grove you will drive across a series of thrust fault slices, each of which is composed of Chilhowee Group rocks that form ridges and valleys underlain by a sheet of Cambrian-age Shady Fm. dolomite (Fig. 8F-9). You will pass the park visitor center at 6 miles from I-81 where maps and information about the park is available. Exposures of the Chilhowee Group occur at the base of a ridge a short distance south of the visitor center. Continue to Sugar Grove. A wide outcrop belt of Shady Fm. dolomite accounts for the broad open valley at Sugar Grove. This dolomite is exposed in roadcuts on the eastern edge of town. Continue for 5 miles south on SR 16 until you reach SR 741 (Homestead Road). (It is easy to miss). Turn right onto SR 741 (Fig. 8F-12).

Stop 1. **Homestead Road (SR 741):** This road connects SR 16 with SR 603, and is located about two miles north of Troutdale.

SR 741 traverses many habitats including open meadows, shrubby grasslands, oak and eastern hardwood forests, and hedgerows. In summer these are prime places to look for sparrows, eastern towhee, house wren, blue-headed vireo, and chestnut-sided warbler. The meadows contain barn and northern rough-winged swallows, wild turkey, and ruffed grouse. Migrant songbirds are numerous in the spring and fall. You might also see white-tailed deer, black bear, red fox, and coyote. A 0.5 mile-long trail off the road leads to Comers Creek Waterfall. At the intersection of SR 741 and 603, turn left on SR 603 and drive toward Troutdale.

Stop 2. **Evidence of Glaciation – Dropstones:** Good exposures of the late-Precambrian Konnarock Fm. are exposed near the intersection of SR 603 (Fairwood Road) and Woodland Lane, 0.5 miles from Troutdale. The Konnarock Fm. is a maroon-colored shale with small quantities of arkose (a sandy sediment containing feldspar). Parts of this rock unit exhibit rhythmic alternation of grain sizes. This suggests that the influx of sediment into the body of water varied systematically (Fig. 8F-5). Several ideas have been advanced to account for this rhythmic layering but the most popular one is that the layers reflect seasonal changes in grain size. Coarser grain sizes are deposited in summer; finer grain sizes accumulate during winter, producing layered sediment. Pieces of granitic rock derived from older Precambrian rocks occur randomly in the fine-grained shale of the Konnarock Fm. (Fig. 8F-5). Most of these pieces do not touch one another and often depress the rhythmically bedded shale and mudstone beneath them. They are interpreted as dropstones, rocks that were frozen into icebergs that drifted out into the body of water and dropped as the ice melted. Dropstones are also present in the outcrops at Stop 6. Another idea that has been suggested is that the sediment was deposited from submarine landslides in which the sediment-laden water moved downslope into deep water where it settled out. Commonly sediment deposited from turbidity currents forms beds that are coarse-grained at the bottom and become finer-grained farther up. This type is sediment layering is called graded bedding.

Fig. 8F-13. Outcrop of probable tillite (left) and the stream located directly across the road from the outcrop of glacial till (right).

Garlic mustard is a common ground cover in moist areas.

Stop 3. **Glacial Tillite:** There is a roadcut located along SR 603 opposite a small waterfall known as Fox Falls, 1.8 miles from Troutdale. Geologists refer to the deposits of rock mixtures that were pushed or carried along in the ice of glaciers and deposited when the ice melted as a tillite. The rock in this exposure lacks the layering that is so prominent in most of the Konnarock Fm. Instead, it is a massive unit composed of rock fragments of many sizes. Some pieces are rounded, suggesting that they were transported in a stream, but many are irregular in shape (Fig. 8F-13). Exactly how this deposit was formed remains unclear. It may have been deposited in contact with the ice, as in a moraine, and hence be a tillite, though some suggest that it may have been deposited by water flowing beneath a glacier.

Stop 4. **Birds at Grindstone Campground:** Grindstone Campground, located along SR 603, is a large and well-equipped campground. Trails from this site connect with the Mt. Rogers and Whispering Waters trails. A short 0.6-mile long trail passes through moist, mixed deciduous woodlands. Here you may see warblers, least flycatchers, wood thrush, and Louisiana water thrush. Many songbirds visit here during migrations. Yellow-bellied sapsuckers nest within the campground. This is an excellent place to look for salamanders including black-bellied, mountain dusky, northern spring, and red-backed salamanders.

Stop 5. **Excellent exposures of the Konnarock Fm:** are along SR 603 (Fig. 8F-5). One is 7.8 miles west of Troutdale on the north side of SR 603. Others are located in the community of Konnarock and farther along, toward the intersection of SR 603 and US 58.

Traveling north along SR 600, north of Whitetop Mtn., you will traverse the rocks that were deposited over half a million yrs. ago on top of the Mt. Rogers volcanic centers, as the Iapetus Ocean advanced across the continent. The first of these is a short distance off SR 603. It has the characteristic maroon color of the Konnarock Fm. seen and described at the last few outcrops on this traverse. As you continue north you will pass through the Chilhowee Group which bears the folding and faulting effects of Paleozoic-age mountain building. Continue across the Chilhowee for about 3 miles until you reach a parking area at Skull's Gap. Look for a runaway truck ramp on the east side of SR 600. The overlook is on the west side of the road from the ramp.

Stop 6a. **Skull's Gap Overlook:** On the trip to Skull's Gap, you will pass through exposures of the Konnarock Fm. followed by Chilhowee Group rocks that are folded into a large synclinal fold. In sequence, you will drive across the Unicoi, Hampton and Erwin Fms. on the southern limb of the fold. Then you will cross the Harpers and Unicoi a second time. Continuing farther north you will

Fig. 8F-14. View west from Skull's Gap Overlook (Stop 6a in Fig. 8F-12). The ridge you are standing on is composed of rocks of the Unicoi Fm. The low rolling country in the foreground is a valley underlain mainly by carbonate rocks of Cambrian and Ordovician-age. Silurian-age sandstones hold up Walkers Mtn. in the distance.

cross a thrust fault (Fig. 8F-10) before entering the limestones and shales of a valley in the Valley and Ridge Province.

Skull's Gap Overlook (Fig. 8F-14) provides an excellent view of the Valley and Ridge Province. The presence of Cambro-Ordovician carbonate rocks, which are soluble and form valleys in humid climates, accounts for the low topography of the valley. Ridges of Silurian-age sandstones hold up Walker Mtn. and Little Mtn., which are visible in the distance.

Skull's Gap Overlook is an excellent place to look for hawks, deer, and the occasional bobcat in the distance. Butterflies are often numerous here during summer months. Note the entrance to a birding and wildlife trail in the parking lot. This trail connects into the Iron Mtn. Trail that extends from Damascus, Va. to SR 650 (Fig. 8F-3).

Beautiful exposures of the Unicoi Fm. are present south of Skull's Gap on both sides of SR 600. The Unicoi consists of quartz sand, granules, and pebbles deposited over 500 million yrs. ago in streams that flowed across what was then a subdued landscape that was eventually submerged under the marine waters of the Iapetus Ocean in which shales would be deposited.

Stop 6b. **A thrust fault** is exposed a short distance (0.1 miles) south of Skull's Gap (Fig. 8F-15). On the west side of the road, light gray exposures of the Unicoi Fm. have been thrust onto a basalt lava

Fig. 8F-15. This thrust fault is located near Skull's Gap on SR 600.

Fig. 8F-16. Exposure of Unicoi marine shales near Skull's Gap on SR 600. (Stop 6b in Fig. 8F-12)

Fig. 8F-17. This large block is part of a landslide that probably took place when these rocks were being deposited underwater. (Stop 6c, Fig. 8F-12)

flow that had formed during the time the Unicoi was deposited. It is likely that this basalt may be similar in age (about 570 Ma) to the Catoctin lavas that are found in Maryland and northern Virginia (Ch. 6A).

Stop 6c. **Submarine landslide:** Continue 2.2 miles to the south on SR 600. A long roadcut passes through shales of the Unicoi that were deposited in a marine environment (Fig. 8F-16). A board fence has been built to keep falling rocks off of the highway. Parking places are available a short distance south of the fence. A remarkably well-exposed submarine landslide is present at the eastern end of this shale outcrop. Large slabs of sandstone, including a large block of rock that rotated as it slipped downslope (Fig. 8F-17), are beautifully exposed here. About 1.2 miles further south on SR 600 is an exposure of nearly vertical Unicoi Fm. sandstones (Fig. 8F-18).

Fig. 8F-18. Vertical beds of the Unicoi Fm. sandstones located along SR 600 about 1.2 miles south of Stop 6c. These layers were nearly horizontal when the sediment was deposited nearly half a billion years ago. Much later, the layers were rotated into a nearly vertical position as the Blue Ridge was pushed—folding and shifting the rocks to the northwest during the Alleghanian Orogeny between 200-300 million years ago.

Fig. 8F-19. Boulder in the Konnarock Fm. at Stop 7.

Stop 7. **Glacial deposits:** From the intersection of SR 600 and SR 603 turn right and continue west on SR 603 that ends where it joins US 58. Large, rounded boulders of granite are present in the Konnarock Fm. about 0.1 miles past the intersection of US 58 and SR 859. An exceptionally large rounded boulder composed of Blue Ridge basement rock (Fig. 8F-19) is present in the Konnarock Fm. at this location. Its origin is uncertain. It could be a large dropstone or something caught up in a moraine. Note that many of the smaller pieces are not in contact with one another. They are surrounded by fine-grained sedimentary rock and do not touch one another. This suggests that they were dropped into fine-grained sediment.

Stop 8. **The Virginia Creeper Trail:** This superb hiking and biking trail was once a rail line. The old rail line was started at Abingdon, Va. in the late 1800s in anticipation that coal, iron, and other resources would be found in the mountains. Construction costs and the failure to make these discoveries led to bankruptcy for the original company.

The next owners extended the line and used it for the movement of timber from the Whitetop Mtn. region. By 1918, the line extended beyond Whitetop to Todd, N.C. As the supply of timber dropped, the depression took its toll on the rail line, but it remained in business until 1977 when the Norfolk and Western Railroad Co. abandoned the line. Flooding in 1977 led to removal of the rails leaving more than 100 tressels standing. The Forest Service obtained much of the right-of-way in Virginia and turned it into a trail. It is one of the most beautiful trails in the country and is heavily used during much of the year. See **www.vacreepertrail.com** for more details.

One of the best entries to the trail is located off of US 58 or at Whitetop Station a short distance west of US 58 on SR 726. If you would like to see one of the trestles from below, drive about half a mile toward Damascus from Stop 7 and turn left on SR 728. This is a single-track road with pullouts. Continue to the end of the road at Creek Junction where you will see

Fig. 8F-20. The Virginia Creeper Trail passes over an old train tressel.

Fig. 8F-21. Elk Garden meadow and forest located where the Appalachian Trail crosses SR 600. A few spruce and oak trees grow on this grassy bald that is grazed by cattle and horses.

one of the trestles on the trail (Fig. 8F-20) near the beautiful Laurel Creek. You can drive over the Virginia Creeper Trail if you take SR 859 (located 0.9 miles west of the intersection of US 58 and 603). (Stops 4, 5, and 6, described in Merschat, et al., 2014 are located along SR 859.)

Stop 9. **Elk Garden:** is located on SR 600 at a gap along the drainage divide between the Tennessee River and the New River, both of which drain into the Gulf of Mexico (Fig. 8F-21). The AT crosses SR 600 at this gap and continues to Mt. Rogers to the east and Whitetop Mtn. to the west. The more localized Elk Garden Trail also passes across SR 600 at this site. Lovely open meadows and balds near the road afford good viewpoints for those interested in watching hawks, ravens, and turkey vultures from hills composed of Whitetop rhyolite lava flows.

To hike to Mt. Rogers from Elk Garden, take the AT from there or the longer trail that begins at the visitor center in Grayson Highlands State Park. The trail leads into a damp spruce forest where mosses and lichens are abundant, forming a mat along the trail. Birds found in or near the forest include the dark-eyed junco, winter wren, veery, hermit thrush, black-capped chickadee, golden-crowned kinglet, red-breasted nuthatch, red crossbill, magnolia warbler, and hairy woodpecker. Many of these birds are generally seen much farther north. The thick, cool, and damp forest is an ideal environment for mosses, mushrooms, and insects that inhabit fallen branches. A great variety of salamanders also make their homes here. The thick forest prevents mountain views. The best viewpoints lie along Wilburn Ridge and close to Elk Garden. The AT crosses SR 600 at Elk Garden. The trail continues very close to the top of Mt. Rogers. Elk Garden is a grassy bald named for elk that once roamed the region. In earlier years elk and other animals may have been responsible for keeping trees and other high vegetation from growing here. Once early settlers arrived, they used the open grassy slopes as pasture. Today herds of cattle and ponies keep the area open.

Close to the junction of the AT and the Mt. Rogers trails, the open grassy bald gives way to a dense growth of evergreen spruce-fir forest typical of the mountain tops throughout the Highlands. The remains of similar forests have been found on recent sedimentary deposits in the Valley and Ridge Province where these deposits are located several thousand feet lower in elevation than modern spruce-fir forests, suggesting that the evergreen forests covered most of this region during the Pleistocene Ice Ages. A side trail that branches off of the AT leads to the top of Mt. Rogers. The AT follows along Wilburn Ridge to the north.

Muskeg (moss-covered bog) covers large parts of the grassy bald near the crest of Whitetop Mtn.

Fig. 8F-22. Grassy bald near the crest of Whitetop Mtn.

Hobblebush (also Moosewood)

Stop 10. **Whitetop Mountain:** For those who do not want to hike the AT from Elk Garden to the top of Whitetop Mtn. it is possible to drive there on a private (but open) road that leaves SR 600 about one mile south of Elk Garden. This is a gravel road marred by numerous potholes, ruts, and rocks. Drive carefully and do not take a vehicle that has low road clearance. This is one of the few places in the Appalachians where you can drive into an ecosystem located above 5,000 ft. (5,344 ft.). On the way up notice that changes occur in the plants found along the road. Most of it is a northern hardwood forest with oaks and beech trees. Mosses, lichens, and a variety of low-growing plants such as garlic mustard cover the ground. As you approach the top most of the hardwoods disappear and are replaced by a large grassy bald (Fig. 8F-22), fringed by spruce trees and scattered hawthorns. Vegetation on the bald includes mountain oatgrass, white-edge sedge, forbs, and Blue Ridge St. Johns wort (Fleming and Coulling, 2001). The thin soil on the bald shows pronounced effects of frost heaving including the elevation of rocks through the moss and grasses. A red-spruce forest covers the top of the mountain. This area provides easy access

Hawthorn trees in the muskeg on Whitetop Mtn.

to highly diverse environments in close proximity to one another. You will likely see, or at least hear, many of the birds on Mt. Rogers.

Stop 11. **Greenstone:** One exposure of greenstone is located along US 58 in a roadcut opposite Mt. Rogers School. Most of the volcanic rocks in the Mt. Rogers area are rhyolite formed by melting of continental crust, but some basaltic intrusions, likely derived from Earth's mantle, rose from great depth as continental rifting took place. This one, which originated as a basalt, was formed when Rodinia began to break up in the late Precambrian. The greenish color came later as the basalt was metamorphosed, and new green minerals, notably chlorite formed. Continue about 4.1 miles east along US 58 and turn left through the entrance to Grayson Highlands State Park.

Stop 12. **Grayson Highlands State Park north of US 58:** This excellent 4,800-acre state park provides a fine campground, museum, horse stables, and numerous trails including one up to Mt. Rogers. The park provides many opportunities to view wood thrush, ovenbirds, and black and white warbler, black-throated blue warbler, black-throated green warbler, Canada and chestnut-sided warblers, scarlet tanager, and rose-breasted grosbeak. Many more birds pass through during migrations in fall and spring. Wildlife living in the area includes black bear, bobcat, red fox, ruffed grouse, deer, and wild turkey. A rare salamander, Weller's salamander, is also found in this area.

Sugarlands viewpoint is on the east side of SR 362 about 0.5 miles from the entrance station. The view to the east is across the Blue Ridge Upland Plateau discussed earlier (Ch. 8B). Blocks of the Mt. Rogers rhyolite showing flow structures (Fig. 8F-8a) were used to build the retaining wall.

Stop 13. **The Rhododendron and Wilburn Ridge Trails:** Both trails go to the top of Mt. Rogers in Grayson Highlands State Park. They begin at a parking lot in Massie Gap and connect with the AT that continues 4.2 miles to the top of Mt. Rogers. Many habitats including open meadows, northern hardwoods, rocky outcrops, rhododendron thickets, sphagnum bogs, grazed pastures, Fraser fir groves, and red spruce forests are present in the park.

Remains of a volcanic ash flow eruption at the Mt. Rogers volcanic center are exposed on the lower part of Wilburn Ridge. From the Massie Gap parking lot, the Wilburn Ridge Trail goes uphill across a meadow and swampy area into open woods. The first outcrops are Whitetop rhyolite. The trail proceeds across zones of volcanic ash (tuff) that was so hot the particles welded together as the ash flowed downslope. Columnar joints, basaltic dikes, and features interpreted as fumaroles are present on this part of Wilburn Ridge. See the description of Stop 9 for the higher portions of the AT.

Stop 14. **Volcanic Conglomerate in the Mt. Rogers Fm:** The park visitor center is located at the end of the road. It offers a museum, information desk, and other facilities. One of the most interesting outcrops in the park is near the southern end of the parking lot below the Buzzard Rock Visitor Center in the picnic area called Wildcat Rocks. These outcrops contain rounded cobbles and boulders of greenstone, quartz, and granitic basement rock that are present in the Mt. Rogers rhyolite. This part of the rhyolite lava contains crystals of feldspar (Fig. 8F-8c). These crystals (called phenocrysts) formed while the molten rock (magma) was slowly cooling at depth in the Earth. When the eruption took place the mush of liquid and crystals came to the surface and solidified so rapidly that no additional crystals had time to form. The resulting rock is called rhyolite porphyry.

Stop 15. **Strongly Deformed Rhyolite:** From the entrance to Grayson Highlands State Park turn west (right) onto US 58 and drive about 0.6 miles. A large loose rock that moved downslope stopped by the road. It contains some of the rocks seen at Stop 14 except that these have been strongly deformed. The fragments in the rhyolite have been flattened and sheared out. They reflect the flattening and shearing that resulted from the plate collisions that took place

Fig. 8F-23. Highly deformed lower Mt. Rogers Fm. rocks that were sheared during Paleozoic mountain building, when plate collisions pushed the rocks of the Mt. Rogers area northwestward.

Fig. 8F-24. Drawn out fragments of rhyolite porphyry in a fine-grained rhyolite matrix, indicating great deformation at a very high temperature.

Fig. 8F-25. Grenville-age Precambrian basement rock near Rugby Rescue Station on US 58.

Fig. 8F-26. This excellent exposure of the Precambrian-age Blue Ridge basement complex at Stop 17 exhibits several low-angle thrust faults formed when the region to the southeast was pushed to the northwest during mid-to-late Paleozoic mountain building.

during the Paleozoic Era. Unmoved outcrops of these highly deformed rocks (Figs. 8F-23, -24) occur on slopes a short distance south of US 58 along Spencer Branch Road (SR 797). They are located on private property. Permission is needed to access them.

Stop 16. **Blue Ridge Basement Outcrop:** Return to US 58 and continue east to the intersection of US 58 and SR 743 opposite the Rugby Rescue Station. This Stop provides an excellent opportunity to see the basement rocks (often mapped as Cranberry gneiss) on which the Mt. Rogers volcanic center was built, and one of the most widespread Precambrian gneisses found in the highland region (Fig. 8F-25). The gneiss has been dated at about 1.2 Ga (billion yrs. old, the age of the Grenville mountain building). Greenstone lies above the gneiss at this locality. Note that these gneisses do not show signs of the Paleozoic deformation seen at Stop 15.

Stop 17. **Fault Zone in the Blue Ridge Basement:** From Volney, continue south on US 58/SR 16. Turn off of US 58 onto SR 93, cross the river and park. This Stop is located on the south side of the New River, east of Mouth of Wilson. At this long outcrop you will see excellent examples of gneiss, augen gneiss, and several zones of highly sheared gneiss (Fig. 8F-26). All are part of the Precambrian-age Blue Ridge basement complex found on the west side of two major faults, the Fries and Gossan-Lead faults, separating the eastern and western sides of the Southern Blue Ridge (Ch. 7). The basement complex, the Mt. Rogers volcanic centers, and rocks of the Chilhowee Group lie west of these faults. The Alligator Back, Ashe, and Lynchburg Group units lie to the south of these major faults. One of the best places to see the Alligator Back is at Doughton Park, which you reach by continuing south on SR 113 and then taking SR 18 to the Parkway. Doughton Park is located a few miles north of the intersection of SR 18 and the Parkway (p. 200). ❖

Chapter 8G

Grandfather Mountain Area and Linville Falls

Above: Fig. 8G-1. Grandfather Mtn.

Location and Access: The Grandfather Mountain area encompasses a vast region surrounding Grandfather Mtn. State Park in North Carolina. This area lies within the Blue Ridge Highlands southwest of Mt. Rogers and northeast of the Great Smoky Mtns., near Parkway milepost 305.

Accommodations: A restaurant is located in the nature museum at the Grandfather Mtn. Stewardship Foundation. Two picnic areas are located inside the attraction park. Camping is permitted along some of the trails in the state park. Motels and more restaurants can be found in nearby towns such as Blowing Rock, Linville, Banner Elk, N.C. and others along US 221. Campground facilities are located at Julian Price Memorial Park and near Linville Falls Visitor Center.

Fig. 8G-2. North American black bear in the zoo at the Grandfather Mtn. The cinnamon brown color is more common in the western U.S. Only 1% of black bears have this color in the east.

Grandfather Mountain

Once a 2,700-acre for-profit nature park, most of the backcountry east of its central attraction became a North Carolina state park in 2009. The attraction is now managed as an eco-tourist destination by the non-profit Grandfather Mtn. Stewardship Foundation. The entrance to the attraction is located on US 221, a short distance west of its intersection with the Parkway near milepost 305. A natural history museum, restaurant, swinging bridge, zoo, theatre, numerous trails, and guided tours are available at the attraction. The zoo includes many of the animals native to this area including black bears, panthers, golden and bald eagles, deer, and otters (Fig. 8G-2).

You may see some of these animals when you take the trails that guide you away from the concentrations of

Fig. 8G-3. Mile High Swinging Bridge at Grandfather Mtn.

Fig. 8G-4. Allegheny sand myrtle growing on the rocks near the swinging bridge.

people. Be sure to pick up a map showing roads and trails at the entrance. Rocky summit plants are easily found on Grandfather Mtn. Some may be seen on the rocks at the swinging bridge north of the visitor center (Figs. 8G-3, -4).

Nature Trails at Grandfather Mtn.

Walking the trails in Grandfather Mtn. State Park is the best way to see the diverse environments and natural communities that are found here. Some of these are easy walks; others are very strenuous and involve the use of ladders and cables. Be sure to obtain a trail map and seek advice from rangers if you plan to explore far from the road. To access the trails from the swinging bridge on Grandfather Mtn., hikers must purchase a ticket to the attraction. However, access is free from two trailheads at the mountain's base: one on the Parkway at milepost 300; and another from SR 105, half a mile north of the intersection with SR 184.

Grandfather Trail is the highest and most difficult of the trails. It is 2.4 miles long and has its head at the Swinging Bridge. The trail runs along the crest of the mountain through a spruce-fir forest that occupies the areas between rocky pinnacles and cliff faces held up by a sandstone rock unit known as the Grandfather Mtn. Fm. Other trails are located on the east and west sides of the ridge on which the Grandfather Trail is located. The Asutsi Trail is the easiest. Only 0.4 miles long, it starts at the Serenity Farm on US 221. The Nuwati Trail, also easy, follows the Boone Fork River and passes along an old logging road through a grove of aspens. The Cragway and Daniel Boone Scout trails have steep sections and are strenuous walks. The Cragway Trail passes through thickets of rhododendron and blueberry.

The Profile Trail, located on the west side of the central ridge, starts on SR 105 in Banner Elk and climbs almost 2,000 ft. to intercept the Grandfather Trail. You will walk through a number of ecological communities with numerous wildflowers during the spring and summer months. The lower part of the trail goes through hardwood forests with rhododendrons along streams. Farther along you pass through several coves. One of these is an acidic cove forest, where the ground is covered by a dense shrub growth with plenty of rhododendrons. Other coves here, known as "rich coves" are more open, with the ground covered by ferns and other herbs that flourish in a nutrient-rich soil. As you climb higher you enter a deciduous hardwood forest with many maples, then a hemlock forest, and eventually a boulderfield forest with yellow birch trees growing up through rocks covered with mosses and ferns. Still higher you enter the spruce-fir forest near the top of the mountain.

Geologic Setting

Geologists use the term "window" to describe places where erosion has cut through overthrust sheets and exposed the rocks below the thrust sheets (Fig. 3-25). In a completely developed window, the rocks below the thrust sheet are surrounded by exposures of the thrust fault as well as the rocks carried on the thrust sheet. The Grandfather Mtn. window encompasses a large area around Grandfather Mtn. Here, the rocks inside the window are more resistant to erosion than most of those that surround them. Consequently, after millions of years of erosion, rock masses exposed inside the window are at higher elevations than those surrounding them. The towns of Blowing Rock and Linville lie within the window, and Boone is located at its northern edge (Fig. 8G-5). The size of this thrust sheet, or in this case thrust *sheets*, and the distance the thrust sheets have moved is difficult to imagine. Thrust sheets are present from Alabama to Maryland; some of them are thousands of feet thick and involved lateral movement

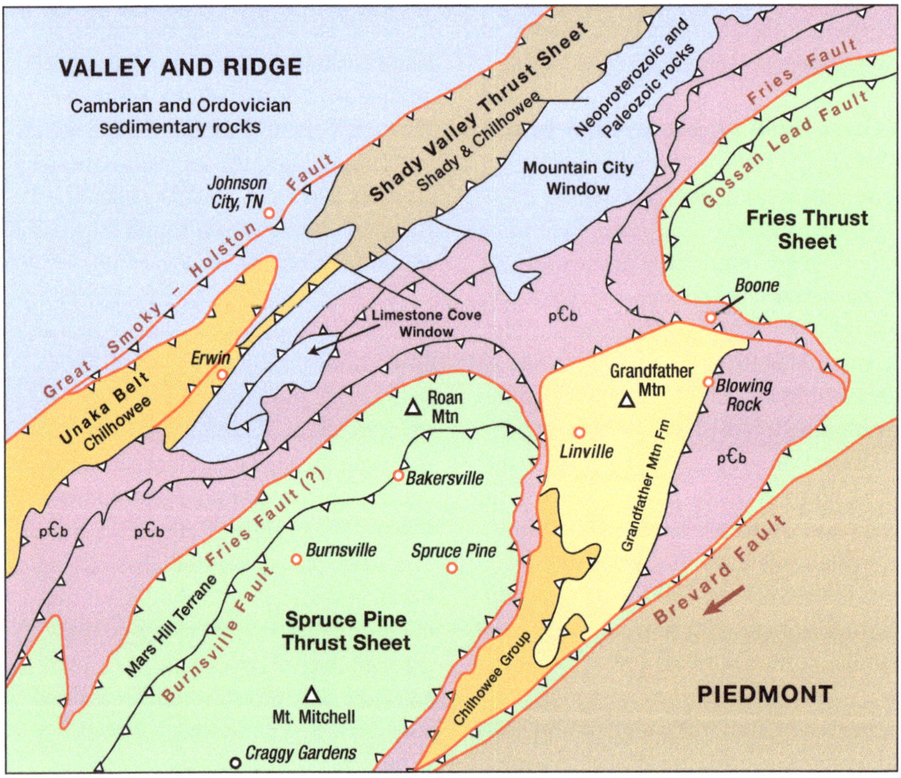

Fig. 8G-5. Geologic sketch map of Grandfather Mtn. and surrounding areas. (Compiled from Hatcher, Merschat and Raymond, 2006; Raymond, Neton, and Cook, 1992; Hatcher and Thomas 1989; and Trupe, et al., 2003)

from southeast to northwest upwards of at least tens of miles. Some think the total displacement may exceed a hundred miles. The emplacement of these sheets is described in Ch. 2.

The Brevard fault zone forms the southeastern boundary of the Grandfather Mtn. window. Both thrusting and strike-slip movements have taken place on this fault. The thrusting took place during mid-Paleozoic mountain building when volcanic island arcs were accreted along the margin of the ancient core of North America (Laurentia). Strike-slip movement occurred during the collision between Gondwana and Laurentia that took place during the Alleghanian Orogeny, late in the Paleozoic Era. During this movement the crust southeast of the Brevard fault shifted toward the southwest relative to that northwest of the fault (Fig. 7-10).

Inside the window, Precambrian basement rocks are exposed over a large area that includes the town of Blowing Rock. These are metamorphic rocks including granitic gneiss and augen gneiss (known as the Elk Park plutonic group). An ancient erosion surface separates this basement complex from a thick sequence of sedimentary and metasedimentary rocks known as the Grandfather Mtn. Fm. (see sidebar). This group of sedimentary rocks is older than the Chilhowee Group (Table 8G-1). It occupies the same stratigraphic position as the Ocoee Group in the Smoky Mtns. and the Mt. Rogers Fm. farther north, but the Grandfather Mtn. Fm. cannot be traced outside the window. Several thrust

Grandfather Mtn. Formation

This is a thick, slightly metamorphosed sequence of conglomerate, sandstone, and siltstone interbedded with basaltic and rhyolitic lava flows. The presence of pink and white potassium feldspar in the conglomerate suggests that the conglomerate was deposited in an area of high relief so rapidly or under such dry conditions that the feldspar did not break down into clay minerals. This unit is 3,000-9,000 yards thick and is only exposed in the Grandfather Mtn. window. It is Late Precambrian in age, and possibly contemporaneous with the Ocoee Supergroup in the Great Smoky Mtns., or the Ashe Fm. The unit is composed of alluvial fan deposits laid down by braided streams carrying sediment derived from surrounding uplands made up of Blue Ridge basement complex rocks. The sedimentary basin may have resembled those of Death Valley in California or those that formed east of the Blue Ridge during the Triassic Period when the Atlantic Ocean began to open (p.16). (Schwab, 1977)

Fig. 8G-6. Large alluvial fan in Death Valley, Calif.

Grandfather Mtn/Linville Falls - NC

Rock Units in the Grandfather Mountain Area (In sequence with the youngest at the top)	
Chilhowee Group	See p. 78
Grandfather Mtn. Group	See sidebar above.
Blue Ridge basement complex	Including the Crossnore Plutonic Group, Elk Park Plutonic Group, Cranberry Gneiss, and Roan Gneiss

Table 8G-1: The various rock layers found on the western side of the Southern Blue Ridge.

Fig. 8G-7. Table Rock Mtn. is composed of sandstones of the Grandfather Mtn. Fm.

sheets are present inside the Grandfather Mtn. window (Fig. 8G-5). One of these, located in the southwestern part of the window consists of the sandstones, shale, and conglomerate rocks of the Chilhowee Group. They have been thrust over the Precambrian basement and the Grandfather Mtn. Fm. They in turn have been overthrust by Precambrian basement rocks in a thrust sheet that surrounds the window and is exposed at Linville Falls. Yet another thrust sheet, known as the Fries thrust sheet carries the Ashe Fm. over the rocks in the window (Fig. 8G-5).

Points of Interest

Stop 1. **View of the Grandfather Mtn. window:** The Grandfather Mtn. window is so large you cannot see all of it from a single place. However, you may see most of it from the top of Grandfather Mtn. For those who do not wish to make such a hike, the Chestoa Viewpoint (near Parkway milepost 320) provides excellent views of much of the window from its western edge. The Catawba River flows in the deep valley directly in front of you. The Linville River is in the next valley to the northeast. The gorges of both of these rivers have cut so deeply into the valley that the rivers are now out of sight. Grandfather Mtn. is on the horizon to the north. The shape of the flat-topped mountain in the distance to your right (Table Rock Mtn.) indicates that the layers of sandstone that compose the mountain are tilted (Fig. 8G-7).

Stop 2. **Deformed pebbles in the Grandfather Mtn. Fm:** At the site called "Split Rock" on park maps, you will find a large boulder (Fig. 8G-8) that has broken along fractures formed during uplift of the region and that was opened further as the boulder rolled downslope to its present position. This rock contains numerous pebbles that have been

Fig. 8G-8. Split Rock

Fig. 8G-9. Pebbles flattened and stretched by the deformation in the Grandfather Mtn. Fm.

Fig. 8G-10. Complex folds formed during the deformation of the rocks at Stop 3.

Fig. 8G-11. Quartz veins in the Grandfather Mtn. Fm. form exotic patterns. This exposure is located at Stop 3.

flattened as a result of the great pressure applied when the thrust sheet was emplaced (Fig. 8G-9). A similar boulder is located about 1.3 miles north of the intersection of US 221 and the Parkway at the Grandfather Mtn. exit. Outcrops of these rocks are present high on Grandfather Mtn.

Stop 3. **Deformation in the Grandfather Mtn. Fm. at Boone, NC:** One of the best exposures of the Grandfather Mtn. Fm. is located on the east side of US 221 opposite Payne Branch Road, 0.7 miles south of the stoplight at Deerfield Road. Park about 0.1 miles west of the outcrop, which extends for about 100 yards along the highway. Most of the outcrop is composed of arkose (sandstone-like rock containing abundant feldspar), but conglomerates and siltstones are also present. All have been metamorphosed to greenschist facies (containing chlorite), which gives the rock a distinct greenish color. The unit is folded into a broad anticlinal structure and shows both graded bedding and cross-bedding that indicate that the layers are right side up. Quartz veins here exhibit unusual patterns (Figs. 8G-10, -11). This exposure inside the Grandfather Mtn. window is located close to the thrust fault that carries Precambrian basement rocks on the thrust sheet covering the younger rocks in the window (Fig. 8G-5). ❖

Linville Falls

Fig. 8G-12. Linville Falls

Fig. 8G-13. Material carried in the swirling current at the top of the main Linville Falls is slowly cutting a pot hole deeper into the rocks. Several potholes are visible in the falls.

Location and Access: Linville Falls is located in The Linville Gorge Wilderness Area in North Carolina. The visitor center is located a short distance southeast of the Blue Ridge Parkway near milepost 316.

Accommodations: A gift shop and campground are located at the park. Fuel, motel rooms and groceries can be obtained at the town of Linville Falls along US 221 a few miles south of the Parkway entrance at milepost 317.4.

The Linville River starts in the Highlands near the southwestern part of the Upland Plateau and west of Grandfather Mtn. at elevations close to 4,000 ft. The river flows southeast across the edge of the Blue Ridge Escarpment and into Lake James at an elevation of 1,300 ft. near the western margin of the Piedmont about 20 miles to the southeast. The waterfalls at Linville Falls (Fig. 8G-12) account for a large part of this remarkably steep gradient. Cascades and waterfalls mark places where the river is cutting its channel downward. In part, this process involves the development of large potholes such as the one that is being formed at the top of Linville Falls. What remains of other potholes are present near the lip of the falls.

Equally remarkable is the pattern of the Linville River (Fig. 8G-14), which for several miles south of the falls flows in a meandering pattern in a steep-sided gorge about 1,500 ft. deep. Most rivers that meander flow across a landscape characterized by low relief. If the region begins to rise, these rivers cut their channels deeper and entrench themselves. That appears to be what happened to the Linville River. If this is correct, the Linville River was meandering across this region before the

Blue Ridge rose relative to the Piedmont. (See the discussion about the Blue Ridge Upland Plateau in Ch. 1).

As the river cut its channel downward, it cut through an upper layer of Precambrian basement rock, through two thrust faults, and into underlying sedimentary rocks of the Chilhowee Group (Stop 2).

Views of the Linville Gorge

Exceptional views of the gorge cut by the Linville River and the waterfall are possible from overlooks located along trails that run downstream on both sides of the gift shop. Views of the gorge and access to trails leading down to the river are also located along roads starting in the town of Linville Falls (Fig. 8G-14). Take SR 183 south out of Linville Falls. The road splits into two roads, SR 1270 and SR 1238. Several places offer views up the gorge including a superb one from Wiseman's View. A number of trailheads to trails leading into the gorge are located along SR 1238, which runs along the edge of the valley for many miles (Fig. 8G-15).

Fig. 8G-15. Trailhead to one of the trails leading into the Linville Gorge.

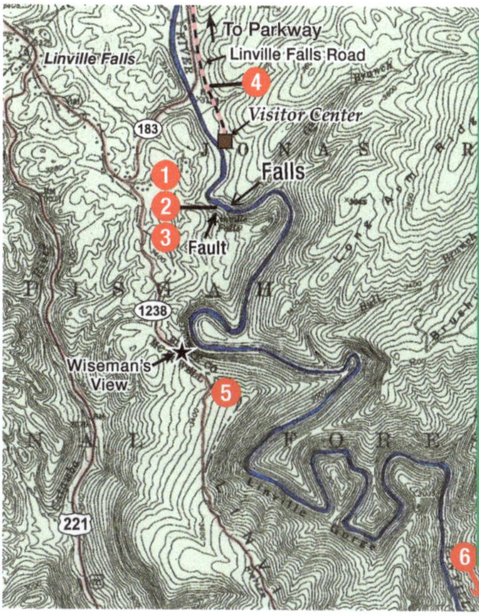

Fig. 8G-14. Topographic map with Stops along the Linville River near Linville Falls. (Note: Stop 6 is beyond the map.) (USGS map, with additions)

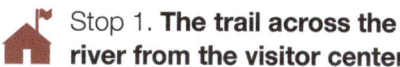

Points of Interest

Stop 1. **The trail across the river from the visitor center:** This trail leads down to the river and is located at the top of the main waterfall. Notice the swirling water in the channel near the top of the waterfall. A pothole is forming here as rocks that are caught in the rotating water abrade and gradually enlarge the bowl-shaped depression. This is one of the important ways that streams cut their channels into even the hardest of rocks. You can see what remains of other potholes where the water passes over the lip of the waterfall.

Stop 2. **The Linville fault:** On the stairs leading down to the falls overlook you cross over a major thrust fault. Precambrian metamorphic rocks (the Linville Falls granite) lie over Cambrian-age sandstones of the Chilhowee Group. The fault separates these two rock types that differ in age by at least half a billion

Fig. 8G-16. The Linville Falls fault zone contains mylonite, formed by shearing and recrystallization of the over- and underlying rocks.

Fig. 8G-17. Broad open folds are present in the Chilhowee sandstones.

years. The original texture of rocks in this zone was destroyed by the crushing and shearing that took place as the thrust sheet, bearing the weight of thousands of feet of rock, moved many miles over the underlying sandstones. The rocks in this zone were transformed into a sheared and recrystallized rock called mylonite (Fig. 8G-16).

Stop 3. **A closer look at the fault zone:** This fault zone can be directly examined a short distance upstream from the overlook, but it is so close to the water level of the river there that it is dangerous to access. To get a good look at this exposure it is necessary to cross the wall on the upstream side of the overlook (but you must first obtain permission from a park ranger to do so). At that viewpoint you will be standing on the bedding plane of one of the Chilhowee sandstone layers. Two types of folds were formed here as the overlying thrust sheet compressed the layered rock below. One set of folds looks like large ripples in the bedding plane. The second set of folds is much smaller with linear striations crossing the folds that indicate the direction of movement of the overlying thrust sheet. The layers of sandstone that you stand on at this viewpoint are part of the Chilhowee Group. Note that they are folded into broad open folds (Fig. 8G-17). Much smaller folds are also present in the sandstone (Fig. 8G-18).

Stop 4. **Precambrian gneisses of the Blue Ridge thrust sheet** have been moved many miles over the underlying Chilhowee sedimentary rocks. The resulting deformation is seen in both the overlying thrust sheet and in the Chilhowee. Zones of intense shearing and dislocated blocks of gneiss, granite, and amphibolite are present in the lower part of the thrust sheet. Although the fault contact zone is thin, a thick zone of deformation is present in the lower part of the thrust sheet. Evidence of this shearing is present in the rock exposures along the road leading from the Parkway into the Linville Falls Visitor Center. From this first overlook it is possible to see isoclinal folds in the far wall of the gorge, and small folds in the layers of sandstone at the overlook. Note the potholes being cut in the stream channel below the viewpoint.

Fig. 8G-18. Small-scale folds are also present in the sandstones at Stop 1.

Stop 5. **Carolina hemlock bluff communities** are visible but inaccessible from the overlooks along the Linville River. Because the slopes are so steep, logging is difficult, and an extensive old-growth forest remains in Linville gorge. Unfortunately, here the woolly adelgid (Fig. 4-8) is attacking both the Carolina hemlock (an uncommon evergreen that likes high south-facing rocky outcrops) and the Canadian (or eastern) hemlock (a more common forest giant that prefers cooler and moister east- or north-facing slopes and coves).

The effects of fire are also clear in the Linville Gorge. Evidence remains of two major wildfires that burned much of the gorge. These were apparently natural fires, but more catastrophic than most in the Blue Ridge. The rugged topography of the gorge may have reduced the number of fires, but once started, the steep slopes make them very difficult to control. Hence, they are more damaging than fires that have burned in most places. Table Rock Mtn. pines are now regenerating in burned areas on the rim farther south along the Wiseman's View Road.

Golden heather (Photo: USF&WS)

Allegheny sand myrtle (Photo: Mark A. Garland, USDA)

Stop 6. **Table Rock Mountain and Shortoff Mountain** are two of only three places in the world where you can find mountain golden heather (Hudsonia montana). If you go into these areas, it is important to stay on the trails. Allegheny sand myrtle is also present on these rocky slopes. Apparently, fires tend to keep the sand-myrtle from overrunning the mountain golden-heather. Because few fires have burned the area in recent years, the mountain golden-heather is doing poorly. Table Rock Mtn. is a favorite place for mountain climbers. It is accessible by taking SR 181 south from the Parkway and SR 1265, a gravel road, to the Table Rock Mtn. trailhead. ❖

Chapter 8H

The Mars Hill Terrane at Roan Mountain State Park

Location and Access: Rocks of the Mars Hill terrane are readily accessible along SR 143 at the Roan Mtn. State Park in Tennessee. You can reach this park via US 19E from Elizabethton, or via US 19 and 19E from the town of Spruce Pine.

Accommodations: Cabins and a campground are available at Roan Mtn. State Park. These are also available at Linville and numerous places along SR 194.

Other Attractions: The Museum of North Carolina Minerals is located at the intersection of SR 226 and the Parkway (near milepost 331) close to Spruce Pine. Shops selling minerals are located near Spruce Pine, which is an important mining site for feldspar and for the occurrence of gem-quality minerals.

Some of the oldest minerals in the Appalachian Mtns. are present in rocks located west of the Grandfather Mtn. window. This area, known as The Mars Hill terrane (Fig. 2-12), extends from Mars Hill near Asheville, N.C. to Roan Mtn., Tenn. The rocks found within this terrane include high-grade granitic gneiss that contains feldspar, quartz, garnet, and pyroxenes (Fig. 8H-1).

Above: Park headquarters at Roan Mtn. State Park. (Photo: Trance Mist†)

In addition, a rock called eclogite, that contains unusually high quantities of iron and magnesium is also present. Zircon minerals in the gneiss have yielded radiometric ages of 1.3, 1.8, and 1.9 Ga (and one of 2.7 Ga). These minerals, much older than the Grenville Orogeny, were weathered out and became incorporated into the gneisses of the Mars Hill terrane, which are Grenville-age and dated at about 1.1 Ga. These ancient gneisses have a complex history. After they initially formed, they were intruded by granite (the Beech granite) about 734 Ma. The magma from which this granite formed may have been generated by the melting of the continental crust as Rodinia was being pulled apart, releasing pressure on the deeper parts of the crust. The granite rose into the gneisses and became part of what would become the Mars Hill terrane. A second major intrusion brought masses of gabbro (the Bakersville gabbro), from deep in the crust or upper mantle into the gneiss of the Mars Hill terrane.

This terrane is part of the Fries thrust sheet (Fig. 2-12), which was accreted onto Laurentia during Paleozoic mountain building. Geologists think that a slice of rock, the eclogite, that originated at a depth of 18-25 miles as part of Earth's mantle became incorporated into the

Fig. 8H-1. Granitic gneiss exposed in the Mars Hill terrane.

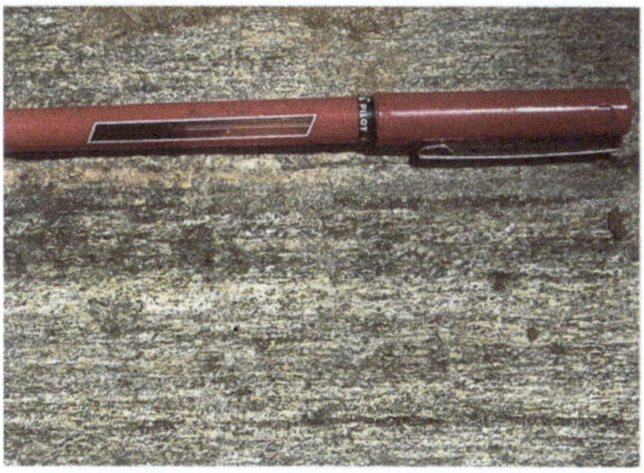

Fig. 8H-2. Sheared gneiss exposed along SR 15 near the entrance to Roan Mtn. State Park.

Fig. 8H-3. Entrance to Peg Leg Iron Mine (above). Collapsed passageway into the mine (below).

Mars Hill terrane. This slice may have been part of a subduction zone that was caught up in mountain building deformation, elevated into ancient mountains and later moved to the northwest and thrust onto the Laurentian (North American) basement.

Points of Interest

Stop 1. **Highly-sheared, Precambrian-age gneiss** (Fig. 8H-2), part of the Precambrian gneiss, is exposed in a roadcut on the north side of US 19E near the Tenn.-N.C. state line east of the entrance to Roan Mtn. State Park.

Stop 2. **Peg Leg Iron Mine Museum and grounds:** From the late 1700s to the 1930s this mine produced iron from lens-shaped bodies in the Precambrian gneiss. The iron here occurred in magnetite (a black iron oxide mineral), titanium-rich magnetite, and hematite (also an iron oxide). The ore-bearing minerals are thought to have originated from fluids that came from deep in Earth's crust or perhaps from the mantle as part of the Bakersville gabbro. The fluids probably moved along a fault zone in the Precambrian-age gneiss. The entrance to the Peg Leg Mine (Fig. 8H-3) is located on a trail that starts at the Roan Mtn. State Park Visitor Center. Although the entrance is well preserved, the mine, which extends about 600 ft. from the opening is flooded.

Stop 3. **The Beech granite**, a 740 Ma old granite that intruded the older gneiss, is exposed in the banks of the Doe River about 1,500 ft. north of Burbank and at the recycling center at Roan, Tenn. (about 0.7 miles west of the entrance to Roan Mtn. State Park).

Fig. 8H-4. Gabbro mass showing beautifully developed joints.

Stop 4. **The Bakersville gabbro** is exposed at a number of outcrops along SR 143. One is located about 0.5 miles from the entrance to SR 143 from US 19E (Fig. 8H-4). Magmas rich in iron and magnesium such as the Bakersville gabbro are thought to have originated in the mantle or lower crust. This one probably came into these gneisses when the ancient continent of Rodinia split apart leading to the formation of the Iapetus Ocean in the late Precambrian (730 Ma). It was later moved as part of a large crustal fragment into its current position during the mountain building that took place in the Paleozoic Era.

Fig. 8H-5. At Carvers Gap you can see a grassy bald on one side of the road and a view of the spruce forest above a northern hardwood forest on the other side.

Stop 5. **Carvers Gap grassy bald:** The Appalachian Trail crosses SR 143 at Carvers Gap (Fig. 8H-5). The gneisses at this gap (known as the Carvers Gap granulite gneiss) are among the oldest rocks found in the Mars Hill terrane. They were metamorphosed at high temperature and pressure. This gneiss is dark brown in color and has a "salt and pepper" texture. The "salt" is composed of plagioclase feldspar and quartz. The darker-colored "pepper" is made up of pyroxene, biotite, and amphibole (hornblende). Unusually large garnets that have weathered out of the host rock, probably an amphibolite interlayered with the gneiss, can be found along the Appalachian Trail about a half-mile west of Carvers Gap.

This is an excellent place to see one of the grassy balds found at high elevations in many places in the Highlands, especially in the Great Smoky Mtns. The origin of these balds is still being debated. Many of the balds are high enough to have been covered by spruce and fir trees. This is obvious at Carvers Gap. Many think the balds were areas originally cleared by American Indians and kept open by set fires. Others postulate that early settlers enlarged balds previously created by natural fires. It has also been suggested that spruce and fir trees covered the bald areas soon after the last glacial advance ended (about 18,000 yrs. ago), but were later killed off by a long warm period that occurred anywhere from 9,000 to 5,000 years ago when temperatures may have been as much as four degrees warmer than at present.

Fig. 8H-6. Outcrop of Newdale dunite (left). Close-up of the Newdale dunite (right). Chemical weathering has discolored the lower section of the Newdale dunite.

A Slice of Earth's Mantle: The Newdale Dunite

Dunite (Fig. 8H-6) (an olivine-rich rock) and eclogite (a rock composed mainly of minerals that make up the upper part of Earth's mantle) both occur in the Mars Hill terrane. Eclogites form at temperatures in the range of 600-800°C and at pressures in the range of 13-17 kilobars. These conditions are found at depths of 18-25 miles. This slice of the mantle probably moved along faults high into the crust during Grenville-age mountain building. Once in the crust the eclogite was in a lower temperature-pressure environment and the minerals underwent adjustments, called retrograde metamorphism that had the effects of changing the mineralogy, producing some hornblende and plagioclase feldspar in the rock. During this process, some of the minerals that formed at higher temperatures and pressures are transformed into minerals associated with lower grades of metamorphism. This is the same metamorphism that caused the original volcanic and sedimentary rocks of the Ashe Fm. to become schists and amphibolites.

Points of Interest

Stop 1. **Newdale dunite:** An excellent and accessible exposure of dunite (Fig. 8H-6) rarely found in the Appalachians is located on SR 80 about half a mile north of its intersection with US 19E. This rock formed under high-grade (granulite facies) metamorphic conditions. Olivine makes up most of the rock with minor amounts of pyroxene and chromite. It is a good example of the ultramafic masses found in the Ashe Fm., and appears to have been emplaced along a fault.

Stop 2. **Eclogite:** Another exposure of these ultramafic rocks, eclogite, occurs near the town of Bakersville, located west of Spruce Pine. At Bakersville, drive 0.6 miles north on Redwood Rd. (SR 1217). The exposure extends several hundred feet along the road, however it is difficult to park near this outcrop. ❖

Chapter 8I

Spruce Pine Mining District

Fig. 8I-1. An open-pit mine at Spruce Pine.

Location and Access: Access to the quarries is limited but some are visible from SR 226, the main east-west artery through Spruce Pine, North Carolina (Fig. 8I-1).

This district, one of the leading producers of feldspar and mica in North America, covers an area that is about 25 miles long and 10 miles wide. Feldspar is used in the manufacture of ceramics. Mica, formerly known as isinglass, is used as an insulator in electronic instruments. Both of these economically important minerals are derived from large pegmatite bodies. These intrusions commonly form during the late stages of the cooling of granitic plutons after most of the magma has crystallized. In this case, the intrusions took place during the early Paleozoic Era when granites intruded the gneisses and schists of the Ashe Fm. A thick residual soil formed from chemical weathering of the underlying rock is up to 50 ft. thick here and covers a large part of this region.

Large crystals of feldspar and mica are major components of the pegmatites, but a number of rare and in some cases valuable minerals such as beryl (emerald), corundum, garnet, amethyst, rare earths, and uranium-bearing minerals are present. Samples of some of these minerals may be purchased from shops in and around Spruce Pine. Owners of several mines allow people to look for minerals in the mine tailings (materials left over after the ore has been removed).

An eastern fence lizard basks in the sun.

Stop 1. **Outcrop of Spruce Pine Pegmatite:** This pegmatite is exposed on both sides of US 19E, about 1.3 miles west of the intersection of US 19E and SR 226 (Fig. 8I-2). You will find good distant views of the main mine workings on SR 226, a short distance north of the intersection with US 19E. The mining is an open-pit operation in which the topsoil is removed to expose the underlying pegmatite. Unfortunately, this leaves scars that remain visible for many miles. ❖

Fig. 8I-2. Close-up of quartz-feldspar pegmatite exposed on SR 226.

Chapter 8J

Mt. Mitchell & Craggy Gardens

Elisha Mitchell, an 18th century geologist. Right: Fig. 8J-1. Mt. Mitchell.

Mt. Mitchell was made a state park in 1915 as a way of controlling timber harvesting on its steep slopes. At 6,684 ft., Mt. Mitchell is the highest mountain in the Appalachians (Figs. 8J-1, -2). It is one of 18 mountains rising over 6,300 ft. in a part of the Blue Ridge known as the Black Mtns. (Fig. 8J-3). The thick forest around Mt. Mitchell is dominated at high elevation by spruce and fir trees that give the slopes a dark green color, hence the name "Black Mtns." On a clear day you can see Grandfather Mtn., Table Rock Mtn., and Hawksbill Mtn. near Linville Falls, 25 miles to the northeast. The large open pit mine at Spruce Pine and Clingmans Dome is nearly 75 miles to the southwest, in Great Smoky Mtns. National Park.

Mt. Mitchell is named for a geologist, Elisha Mitchell who taught at the University of North Carolina and explored the Black Mtns. in the late 18th century. He calculated the height of the mountain and later died when he slipped and fell at a waterfall now named in his honor.

Location and Access: The only access to North Carolina's Mt. Mitchell State Park and Craggy Gardens is via the Parkway. SR 128 leads from the Parkway near milepost 354 into Mt. Mitchell State Park. Craggy Gardens Visitor Center is located on the Parkway near milepost 364. Access depends on weather conditions.

Accommodations: A restaurant and picnic tables are at Mt. Mitchell. Other accommodations can be found in Asheville, Boone, Blowing Rock, Linville, and Little Switzerland, N.C. Crabtree Falls Campground is located about 15 miles north of Mt. Mitchell.

Trails: Obtain a trail guide at the visitor center near the top of Mt. Mitchell or refer to their website. The Craggy Gardens, Craggy Pinnacle, and Douglas Falls trails start near the visitor center at Craggy Gardens.

Outstanding Features: Mt. Mitchell is famous for its severe weather conditions. Craggy Gardens contains one of the most diverse plant communities in the Appalachians.

Nearby Attractions: Visit the small museums and viewpoints at these two sites.

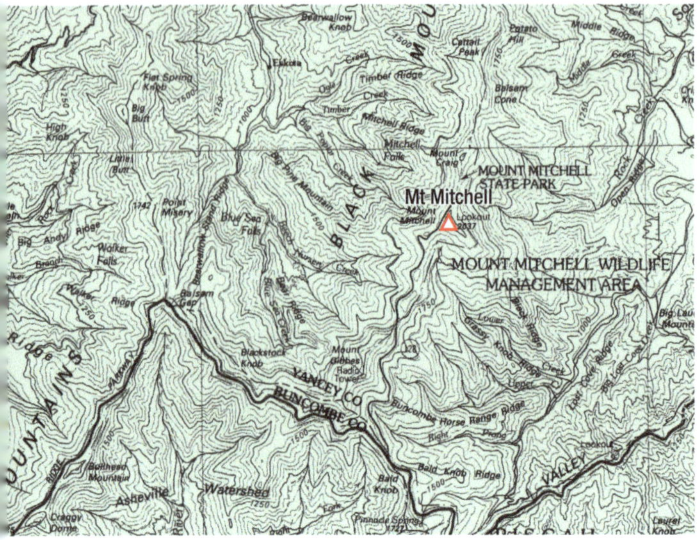

Fig. 8J-2. Topographic map of Mt. Mitchell. (USGS)

tures make the high elevations around Mt. Mitchell prone to the development of fog and dew on vegetation.

The conifer forests in the Black Mtns. are showing signs of regeneration after a die-off caused by the combined effects of air pollution, acid rain, and the balsam woolly adelgid (Figs. 8J-4, -5) in the 1950s and 60s. A brown layer of concentrated air pollution located near the elevation of the mountaintop is still visible at times. The mountains have also suffered as a result of clear-cut logging and fires that subsequently swept through the slash produced by the cutting. The clear-cutting stopped when the area was declared a state park. The steep slopes are also prone to the development of landslides that leave scars that are visible from Parkway overlooks. New growth of heartleaf paper birch trees is present on some of the landslide scars. Although red spruce and Fraser firs

Ecology on Mt. Mitchell

Meteorological records have long been kept at an observatory near the top of Mt. Mitchell. The summers here are mild and winters are extremely cold. The cold temperature record occurred in January of 1985, when thermometers fell to -34°F. The warmest temperature record from 1972-82 was 81°F in Aug). Rainfall averages about 74 in. per year. Winds of 178 miles per hour are the record to date. High humidity and cool tempera-

Fig. 8J-3. View across part of the Black Mtns.

Fig. 8J-4. Park headquarters with dead spruce trees on the mountain above.

dominate the high elevations, many other types of trees, including American chestnuts, oaks, and hickories are common below about 4,500 ft. and rhododendron thickets cover many of the steep slopes, especially those close to streams. Young red spruce, fire cherry, yellow birch, mountain ash, and mountain maple are gradually covering the areas formerly populated by Fraser fir and red spruce. The Black Mtn. forest is considered by some to be the most diverse forest in the Southern Appalachians. Some rare plants (including Appalachian avens and deergrass), once abundant in the alpine tundra that covered this area earlier in the Ice Age, still grow on the summits in the Black Mtns.

In addition to plants, 91 species of birds have been reported on Mt. Mitchell. The bird population here is similar to that found in the colder Canadian and New England mountains. Wrens, juncos, red crossbills, golden-crowned kinglets, and peregrine falcons are often seen from the observation tower on the mountaintop.

Fig. 8J-5. Spruce trees were knocked over by high winds here.

Geologic Setting

Both Mt. Mitchell and Craggy Gardens are located on the Spruce Pine thrust sheet (Fig. 8G-5), a southern equivalent of the Fries and Gossan-Lead thrust sheets seen north of the Grandfather Mtn. window. The rock carried on this sheet is part of the Ashe Fm., and probably once resembled the sediments and volcanics that initially formed late in the Precambrian Era in an island arc situated off the coast of Laurentia. Later, in the Paleozoic Era, higher temperatures and pressure altered the rocks on this part of the thrust sheet, producing metamorphic rocks of higher grade. The original basalts and volcanic ashes were altered to amphibolites, and what had been sediments were changed to quartzites, schists and gneisses. These are now well exposed along the Parkway near Craggy Gardens. This alteration took place as the ocean that had separated the island arc from Laurentia slowly closed, lowering the rocks to great depth where, subjected to high temperature and pressure, they transformed into metamorphic rocks. Later, collisions between the island arcs and the continent took place, and the rocks were folded and moved in thrust sheets to the northwest, onto the Laurentian continent. As in other accreted terranes, this one is thought to have moved a long distance, perhaps as much as 100 miles. The timing of the movement is still debated, but recent studies support the idea that these thrust sheets were subjected to more than one period of metamorphism. They appear to be part of the highest, in the great pile of thrust sheets found in the Highlands region, and probably were last moved during the Alleghanian Orogeny when the southern Appalachian Mtns. rose late in the Paleozoic Era. ❖

Craggy Gardens

Fig. 8J-6. Mountain bald at Craggy Gardens.

Fig. 8J-7. Scar left after soil slipped off a steep slope near Craggy Gardens.

The heath balds at Craggy Gardens make this an especially interesting and accessible place to study plants that prosper at high elevations where spruce-fir forests are absent (Figs. 8J-6, -7, -8). It is unclear why trees do not cover these areas. (See discussions on p. 69 and p. 199.) Some of the balds seen from Craggy Gardens appear to have formed on steep slopes where the soil cover may have slipped off the underlying rock. It has been suggested that the trees succumbed to higher temperatures during the interglacial periods of the Ice Ages, or alternatively, that fires swept up the slopes killing the trees. Trails into the heath that start near the small visitor center located on the Parkway (near milepost 364.5) provide a rare opportunity to stroll through the heath. Occasionally the trails are closed to help restore areas where the plants have been trampled.

Fig. 8J-8. Trail through a rhododendron thicket at Craggy Gardens.

Fig. 8J-9. Pegmatite dike by the Parkway near Craggy Gardens.

Fig. 8J-10. Amphibolite mass cut by veins of quartz.

Rocks that make up part of the Spruce Pine thrust sheet are exposed along the Parkway about a hundred yards south of the visitor center (Figs. 8J-9, -10, -11, -12). This is an especially good place to examine a thick pegmatite dike that cuts the dark, nearly black amphibolites and quartz veins that cut both the pegmatite and the amphibolites. Like most pegmatites, those found in this area are composed of large crystals of quartz and feldspar with micas but a number of other minerals including some rare precious minerals have been found in these pegmatites, notably at Spruce Pine. ❖

Fig. 8J-11. Metamorphosed quartzite in the Ashe Fm. at Craggy Gardens.

Catawba rhododendrons peak at Craggy Gardens in early June.

Fig. 8J-12. Folded quartz pegmatite near Craggy Gardens.

Chapter 8K

Precambrian Basement Rocks on I-26

Location and Access: From Asheville, N.C., drive north on I-26 toward Erwin, Tenn. to see the roadcuts exposing these rocks.

One of the best places to see the Precambrian basement rocks deformed at high temperature and pressure is along I-26 between Asheville, N.C. and Erwin, Tenn. (Fig. 8K-1). This exposure is so remarkable that the highway dept. built an overlook on the south side of I-26 to provide a view of the eastern end of this long outcrop. It is of special interest for revealing an extraordinary view of the complexity of the rock structures present in the late Precambrian metamorphic-igneous terranes that comprise so much of the rocks in the core of the Blue Ridge. These rocks include granitic gneisses that have a pinkish gray color. White quartz veins and light-colored pegmatites cut through them. Around 734 Ma, large gabbro bodies intruded the gneisses. These are the black to dark gray masses that have been altered to amphibolites. They are seen at the southern end of the exposure (Fig. 8K-2). In some places, it is clear that the rocks were so hot and mobile that they mixed together. They were close to the temperature and pressure conditions needed to melt them (Fig. 8K-3). Geologists are still trying to decipher the history of these rocks—how they were affected during the initial mountain building at the end of the Precambrian and how they were changed by the subsequent mountain building episodes that took place during the Paleozoic. (See Labotka and Hatcher (2006) for more details about this outcrop.) ❖

Fig. 8K-1. Folded gneisses indicate the rock was once hot enough and under sufficient pressure to cause it to be highly ductile (plastic). The vertical streaks are drill holes made during road construction.

Fig. 8K-2. The contact between the large mass of gabbro (dark) and the gneisses (light) indicates the gabbro was intruded into the gneiss. During the intrusion the gabbro cut across the layering in the gneiss. (The vertical lines are drill holes.)

Fig. 8K-3. Amphibolite (dark) and gneiss (light) were broken up and mixed at high temperatures, forming a migmatite.

Chapter 8L

Blue Ridge Parkway - Asheville to Oconaluftee

Geologic Setting

East of the entrance to the Great Smoky Mtns. Natl. Park at Oconaluftee, the Parkway passes through the eastern part of the Ocoee Supergroup and into an area where the rocks have been subjected to multiple deformations under conditions of high-grade metamorphism (Ch. 9). The complex geological history for this central part of the Blue Ridge is still in a relatively early stage of investigation. The 1985 state map of North Carolina shows the Parkway crossing an area near Balsam, N.C. composed of Precambrian-age gneisses and amphibolites, biotite gneiss, and migmatite (mixed gneiss and igneous rocks). Beyond that most of the region is underlain by muscovite-biotite gneiss, amphibolites, hornblende gneiss, and mica schists that are interpreted as part of the Ashe/Tallulah Falls Fms., which lie on the east side of the Blue Ridge. This whole region was subsequently subdivided into several terranes (Fig. 2-12) by Hatcher and described in greater detail in guidebooks by Merschat, et al. (2012), and Carter, Merschat, and Wilson (2001). This second guidebook also provides detailed geological information about most of the overlooks along the Parkway.

Location and Access: This section of the Parkway in North Carolina may be accessed from I-40 on the north side of Asheville; from US 25 on the east side of Asheville; from NC 191 (near I-26) south of Asheville; and also from US 276, US 23/74, and US 19.

Accommodations: The only accommodations on this section of the Parkway (a motel, restaurant and campground) are located at Mount Pisgah Visitor Center (about 20 miles south of Asheville). Other convenient facilities are located at Asheville, Waynesville, and on the Cherokee Indian Reservation at the southern entrance to the Parkway.

Nearby Attractions: The Folk Art Visitor Center and the headquarters for the Blue Ridge Parkway are located north of Asheville, near the I-40 entrance to the Parkway. The Biltmore Estate may be reached southeast of Asheville from the US 25 entrance to the Parkway. The NC 191 entrance is close to the North Carolina Arboretum. The Cradle of Forestry in America Visitor Center (U.S. Forest Service) is located a few miles east of the Parkway on US 276.

Above: The Highlands southwest of the highest point along the Blue Ridge Parkway.

Fig. 8L-1. Roads and overlooks along the Parkway south of Asheville, N.C. (NPS map, with additions).

Points of Interest

along the Blue Ridge Parkway from Asheville to Oconaluftee

This section of the guide follows the Parkway from Asheville to Oconaluftee, with a side trip from the Parkway to Brevard, N.C. along US 276 (Fig. 8L-1).

The French Broad River at Asheville, N.C.

The drop in elevation as one approaches Asheville on the Parkway is striking. The road descends from elevations over 5,000 ft. into a large basin with elevations slightly above 2,000 ft. This basin is unusual in the Blue Ridge because higher mountains surround it. The area appears to be the result of erosion by streams. It is not obviously related to anything in the composition or structure of the bedrock, which is part of the Ashe/Talullah Falls metamorphic suite. The French Broad River and its tributaries drain this basin. The French Broad is one of the few streams flowing from east to west that cuts completely across the Blue Ridge. Its headwaters lie in the Piedmont and along the Brevard fault zone where movement along the fault has weakened the rocks and made them more susceptible to erosion and because of this, the Brevard zone shows up clearly on maps and in aerial photographs (Fig. 7-1). South of Asheville the French Broad River turns north and begins to flow to the west across the Blue Ridge into the Valley and Ridge where it joins the Holston River. The two form the Tennessee River, a tributary of the Mississippi River. Consequently, sediment eroded from the Piedmont finds its way into the Mississippi Delta. The Parkway crosses the French Broad River in the southern part of Asheville and affords a good of it view at an overlook at milepost 393.8.

The rocks exposed along the Parkway demonstrate the complexity of this region. It has been subjected to strong deformation several times during Paleozoic orogenies. The high-grade of the metamorphic rocks, some of which are sillimanite or granulite facies, along with radiometric dating of rocks in this area, confirm that they came to this level in the crust from great depth and that their orogenic history is complex (Figs. 8L-2a-b). Excellent exposures of some of the resulting features are present at various overlooks along the Parkway. These are listed in sequence from Asheville to the entrance to the Great Smoky Mtns. National Park.

Haw Creek Valley Overlook: (mp 380) The stream valley seen from this overlook is formed by a tributary of the Swannanoa River, which flows into the French Broad River. A long outcrop composed of quartz and biotite-rich gneiss and metamorphosed greywacke forms the border of the Parkway across from the overlook. These rocks are currently mapped as part of the Ashe Fm.

French Broad Overlook: (mp 393.8) Outcrops of gneiss opposite this overlook are part of the Ashe/Tallulah Falls metamorphic suite. A few purple colored garnet crystals and many quartz veins and pods are present. The mica-rich layers of the gneiss also exhibit many small crenulations formed when the region was subjected to strong deformation during the episodes of compression that accompanied Paleozoic mountain building in this region.

Fig. 8L-2a. Migmatites developed in the Ashe/ Tallulah Falls metamorphic complex exposed between Parkway mileposts 429 and 430.

Fig. 8L-2b. Exposure of the Ashe/Tallulah Falls metamorphic complex exposed along US 276 several miles west of Brevard, N.C.

Walnut Cover Overlook: (mp 396.4) The rocks exposed here are similar to the gneisses seen at the French Broad Overlook described above.

Pisga Inn: (mp 408.5) This is the only place that offers overnight accommodations along the Parkway between Asheville and Oconaluftee.

Pink Bed Overlook: (mp 410.3) The large outcrop here is composed of mica-gneiss similar to that seen at French Broad Overlook. *(Continued on p. 285)*

A Side Trip to Brevard Along US 276

A trip to the Brevard fault zone at Brevard N.C. along one of the most beautiful routes in the Blue Ridge, US 276, is worthwhile. Leave the Parkway at Wagon Road Gap near milepost 412.

Stop 1. **Cradle of Forestry in America Historic Site:** This was the site of the first school of forestry in America. Known as the Biltmore Forest School it was founded in 1898 by Dr. Carl Schenck as a way to educate the public and especially loggers whose techniques had left many of the forests in the Appalachians devastated. This historic site includes a vast forest covering about 6,500 acres that abounds in wildflowers, birds, secluded woods, creeks with cascades, and paved trails that make it accessible to everyone. Many old buildings have been restored or rebuilt and a number of interactive exhibits provide insights into older and modern forest management techniques. The visitor center contains exceptionally well-designed displays as well as a place where you can find lunch during the spring, summer, and fall seasons when the Cradle of Forestry is open. It is also used for a variety of conservation, natural history, and music programs.

Fig. 8L-3. Looking Glass Falls near US 276.

Stop 2. **Looking Glass Falls:** This beautiful waterfall (Fig. 8L-3) visible from the road is a popular place to see knickpoints, places where stream erosion cuts a valley as it undermines the rocks that underlie the valley upstream.

Several hundred yards downstream you will encounter a near vertical rock wall (Fig. 8L-4) formed where the granitic rocks that make up Looking Glass Mtn. were broken by a long joint or fracture that formed as the massive granitic rocks cracked. The fractures most likely formed during deformation as the region warped while it was uplifted.

Fig. 8L-4. Joint formed in the granite pluton that makes up Looking Glass Mtn.

Stop 3. **Looking Glass Rock:**
This mountain is visible from several overlooks along the Parkway near mile-post 417 (Fig. 8L-5). From these over-looks you see a mountain that closely resembles Stone Mtn. The two share much of their geologic history. Both are Devonian-age (390 Ma old) plutons composed of granitic rocks (diorite to granodiorite) that were intruded into the older rocks of the Blue Ridge, then cooled slowly deep in the crust before ultimately being exposed as erosion removed the overlying rock. Once ex-posed, the layers of the mass spalled-off through a process known as exfoliation. This left the sides of the mass almost completely bare of vegetation.

If you would like to climb the mountain and witness the spectacular views from its crest, turn west from US 276 and drive to a parking lot, which is at the head of a trail that leads to the top.

⚠ At the top be very careful if you venture close to the edge. A fall here would likely be fatal.

Stop 4. **The Brevard fault:**
The Brevard fault is recognized as the eastern boundary of the southern section of the Blue Ridge (Fig. 7-1). It has been traced from the coastal plain in Georgia to central Virginia where another fault, the Bowens Creek fault, continues to the northeast. These are thrust faults that formed during the Paleozoic orogenic deformation that culminated in the Alleghanian Orogeny when the most prominent structural features of the Southern Appalachians formed. The origin and significance of the intensely deformed rocks of the Brevard zone have long been debated by geologists. It marks one of the most important thrust faults in the Appalachians. It is steeply inclined to the southeast and continues at depth beneath the eastern part of the continent. At one time it was thought

Fig. 8L-5. View of Looking Glass Rock from the Parkway near milepost 418.

Fig. 8L-6. Highly deformed rocks in the Brevard fault zone at Brevard, N.C. Note that in both of these outcrops the feldspar crystals have been flattened and drawn out as a result of the shearing that took place in the fault zone under extremely high pressure.

to be the contact between the North American and Eurasian plates but hat idea thas been given up because the rocks on both sides of the fault are so similar. However, it is still a fault on which great lateral movements have taken place. The Brevard fault zone is often a mile or more wide and contains rock structures indicating both a southeast to northwest as well as a northeast to southwest movement. Most geologists now agree that movement on the Brevard fault involved an oblique slip of the southeastern side toward the southwest (Fig. 7-10). The Brevard fault differs from many other faults in that left lateral strike-slip movement as well as a major displacement to the northwest has taken place. It appears that the whole region southeast of the Brevard fault (called Gondwana, which included Africa) was moving counter-clockwise and toward the northwest relative to the central part of North America, as the collision between Gondwana and Laurentia took place near the end of the Paleozoic Era (Fig. 2-14).

Because the rocks along the fault have been sheared and weakened by erosion, the fault zone shows up clearly on many satellite images as a valley, yet as a result of that erosion outcrops of rocks in the zone are not well exposed. One exposure is located close to the town of Brevard. To reach it, drive about one mile east on US 276 from the courthouse located on Main St. Turn right onto Galimore Road, continue to Sugarloaf Road (SR 1119) until you cross a one-lane bridge. The outcrops are located about 0.2 miles along that road. The rocks exposed there are strongly sheared augen gneisses (Fig. 8L-6). Another very different exposure is located west of Rosman, N.C. Turn onto Frozen Creek Road about a mile west of Rosman, then turn right onto Windy Hollow Lane and park near an old quarry. The rocks in the fault zone here are highly deformed phyllites called phyllonite that is interlayered with marble (see Merschat, et al., 2012 for more details). ❖

East Fork Overlook: (mp 418.3) The outcrop at this overlook is an excellent example of a migmatite, a mixture of quartz-feldspar veins in a dark biotite-rich gneiss. Deformation of the rock at a temperature close to the melting point leads to this type of feature (Fig. 8L-2a-b).

Graveyard Fields Overlook: (mp 418.8) The name of this overlook comes from the resemblance of the landscape here to a graveyard with monument-like spruce tree stumps left by a forest fire that swept through the area in 1925. This event produced scenery that is somewhat different from what you see along most of the Parkway. The stream that flows down the valley, the Yellowstone Prong of the Pigeon River gets its name from the yellow color of limonite (an iron oxide mineral also known as rust) produced by oxidation of iron-bearing minerals in the rock. The limonite becomes incorporated into clay that forms as a result of weathering of feldspar minerals in the rocks, and in turn creates yellow-stained soil. Outcrops exposed in the creek have been polished as a result of being abraded by rocks moving downstream with the current. As a result we find excellent exposures of small-scale folds developed in the migmatitic (mixed) gneiss, part of the highly altered Ashe and Talullah Falls fms. (p. 195-6).

Two views of the valley reveal a prominent change in its shape. The image taken down-valley from the overlook (Fig. 8L-7a) shows a deep v-shaped valley. Above the overlook, and especially visible from the Black Balsam Knob Road, you see a broad valley and a stream flowing down a low slope (Fig. 8L-7b). These two sections of the valley are connected by a series of cascades and a waterfall that provide a good example of how a stream deepens its valley and the adjacent landscape as it gradually cuts its channel lower. This lowering of the channel works its way upstream (see the description in Ch. 3). A 2.3 miles long loop trail from the overlook takes you down to the upper part of the cascade and waterfalls.

Beech Gap: (mp 422) Superb views of the vast expanse of the highland mountains to the southwest occur between Beech Gap and the highest point on the Parkway at milepost 431 (Fig. 8L-8).

Fig. 8L-7a. View down the drainage of the Yellowstone River from a road labeled Balsum Spring Gap. Note the low slope of the stream valley.

Fig. 8L-7b. View down the Yellowstone River valley near Graveyard Fields Overlook. Note the steep slope of the stream in this section. Waterfalls and cascades mark the transition of the low to the steep sloping sections of the stream.

Fig. 8L-8. Southwest view across the Nantahala Natl. Forest in the Highlands. Photo was taken near the highest point on the Parkway.

Highest Point of Parkway: (mp 431.4) At an elevation of 6,410 ft., this is the highest point on the Parkway. This is an excellent place to pause and reflect on what we see along this spectacular section of the Parkway. The outcrops show how the region was deformed at great depth in the Earth during the Paleozoic Era only to be uplifted to a high mountain range, one much higher than the present mountains as a result of repeated continental collisions. These uplifts were followed by millions of years of erosion that ultimately created a landscape that even today is being affected by more recent upward warping of the region that has caused streams to cut it down more rapidly. The evolution of the Blue Ridge is an ongoing process, one that may appear unchanging in terms of human lifespans, but that eventually alters the landscape.

Highly deformed gneisses and schists, generally known as migmatites are beautifully exposed at a number of overlooks along this part of the Parkway near Herrin Knob (milepost 424.4), Carney Fork (milepost 428), and Cowee Mtn. (milepost 430.7), which is one of the most popular places to take pictures showing the vast extent of the highland mountains to the southwest (Fig. 8L-8).

Tallulah Falls Fm: (mp 422) Excellent exposures of the Tallulah Falls Fm. occur at the intersection of SR 215 with the Parkway near milepost 422. You may collect samples outside the Parkway boundaries. Exit the Parkway to see the exposures south of the overpass (Fig. 8L-9). At this locality the Tallulah Falls Fm. consists of a chlorite-muscovite schist and metamorphosed greywacke containing large (>1 cm) black crystals of magnetite. At about 150 miles south of the overpass, look for thin layers of tourmaline schists (the tourmaline occurs as black prismatic crystals that are nearly triangular in cross-section) interlayered with quartz (Merschat, et al., 2012).

Fig. 8L-9. Exposure of the Ashe/Tallulah Falls metamorphic complex exposed at the intersection of the Parkway and SR 215.

Fig. 8L-10. A large recumbent fold in the Ashe/ Tallulah Falls Fm. at Cove Field Ridge Overlook.

Cove Field Ridge Overlook: (mp 439.4) The outcrop across the Parkway appears to have the form of a large recumbent anticline (with an axial plane that is subhorizontal) that formed as the rocks moved from southeast to northwest (Fig. 8L-10). The rocks are metamorphosed greywacke, a sandstone mixture containing feldspar and some dark minerals such as biotite and hornblende. This rock unit is part of a thick section of rock widely exposed in Great Smoky Mtns. National Park and bears the name Great Smoky Group. A number of intrusions cut across this exposure.

Standing Rock Overlook: (mp 441.4) The large block of rock at the west end of this overlook (Fig. 8L-11) fell down the mountainside and settled in this position. The rock is composed of complexly folded gneiss in which some dark layers have high concentrations of biotite and hornblende interlayered with layers rich in quartz and feldspar.

Balsam Gap: (mp 443.1) US 23/74 passes under the Parkway at this gap. Accommodations are available in Waynesville, N.C., eight miles to the northeast.

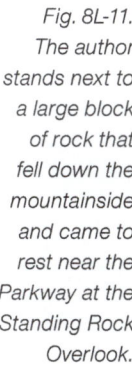

Fig. 8L-11. The author stands next to a large block of rock that fell down the mountainside and came to rest near the Parkway at the Standing Rock Overlook.

Hayesville Fault: (mp 446.2)
This important thrust fault, which has been traced all the way from the Grandfather Mtn. window, carries gneisses of the eastern Blue Ridge over rocks of the Great Smoky Group. The gneisses are exposed at Woodfin Valley Overlook and metagraywackes are exposed at the Woodfin Cascades Overlook at milepost 446.7. The fault is not exposed, but must lie between these two overlooks.

The USGS has published a geologic map of the Great Smoky Mtns. National Park in cooperation with the Great Smoky Mtns. Association. In addition to the park itself, the map covers the Parkway from its intersection with US 23/74 to Oconaluftee where it enters the park. This section shows the Hayesville and other faults in this western section of the Parkway.

Scott Creek Overlook: (mp 448.3)
Outcrop of amphibolite in gneiss.

Pegmatites along the Parkway: (mp 449) Excellent examples of large pegmatites, rocks produced from the last liquid residue to crystallize from magma, are exposed here.

Yellow Face Overlook: (mp 450.2)
A large pegmatite, nearly white in color cuts across a darker colored rock (a metamorphosed graywacke) on the rock exposure at this overlook.

Woolyback Overlook: (mp 452. 3)
Complex folds are nicely exposed in the cliffs across from this overlook (Figs. 8L-12a-b). The red stain on the rocks is an example of the effects of the oxidation of iron-bearing minerals in the rock.

Greenbrier Fault: (mp 455-456)
This major fault crosses the Parkway here. It is not exposed on the Parkway, but can be seen off of it. Precambrian gneisses of the basement complex are thrust onto rocks of the Ocoee Supergroup. Between this point and the intersection of the Parkway with US 441, most of the rocks exposed at overlooks are metasedimentary rocks of the Ocoee Supergroup, described in the following section.

Fig. 8L-12a. Folds in rocks of the Great Smoky Group at Wooly Back Overlook (above). In this photo the camera looks from the side, across the axis of the fold (roughly perpendicular to the axis).

Fig. 8L-12b. Right: In this photo the camera looks directly into the axis of the fold (parallel to the axis).

Fig. 8L-13. Dense deciduous forest grows along the French Broad River at Asheville. Note the change in vegetation from the edge of the river to farther upslope.

Fig. 8L-14. October leaf color reveals the variety of deciduous trees to be found at 4,000-5,000 ft. between the Parkway's highest point and Balsam Gap.

Ecological Setting

Asheville is located in a topographic basin surrounded by the Blue Ridge Highlands. If you take the Parkway south you will rise from elevations of 2,000 to 6,000 ft. As you rise from the level of the French Broad River at Asheville to the highest point along the Parkway at milepost 431 you'll see a dramatic change in ecologic conditions most obviously revealed by the change in the forests. At the level of the river pines, red and chestnut oaks, and hickories are present in a mixed forest of deciduous trees (Fig. 8L-13). As you drive higher the pines disappear and a northern hardwood forest dominates (Fig. 8L-14). Above 5,000 ft. an increasing number of spruce and fir trees make up the forest until they become the dominant species (Fig. 8L-15) (see Chs. 4 and 7 for more details).

Similar changes in the ecosystem may be seen driving from Brevard to the Parkway up US 276 along a route followed by monarch butterflies in September during their migration, or from Balsam Gap near Waynesville to the highest point on the Parkway. ❖

Fig. 8L-15. Spruce forest dominates the ecosystem at the elevation of the Devil's Courthouse near milepost 424. A few oak trees are here but rapidly increase in numbers downslope. The parking area at this overlook is at about 5,400 ft.

Chapter 8M

Great Smoky Mountains National Park

Above: Fig. 8M-1. View of the Great Smoky Mtns. from Chilhowee Mtn.

Location and Access: This national park lies along the boundary between Tenn. and N.C. From I-81 or I-40 go to Gatlinburg or Maryville, Tenn. The Sugarlands Visitor Center, located on US 441, a few miles southwest of Gatlinburg, is the main administrative center for the park. From Maryville, take US 321 toward Townsend. The Stops in this chapter begin west of the park on US 321. The eastern entrance to the park is located where US 441 joins the Blue Ridge Parkway at Oconaluftee Visitor Center. US 441 is the only highway through the park (Fig. 8M-8).

The Little River Road, SR 73 is a scenic route connecting the Sugarlands Visitor Center south of Gatlinburg with Townsend and with the Lauren Creek Road that leads into Cades Cove. SR 73 continues from Townsend to Maryville.

The Foothills Parkway runs along the western side of the park and connects with US 129 that follows the Little Tennessee River and the southern border of the park to the Fontana Dam.

Accommodations: There are campgrounds and restaurants in the park. Accommodations of all sorts abound in towns surrounding the park.

Trails: Refer to the National Geographic trail map and/or to one of the many trail guidebooks before starting out on long hikes, camping trips, or if hiking during late fall, winter, or early spring. Weather conditions can make trails dangerous.

Nearby Attractions: The Cherokee Indian Reservation is located at the eastern park entrance. Pigeon Forge, Gatlinburg, and Cades Cove attract throngs of tourists on the western side of the park.

The Great Smoky Mountains National Park covers a large part of an area known as the Great Smoky Mtns. located in the Highlands of North Carolina and Tennessee. This park has more visitors than any other national park. Many go to visit the great variety of tourist attractions and playgrounds at Gatlinburg, Pigeon Forge, and other centers around the park. Others go to see the spectacular landscape views, the great variety of birds, the thousands of plant species, and hundreds of animals species in the park, or to learn about the geological history that led to the creation of this extraordinary natural environment. The park is an important destination for all who want to learn about the southern part of the Blue Ridge.

History

Concern about the amount of logging that was taking place in the Blue Ridge led people in Tennessee (notably Mrs. W. P. Davis of Knoxville) and North Carolina to work toward establishment of a park in the area east of Knoxville and west of Asheville. These groups raised much of the money needed to purchase land for the park. The Rockefeller Foundation contributed $5 million and Congress gave $1 million when it authorized the park in 1926. The park was established in 1934, about the same time the Shenandoah National Park was being created near the northern end of the Blue Ridge. Soon after, a decision was made to connect the two parks with the Blue Ridge Parkway that loosely followed the Appalachian Trail along the high ridges. The Civilian Conservation Corps constructed much of the Parkway during the Great Depression of the 1930s, but the final section of the 469-mile long road around Grandfather Mtn. was not completed until 1983.

Champions of Wilderness: *Horace Kephart started out as an expert on early western explorations and was head of the St. Louis Mercantile Library. After a failed marriage he took refuge in the Great Smoky Mtns. wilderness, writing books on camping and highland culture. He became worried that the Smokies were being destroyed by clearcutting and began to write articles for the cause of saving the mountains as a national park. He was assisted in his efforts by George Masa. Born in Osaka, Japan, Masa visited the U.S. while studying mining in Asheville. He fell in love with the area and stayed to work at a local hotel where he learned photography. The two became friends and worked for many years together on promotional materials and maps for the proposed park and the Appalachian Trail. John D. Rockefeller Jr. donated his $5 milliion for park lands after seeing Masa's breathtaking photographs.*

Above: Photograph by George Masa, circa 1928-33. (Courtesy of Pack Memorial Library, Asheville, N.C.) Right: George Masa at work while writer Horace Kephart surveys the horizon. (Photo courtesy of George Ellison)

Fig. 8M-2. High levels of water vapor in the Great Smoky Mtns. causes fogs and clouds, inspiring the name "Smoky." These conditions are common throughout the highland region.

Ecological Setting

Local environments in the Great Smoky Mtns. are greatly influenced by the exaggerated topographic relief with elevations ranging from 875 ft. to 6,643 ft. From the base of the mountains to the top of the highest peaks, climate variations are similar to those found while traveling from Tennessee to eastern Canada, with plants and animals that are typical in the southern U.S. found at lower elevations and those common to the northern states found at higher elevations. Rainfall also varies with elevation, averaging from 55 in. per year in the valleys to 85 in. on the high peaks, feeding the roaring creeks and waterfalls and creating a moist environment that gives rise to the famous "smoky" mountain mists and an abundance of plant growth (Fig. 8M-2).

Plants

With the exception of the grassy and heath balds (the treeless areas found at mid to high elevations), 95% of the park is forested. Considered to be the most species-rich temperate forests in the eastern U.S., they have the largest stands of old growth forest in the Southern Appalachians—25% is old growth. The dramatic variation in elevation provides the setting for five different types of forests with 130 different tree species—roughly as many as in all of Europe. A description of the five types of forests found in the park follows (see Ch. 4 for more details).

Spruce-Fir Forests with red spruce and Fraser fir dominating the canopy have an understory of hobblebush and blackberries covering most of the high (over 4,500 ft.) mountaintops. These forests are present on the highest peaks that receive over 80 in. of rain each year and endure a colder climate similar to Maine and Canada due to the higher elevation.

Northern Hardwood Forests with American beech, yellow birch, and maples, prominent on the mid to upper mountain slopes (from 3,500-5,000 ft.). With their dramatic fall colors, these forests are similar to those found in the northeastern U.S.

Cove Hardwood Forests contain the greatest diversity of tree and shrub species, some with over 40 species. Most of these coves are sheltered from exposure to the sun and have thick, moist, fertile soil. They are commonly inhabited by magnolia, dogwood, basswood, and silverbells.

Hemlock Forests are still present in a few places even though they have been badly depleted and damaged by the woolly adelgid. Hemlocks prefer valleys with shady slopes, moist rich soils, and streams. Most are found below 4,000 ft. The largest of these forests are located in the southeastern corner of the park and areas west of Dellwood, N.C.

Cove hardwood forest along the Baxter Trail. (Photo: Miguel Vieria†)

Fire pink can often be found at the forest edge. (Photo: Jason Hollinger†)

Pine & Oak Forests are found on dry slopes and ridges exposed to the sun. These contain red, black, and chestnut oaks and table mountain, pitch, and white pines, as well as hickory trees. Because fires are needed to promote regeneration, the NPS may sometimes conduct controlled burns.

Air pollution is a serious concern in the park. The high mountains intercept the air coming in from the west or southwest where coal-fired power plants, heavy traffic, and a number of industries emit sulfur and nitrogen oxides that produce rain with an average pH of 4.5, sometimes 2.0—a strong enough acid to damage conifers in the spruce-fir forests and inhibit nutrient uptake (see pH chart on p. 64).

Pollution and woolly adelgid have damaged trees at Clingmans Dome.

Wildflowers Ecologists estimate that more than 1,500 flowering plants and 4,000 non-flowering plant species are present in the park, including 400 that are rare and over 200 that are endemic. Known to some as "The Wildflower Natl. Park," a plethora of blooming flowers grow here. Ephemerals such as bloodroot, crested dwarf iris, and an array of orchids bloom in the deciduous forests from mid-April to mid-May before tree leaves shade the forest floor. Later azaleas, mountain laurel, rhododendrons and other heath shrubs flower profusely, particularly on high elevation balds from mid-June to mid-July. The summer brings Turk's cap lilies, bee-balm and red cardinal flowers. These are followed by autumn goldenrod, sunflowers, coneflowers, and asters. There are also 450 species of non-flowering mosses, liverworts, and ferns in the park.

Animals

With such diverse forests, it is not surprising that the Great Smoky Mtns. are home to a great variety of animals, including notably, the symbol of the Smokies, the American black bear. The mascot

Black bear (USF&WS)

Fig. 8M-3. Black-chinned red salamanders (above left) breathe through their skin and are fairly common in the park at elevations up to 3,000 ft. (Photo: Athene Cuniculariaϯ) Eastern hellbenders (above right) are the largest salamanders in the park, growing up to 28 in., and live up to 25 yrs. in the cool, well-oxygenated streams of the Smokies. Intolerant of pollution, their populations in the Appalachians have waned everywhere except for in this well-protected park. (Photo: J.D. Kleopfer, Va. Dept. of Game and Inland Fisheries)

Little brown bat. Caves are now closed to the public as a precaution against the spread of white nose syndrome in bats. (Photo: Marvin Moriarty, USF&WS)

"Smokey the Bear," created in 1944 to warn the public about the dangers of forest fires is actually unrelated to the Great Smoky Mtns. Black bears in the park number around 1,500 and can best be spotted foraging in fields.

Mammals

Most commonly seen of the other 65 species of mammals in the park are white-tailed deer, groundhogs, chipmunks, squirrels, raccoons, opossums, skunks and bats. The deer mouse and white-footed mouse are the most common mammals here. Many beaver dams can be seen along the lower parts of creeks in the west and southwestern parts of the park. Also abundant is the eastern cottontail rabbit. Nocturnal coyotes, red foxes, and gray foxes live here but are rarely seen. The reclusive bobcat is likewise shy of humans. Although sightings of mountain lions by visitors have been reported, there has been no evidence of their presence (droppings or tracks) for about three decades. Another nocturnal creature, the Carolina northern flying squirrel is active at night in the conifer forests at higher elevations. Bison and grey wolves were once native but no longer exist here. Around 50 elk, the largest mammals here (roughly 700 lbs.) were reintroduced in 2001 and 2002. They can be seen at a distance during the early morning or late evening in the fields of the Cataloochee Valley. An attempt to reintroduce red wolves proved unsuccessful due to low reproduction and high pup mortality. However, northern river otters, reintroduced in the 1990s have successfully taken hold.

Birds

240 species of birds have been reported in the park. 60 of these are year-long residents. Around 120 species breed here, including 52 that fly in from South America. Others stop and forage in the area during seasonal migrations.

Amphibians & Reptiles

Known as the "salamander capital of the world," 31 species of moisture-loving salamanders can be found here, 24 of which are lungless (breathe through their skin) (Fig. 8M-3). There are also 13 species of frogs and toads. Among the reptiles, there are 8 species of turtles, 9 species of lizards and 23 species of snakes. Of the snakes, only two varieties, the northern copperhead and timber rattlesnake are poisonous. Happily, there have been no snakebite fatalities in the history of the park.

Geologic Setting

The park provides exceptional opportunities to examine the operation of the natural processes that are currently shaping the landscape of the Blue Ridge, notably rock weathering, downslope movement of materials, and erosion facilitated by frost action and the work of streams.

Answers to some of the most intriguing questions about the history of the Blue Ridge are revealed in and close to The Great Smoky Mtns. National Park in places of exceptional scenic beauty. Here we see clear evidence of the northwestward transport of the Blue Ridge, also that this movement took place late in the Paleozoic Era, as well as evidence that a huge amount of sediment accumulated in this region of the Blue Ridge late in the Precambrian, and evidence that the region was subjected to temperatures high enough to metamorphose older rocks during the middle of the Paleozoic Era. The roadside guide provides detailed information about Stops of special geological significance.

A few miles northwest of the park a long ridge, named Chilhowee Mtn. (Fig. 8M-4) rises above a valley floored by Paleozoic limestones, shales, and sandstones. Rocks of the Chilhowee Group crop out along the northwestern edge of this mountain. They lie on a thrust sheet above the major thrust fault named the Great Smoky fault that is exposed along the edge of the mountain. It is one of several major thrust faults that cut through rocks of Mississippian-age proving that the movement was post-Mississippian (Fig. 2-9). The Great Smoky fault continues many miles along the northwestern edge of the Blue Ridge both to the north-east and to the southwest.

Many thousands of feet, some estimate as much as a thickness of 9 miles of sandstone, shale, and their metamorphic equivalents, known as the Ocoee Supergroup, lie beneath the Chilhowee Group. This accumulation of sands, silts, and muds filled a huge basin that

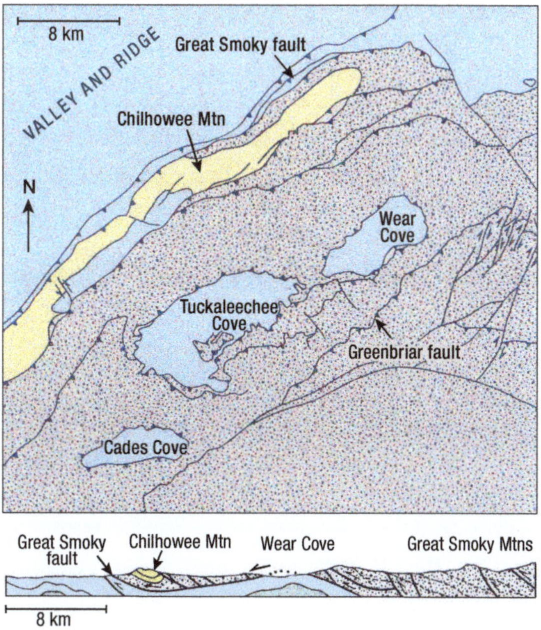

Fig. 8M-4. Geologic sketch map of the northwestern section of the Great Smoky Mtns. Natl. Park. Chilhowee Mtn. is shown in yellow. Precambrian-age metamorphic rocks (speckled lavender) have been thrust to the northwest over Paleozoic carbonate units (blue) that crop out in "windows" eroded through the thrust sheets. (After King, Neuman, and Hadley, 1968. More detailed maps of parts of this area can be found in Hatcher, et al., 2004.)

Rock Units in the Great Smoky Mtns. (In sequence with the youngest at the top)	
Chilhowee Group	Helenmode Fm., Hesse quartzite, Murray shale, Nebo quartzite Nichols shale, and the Cochran Fm. (Note that the units making up the Chilhowee Group in this area are different from those found in the Northern Blue Ridge, and described on p. 78.)
Ocoee Supergroup	**Walden Creek Group:** Sandsuck Fm., Wilhite Fm., Shields Fm., and Licklog Fm.
	Great Smoky Group: Unnamed sandstone, Anakeesta Fm., Thunderhead sandstone, and Elkmont sandstone. (Fig. 8M-6)
	Snowbird Group: Roaring Fork sandstone, Longarm quartzite., and Wading Branch Fm. (Fig. 8M-6)
~~ A major unconformity ~~~~~~~~~~~~~~~~~~~~~~~~~	
Blue Ridge basement complex	

Table 8M-1: The Chilhowee Group and Ocoee Supergroup are composed of a number of different formations and groups of rock units.

formed as Rodinia was breaking apart in the late Precambrian (Fig. 8M-5). It was a rift basin similar to the ones that formed in the Northern Blue Ridge about the same time (Fig. 5-6). The quantity of sediment in this rift basin indicates that land areas, almost certainly mountains containing sources of quartz and clay, were nearby (Fig. 8M-6). The orientation of cross-beds that formed in streams that carried the sediment suggests that these highlands were located off to the southeast of this basin. These sediments fed into the basin for hundreds of thousands—more likely millions of years. Several hundred million years later, the collision between Laurentia and Gondwana (p. 30) took place. At that time (late in the Paleozoic Era), thrust faults sliced through the Ocoee and Chilhowee creating huge slabs of rock that moved to the northwest as the mountains rose. Some of these faults, certainly the Smoky Mtn. thrust, were nearly horizontal. This is clearly shown in several of the windows located along the western edge of

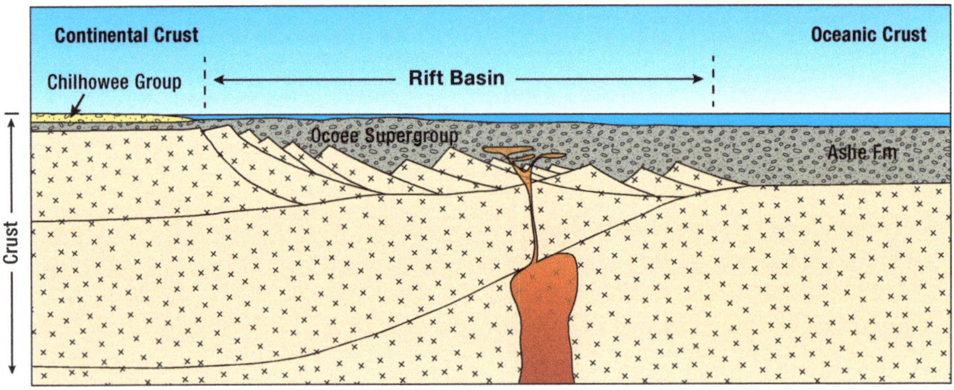

Fig. 8M-5. Schematic cross-section of a rift basin of the type in which the Ocoee Supergroup may have accumulated.

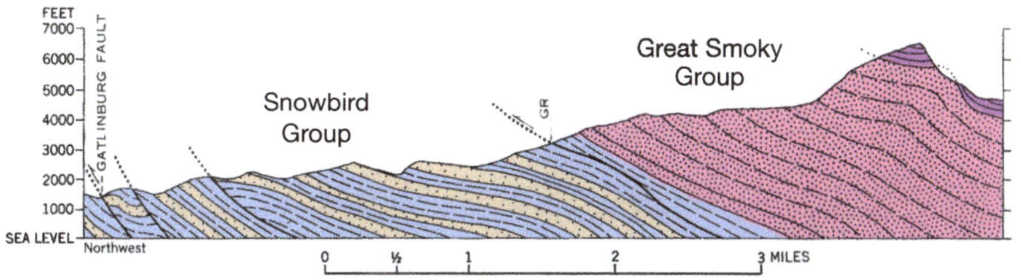

Fig. 8M-6. Cross-section of the Great Smoky Mtns. near US 441 showing a great thickness of sandstone in the Snowbird and Great Smoky groups. (From King, Neuman, and Hadley, 1968)

the park at Cades, Tuckaleechee, and Wear coves. Ordovician-age limestones are present in each of these topographic depressions referred to as coves. They are surrounded by a more resistant part of the Ocoee Supergroup and separated from the Ocoee by the Smoky Mtn. thrust (Fig. 8M-4).

Much older Precambrian basement rocks, gneisses, schists, and granitic rocks that lie unconformably below the Ocoee, are exposed along roads on the southeast side of the park in North Carolina at Bryson City, Ela, and to the north of Cherokee and Oconaluftee.

Events in the Geologic History of the Great Smoky Mountains	
After the Paleozoic (~200 Ma/Million years ago to present)	Streams and mass movements downslope slowly reduced the elevation of the high mountains. This has been accompanied by some slow uplift of the region described earlier (see Ch. 7).
Late Paleozoic (~265 to 200 Ma)	The modern Atlantic Ocean began to form as Pangea broke apart and as Gondwana (Africa) and North America began to drift apart. The Appalachians began to subside as the continental crust was stretched.
Late Paleozoic (~280 Ma)	The Alleghanian Orogeny during which Africa collided with North America affected this area. The older deformed rocks were moved more than 100 miles to the northwest over the Great Smoky fault system. The Great Smoky Mtns. probably reached their maximum height during this time, and may have been comparable to the Alps or even the Himalayas.
Mid-Paleozoic (~354 to 341 Ma)	The region was subjected to another phase of orogenic deformation, the Acadian Orogeny. Early formed faults were folded and the whole region was moved to the northwest over new thrust faults.
Late Precambrian to early Paleozoic (~700 to 440 Ma)	The region was subjected to deformation during the Taconic Orogeny. Folds and thrust faults formed as the region to the east was uplifted and moved toward the northwest. At depth (9 miles) metamorphic conditions reached garnet grade and rocks that contained magnesium and iron (e.g. basalt) were transformed into amphibolite.
Precambrian (~930 to 700 Ma)	Sediments that would become the Thunderhead sandstone were deposited in deep water (Fig. 8M-6). Carbon-rich muds containing pyrite (the Anakeesta Fm.) were deposited in a basin with poor circulation resulting in oxygen depletion. The presence of turbidites, boulders beds, slump, and other types of soft-sediment deformation, and rip-up clasts indicate the units were deposited in tectonically active areas. The activity probably resulted from the rifting of Laurentia as the continent of Rodinia split apart. Great thicknesses of sediment were deposited.

Table 8M-2. Events in the geological history of the Great Smoky Mtns. (Southworth, Schultz, and Denny, 2006)

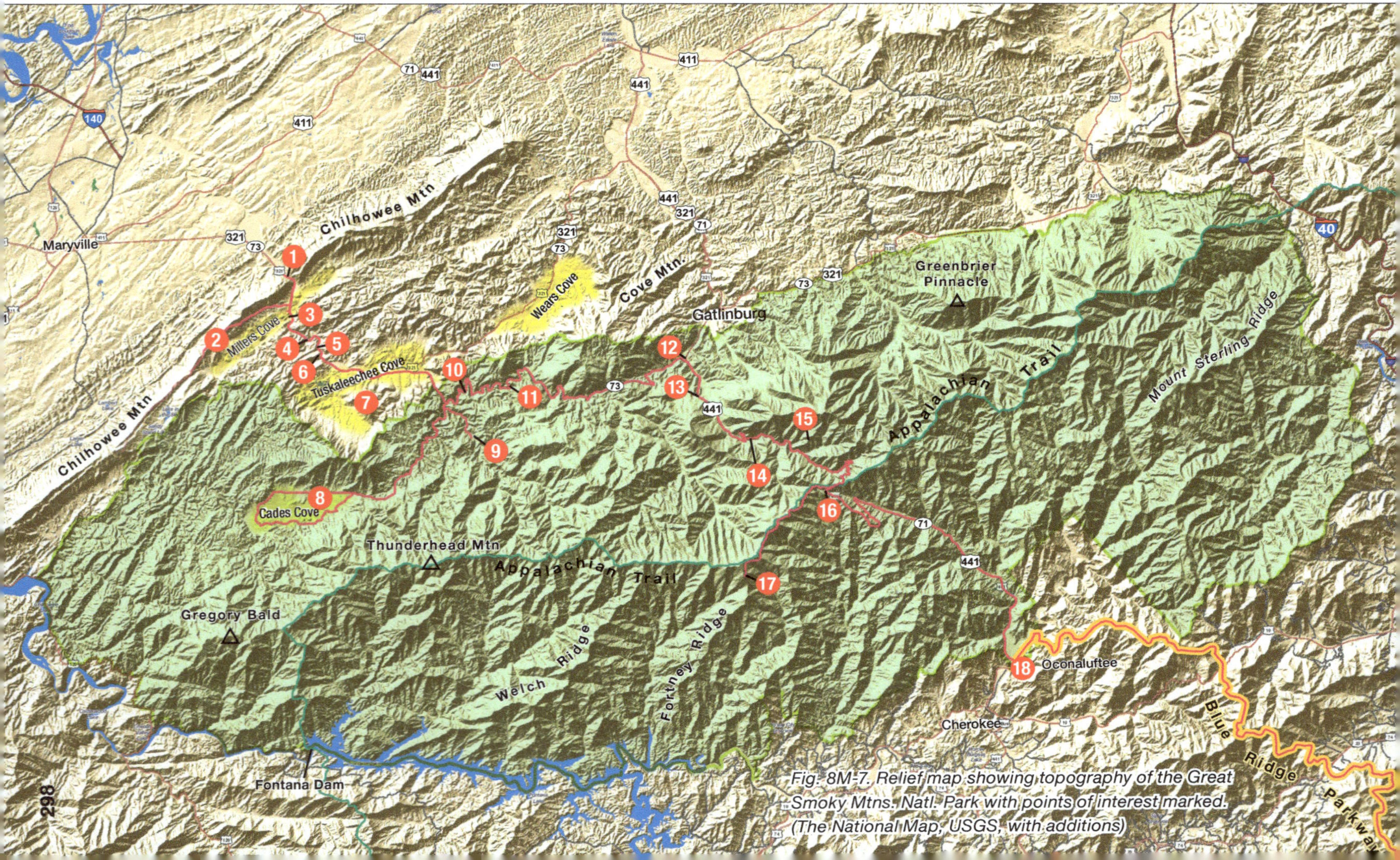

Fig. 8M-7. Relief map showing topography of the Great Smoky Mtns. Natl. Park with points of interest marked. (The National Map, USGS, with additions)

Points of Interest

Fig. 8M-8. Map of Gr. Smoky Mtns. Natl. Park. (NPS map, with additions)

1 - The Great Smoky fault
2 - Chilhowee Mtn.
3 - Millers Cove
4 - Wilhite cleavage
5 - Faults in the Wilhite Fm.
6 - Upper Great Smoky Group
7 - Tuckalechee, Cades, and Wear coves
8 - Cades Cove
9 - Gr. Smoky Institute at Tremont
10 - Little River
11 - The Sinks
12 - Sugarlands Visitor Center
13 - The Great Smoky Group
14 - Chimney Tops Overlook
15 - Alum Cave Bluffs
16 - Newfound Gap Overlook
17 - Clingmans Dome
18 - Eastern park entrance and Cherokee, N.C.

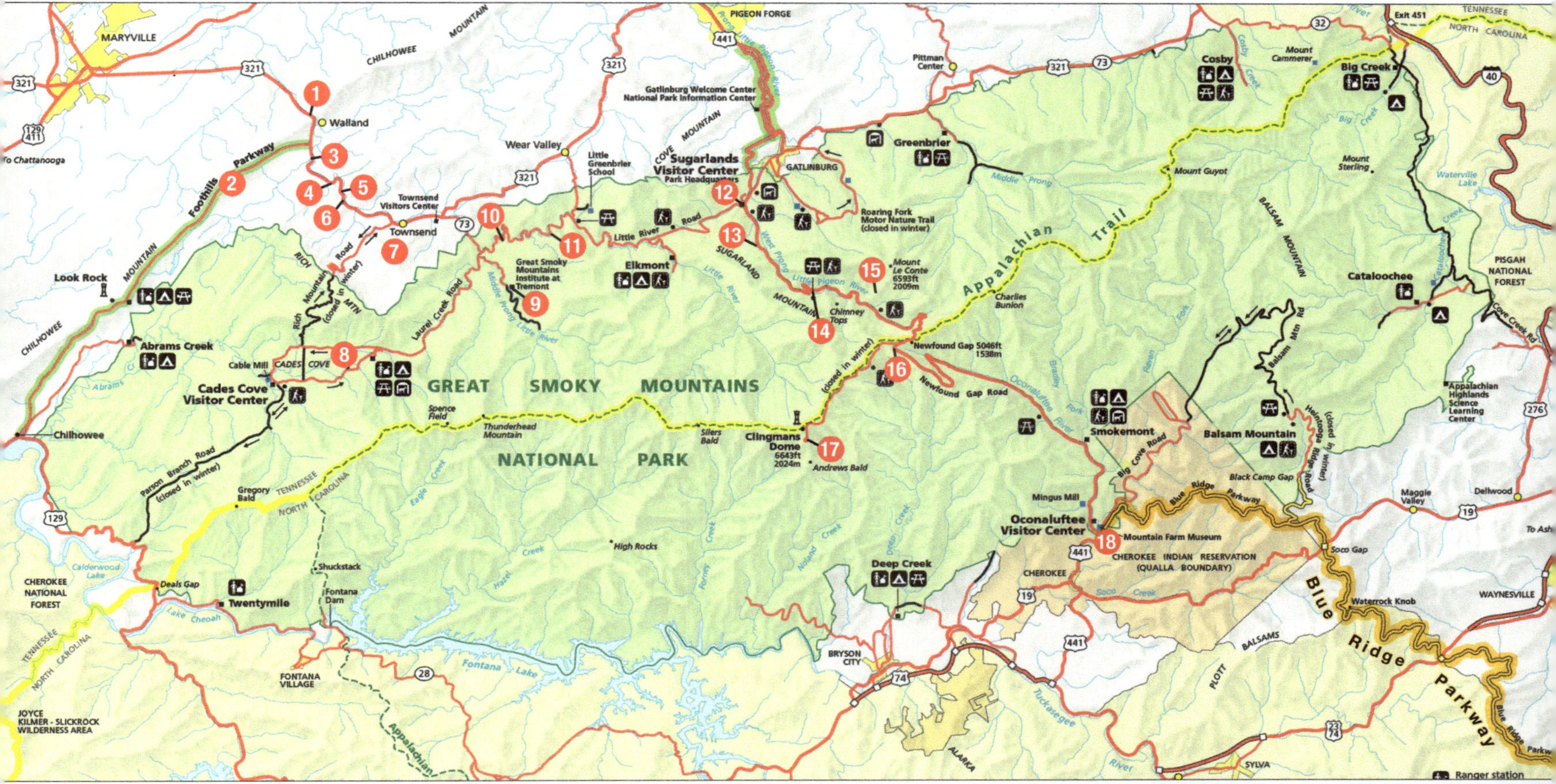

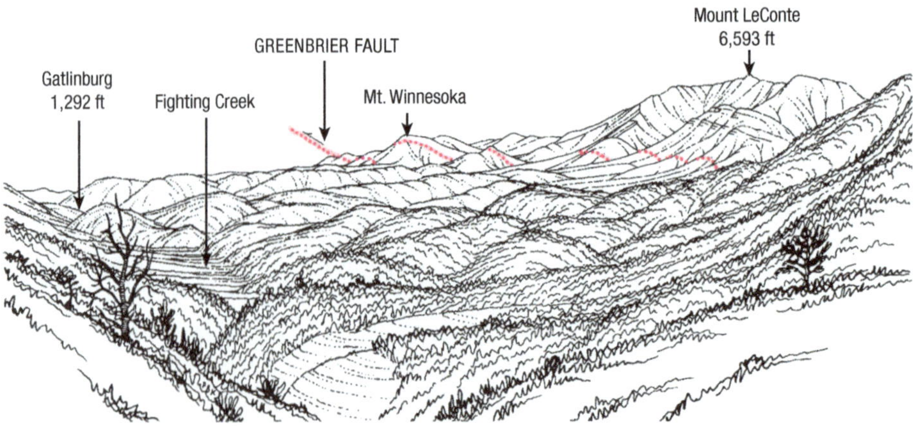

Fig. 8M-9. Sketch of the landscape of the western flank of the Great Smoky Mtns. from Gatlinburg to Mt. LeConte. (From King, Neuman, and Hadley, 1968)

Points of Interest
in Great Smoky Mtns. National Park

Visitors to the park most commonly approach it from the west, so the traverse suggested here begins in the northwest along US 321 from Maryville, Tenn.; has a side trip along Chilhowee Mtn.; continues on US 321 through Tuckaleechee Cove; takes a side trip along Laurel Creek Rd. into Cades Cove; takes another side trip to the Great Smoky Institute at Tremont; continues along Little River Road to the Sugarlands Visitor Center; and proceeds along US 441 to the junction with the Blue Ridge Parkway at Oconaluftee Visitor Center (Fig. 8M-8, -9).

Stop 1. **The Great Smoky Fault Northwest of Chilhowee Mountain:** Stop along US 321 west of the small town of Walland directly opposite a sign labeled "Bike Route 21/73." This Stop is located where the Great Smoky thrust fault carries lower Cambrian Chilhowee rocks over Ordovician-age rocks of the Sevier shale. A long concrete retaining wall separates the Sevier shale from the Chilhowee Fm. The thrust fault is exposed at the southeastern end of the retaining wall. Folds with steep plunges, evidence of brittle deformation, and sediments formed by submarine landslides (turbidites) showing graded bedding are among the interesting geological features exposed in the Sevier shale.

Fig. 8M-10. View across the large valley west of Chilhowee Mtn. that is underlain by Paleozoic-age limestones, shales, and sandstones.

Fig. 8M-11. Quartzites of the Hess Fm. (an equivalent of the Erwin farther north) are exposed near the entrance to the Foothills Parkway.

Fig. 8M-12. Chilhowee shales exposed along the Foothills Parkway.

Fig. 8M-13. View of Millers Cove Valley from the Foothills Parkway. This valley is underlain by carbonate rocks and is bounded by a thrust fault located near the far edge of the field.

Stop 2. **Chilhowee Mountain:** A prominent sign indicates the entrance to the Foothills Parkway along US 321. Many Stops along this road provide excellent views of the Valley and Ridge Province to the northwest (Fig. 8M-10), the Millers Cove Valley immediately southeast, and to the high peaks of the Great Smoky Mtns. farther to the southeast. Exposures of the quartzite (Hess Fm.) (Fig. 8M-11) of the upper part of the Chilhowee Group occur at many places along the Parkway. In a few places, sections of the mountainside, sometimes including part of the road, have slumped down the steep slopes that lead into Miller Cove. Excellent exposures of the shale units in the Chilhowee Group (Fig 8M-12) are present along the Parkway. Return to the Foothills Parkway entrance, turn right and continue along 321 toward Townsend.

Stop 3. **Millers Cove:** A thick section of dolomite, the Shady Fm., underlies Millers Cove. The flat ground (Fig. 8M-13) formed where the carbonate rocks are located. These dolomites have been much less resistant to erosion than the quartzites (Hess Fm.) that hold up Chilhowee Mtn. A similar topography is present in Cades Cove as well as in the other coves. A rare exposure of the Shady Fm. is present a short distance east of West Millers Cove Road. Another major thrust fault forms the southeastern edge of Millers Cove. This thrust brings the upper part of the Ocoee Group (the Wilhite Fm.) onto the Shady Fm. The thrust is not exposed. An outcrop of the Shady is exposed along US 321 about 0.3 miles from the Foothills Parkway entrance.

Fig. 8M-14. This fold in the Wilhite Fm. shows beautifully developed cleavage that is parallel with the axial plane of the fold.

Fig. 8M-15. These types of small-scale folds often form in thin-bedded rocks. These are in the Wilhite Fm.

Stop 4. **Wilhite Cleavage:** At the intersection of US 321 with Fencerail Gap Rd., park along Fencerail Gap Rd. Exposures along US 321 at this location provide an excellent opportunity to examine the slate of the Wilhite Fm. with its beautifully developed slaty cleavage that follows the folds (Fig. 8M-14). Parts of the Wilhite Fm. composed of thin-bedded units exhibit small-scale folds (Fig. 8M-15). Although the Wilhite Fm. is Precambrian in age radiometric dating of the unit indicates that it is only 460 million years old (Ordovician-age). This clearly indicates that the older rocks were metamorphosed in the Ordovician Period, a sign of tectonic activity in the region at that time.

Stop 5. **Faults in the Wilhite Fm:** Three miles from the Foothills Parkway entrance, park on the east side of the road in a large pullout. Excellent exposures here show a subhorizontal thrust fault that has been offset by a high-angle normal fault (Fig. 8M-16).

Fig. 8M-16. In this outcrop of the Wilhite Fm., a nearly horizontal thrust fault is cut by a younger, steeply inclined (dipping) normal fault.

Fig. 8M-17. Large pieces of black shale were ripped up from the bottom of a stream channel by currents that were also strong enough to move the large quartz pebbles. Good exposures of channels are present at this locality.

Fig. 8M-18. Cades Cove (Photo: Anthony Chavez†)

Stop 6. **Upper Great Smoky Group:** At 0.3-miles past Stop 5 park on the west side of the road. Pebble conglomerates, ancient stream channels, massive conglomerate beds, and graded beds are exposed at this locality (Fig. 8M-17).

Stop 7. **Tuckaleechee, Cades, and Wear Coves:** (Fig 8M-18) Nearly-flat ground underlain by fertile soil sets these three areas apart from the surrounding mountains. Bedrock in the coves is composed of Cambrian-age Shady Fm. carbonate rocks. They are surrounded by mountains composed of older metamorphic rocks of the Ocoee Supergroup that were thrust over the carbonate rocks. Erosion has cut through the thrust sheet exposing the rock beneath it in windows. The presence of limestone and dolomite in Tuckaleechee, Wear, and Cades coves accounts for the caves found in these areas. As with most other caves, they form from the dissolving of carbonate rocks by water that infiltrates into the ground and slowly dissolves the rock as it moves through pore spaces, especially fractures.

The thrust fault that surrounds these coves is poorly exposed, but can be seen in Tuckaleechee Cove along the road located on the north side of the creek, across the road from the Back Porch Restaurant. This fault is east of a limestone quarry on this road. (Parking is very limited here.)

Stop 8. **Cades Cove:** (Figs. 8M-4, -18) Cades Cove can be accessed by taking SR 73 southeast from Townsend to its intersection with Laurel Creek Road. Continue on Laurel Creek Road to Cades Cove, a popular tourist site in the national park. The presence of limestones of Ordovician-age in the floor of the cove (Fig. 8M-19) accounts for its low elevation relative to the surrounding mountains, and for the presence of caves in Cades and Tuckaleechee coves. These coves were among

Fig. 8M-19. The limestone that underlies Cades Cove has weathered to form a productive soil, which made this an excellent place for farming. The red line indicates the road that traces the edge of the cove and is close to the trace of the thrust fault that outlines the area underlain by limestone. (NPS)

The Cable Mill is a working grist mill in the historic area at Cades Cove. (Photo: Chris Totsky†)

White-tailed deer buck with full set of antlers. The open fields at Cades Cove offer good views of wildlife.

the few places settlers could sustain communities throughout the year. Weathering of the carbonate rocks produced a soil of good quality and made this a prime agricultural site for early settlers. American Indians may have been initially responsible for clearing the land. In any case, once it had been cleared it was kept open for cultivation and grazing of cattle. Grassy bald areas high in the mountains are the only other areas where grass-covered openings are found.

Accounts from early settlers make it clear that **grassy balds** were used for the grazing of cattle during summer months. So little land in the region is suitable for cultivation that settlers planted the rich soil in Cades and other coves and moved their cattle up the mountains to the grassy balds after the snow cover melted. Since the park was established in 1934 these grassy balds are no longer being grazed and so are now being invaded by blueberries and beech saplings. How these balds originated is not clear. It seems unlikely that they are natural features of the mountains and were probably created by the clearing of land. Some think they may have been areas that were originally burned by forest fires and later cleared and enlarged by settlers. Once established, the grassy balds became valuable assets. Several of these balds are still partially open and visited by hikers (Fig. 8M-20). A strenuous 9-mile hike on the Gregory Ridge Trail will take you to Gregory Bald. It is famous for the large number of azaleas that grow there. A trail starting at the visitor center in Cades Cove leads to balds at Gregory, Parson, and Russell fields.

Fig. 8M-20. Hemphill Bald is a grassy bald that is located along the Cataloochee Divide Trail. (Photo: Jeff Clark, jeff@internetbrothers.com)

From Cades Cove follow Laurel Creek Road and return to SR 73, which is known as the Little River Road.

Chestnut Top Trail: A spectacular variety of wildflowers can be found within the first half mile of this steadily rising trail that continues for 4.4 miles. The

Hairy beard-tongue blossoms (Photo: Fritz Flohr Reynolds†)

Bloodroot blossom

trailhead is located on US 73 about 100 yards north of the intersection with Laurel Creek Road. The trailhead is across the street from the parking lot. Among the early spring blossoms you will find yellow trillium, bloodroot, hepatica, and violets. Later, Jack-in-the-pulpit, white trillium, bishop's cap, purple phacelia, fire pink, star chickweed, and wild stonecrop appear. In the late spring/early summer hairy beard-tongue, rattlesnake hawkweed, and squawroot grow here. A mile up the trail the slopes become drier and there you will find laurel, pines, and oaks.

Stop 9. **The Great Smoky Institute at Tremont**, an educational center that places emphasis on the ecology and environment of the Great Smoky Mtns., is located in the park southeast of Townsend. To reach the Institute, take US 73 southeast into the park. Turn right at the junction of US 73 and Laurel Creek Road, which leads into Cades Cove. About a quarter of a mile along this road turn left onto Tremont Road (a gravel road). The Institute is about two miles southeast on this road. For those interested in a hike of moderate difficulty, continue to the end of Tremont Road and take the Middle Prong Trail that continues to Lynn Camp Prong Falls. The trail continues up steeper grades into the mountains. A second education center, the Appalachian Highland Science Learning Center has recently been established near the southeast corner of the park. Directions can be obtained from their website.

Stop 10. **Little River** is a beautiful stream (Fig. 8M-21) that meanders through mountains composed of various rock units that are parts of the Ocoee Supergroup (Table 8M-1). These units are folded and thrust faulted. A pullout called "The Sinks" is one of the best places to see this deformation.

Fig. 8M-21. View of the Little River.

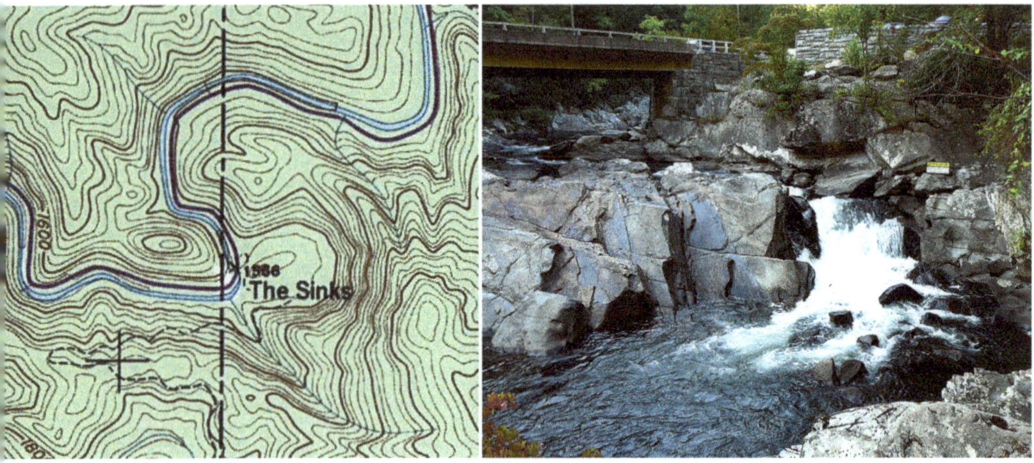

Fig. 8M-22. Topographic map of The Sinks area (left). (USGS) The stream has cut off a meander loop. This small waterfall is located where the Little River cuts through the meander loop (right).

Stop 11. **The Sinks:** Located near a bridge that crosses the Little River, "The Sinks" get their name from a drop in the slope of the stream (Fig. 8M-22). This was not caused by the effects of solution that produce sinkholes in carbonate rocks. Instead, the drop resulted from the cutting off of a meander loop. From the parking area at this pullout you can walk the Meigs Creek Trail that follows the previous course of the Little River.

All of the rocks at The Sinks are sandstones, part of an Ocoee unit known as the Thunderhead sandstone. The structure at this locality is deceptively complex. **Graded bedding** (in which larger and hence heavier rock grains tend to settle at the bottom of a sedimentary rock layer) in the sandstone (Fig. 8M-23) indicates that the Thunderhead sandstone has been turned upside down locally (not shown). Regional mapping in this area shows that the Thunderhead is part of a klippe (p. 53) that has been folded along the edge of another major thrust fault, the Greenbrier thrust fault that carried older Ocoee rocks (the Great Smoky Group) over younger Ocoee units (the Snowbird Grp.).

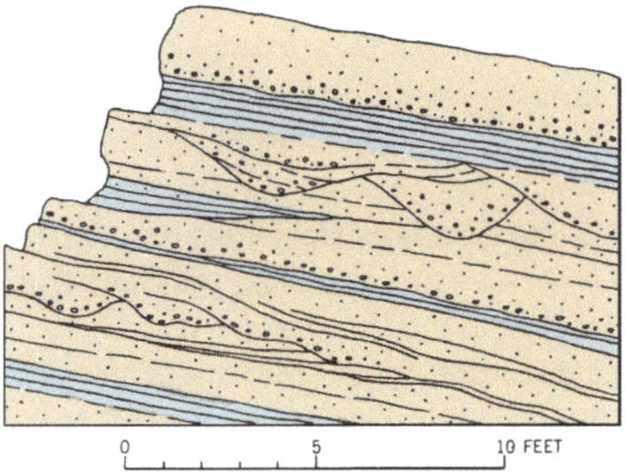

Fig. 8M-23. Channels, cross-bedding, and graded bedding are present in the Thunderhead sandstone, which is exposed at The Sinks and along US 441 south of Gatlinburg. (From King, Newman, and Hadley, 1968)

Stop 12. **Sugarlands Visitor Center:** The center is located at the intersection of Little River Road (SR 73) and US 441. Here you can visit the natural history museum, see a free 20-minute film about the park, speak with rangers and pick up maps or backcountry permits. Also on the premises are a bookstore, shop, restrooms, and soda machines.

Stop 13. **The Great Smoky Group:** Turn right at the intersection of the Little

River Road (SR 73) and US 441. For most of the distance from this intersection to Oconaluftee you will drive across the rock units of the Great Smoky Group, notably the Thunderhead sandstone and the Anakeesta Fm.

Stop 14. **Chimney Tops Overlook on US 441:** This overlook provides an excellent place to have a close look at the Thunderhead sandstone and a distant look at the Anakeesta Fm. which overlies the Thunderhead, and makes up the steep-sided craggy pinnacles called the Chimneys. Graded bedding and channeling is present in the feldspathic sandstone of the Thunderhead sandstone (Fig. 8M-26), exposed in roadcuts near this overlook.

An exceptional site for spring wildflowers is the easy 0.7-mile **Cove Hardwood Nature Trail**, a loop that starts at the Chimneys Picnic Area one mile before the overlook. The trail passes through old growth forests. Pick up the interpretive pamphlet at the trailhead.

Stop 15. **Alum Cave Bluffs:** The entrance to the Alum Cave Bluffs is located about eight miles south of Sugarlands Visitor Center along US 441. The trail is about 2.5 miles long with a climb of about 1,300 ft. The bluffs are composed of the Anakeesta Fm., which formed from mud and silt deposited in a deep marine basin. The presence of pyrite in the sediment indicates that the environment in which it formed contained very little oxygen. These rocks were folded and moved to the northwest during two periods of mountain building. One took place early in the Paleozoic Era and was followed by deformation during the late Paleozoic Alleghanian Orogeny. A number of unusual salt and sulphate minerals are present along the bluffs. This trail leads to the top of Mt. LeConte. An adjoining trail on the northwest side of the mountain passes by the 80 ft. high Rainbow Falls, one of the park's most popular waterfalls. It flows over the Thunderhead Fm.

Stop 16. **Newfound Gap Overlook:** The Anakeesta Fm. (Fig. 8M-24), the upper unit in the Great Smoky Group is exposed at this overlook. The unit is composed of slate, phyllite, or schist rocks that were metamorphosed from dark silty and clay-rich sedimentary rocks.

Fig. 8M-24. The Anakeesta Fm. seen here at Newfound Gap shows well-developed cleavage.

Fig. 8M-25. Small folds (called kinks) are beautifully developed in the Anakeesta Fm. phyllite.

Red berries are a prominent feature of the mountain ash tree during fall. These were found at Clingmans Dome.

Parts of the formation have an orange color caused by the oxidation of the iron sulphide mineral pyrite. Large boudins (sausage-shapes bodies formed by the stretching of the rocks) of sandstone are present in the slates and phyllites exposed at this Stop. Also note the small-scale kinks in the phylites (Fig. 8M-25). The lithologies of the Anakeesta and Thunderhead interfinger in other parts of the Great Smoky Mtns., and in some places the Anakeesta rocks are completely missing. The Anakeesta often has reddish stains produced by the oxidation of the mineral pyrite (an iron sulphide). The presence of pyrite in the rock indicates that it was deposited in a body of water containing little oxygen. This is often found in deep basins where the water does not circulate.

Hikers may want to walk the section of the **Appalachian Trail** from Newfound Gap to Clingmans Dome. You will pass through typical southern spruce-fir forests. The understory here is largely composed of rhododendron, blackberry, mountain cranberry, and witch-hobble. Ferns and several hundred varieties of mosses reflect the large amounts of moisture found in the high parts of the mountains.

Stop 17. **Clingmans Dome:** From Newfound Gap take the road leading to Clingmans Dome. The tower on Clingmans Dome affords sweeping views of the Great Smoky Mtns. On clear days you can see parts of seven states from the observation tower. The dome is underlain by the Thunderhead sandstone, a thick-bedded, coarse-to-fine grained metamorphosed sandstone and conglomerate (Fig. 8M-26). A close look

Fig. 8M-26. Massive outcrops of the Thunderhead sandstone at Clingmans Dome (left). A close-up of the Thunderhead sandstone (right) reveals blue quartz, a sign of metamorphism during which rutile minerals grow within the quartz grains.

Eastern American red fox with its thick winter coat. They are usually solitary hunters preferring small mammals such as rabbits but will also eat birds, fish, insects, fruits and berries. Their excellent eyesight is similar to cats, with vertically slit pupils. They tend to live in edge habitats with a diverse food supply and will burrow 1-8 ft. into hillsides, ditches and near tree trunks to create a den for their kits during breeding season.

Raccoon

at it will reveal blue quartz, a sign that the quartz has been subjected to high temperatures. A few thin beds of schist are present in the Thunderhead sandstone. These are deeply weathered producing a thick soil, called saprolite.

During the glacial advances, ice was present on the high peaks in the Great Smoky Mtns. At that time the accumulations of talus now found at these high elevations probably contained ice and moved as "rock glaciers." Fields of boulders composed of the Thunderhead sandstone are present in steep valleys on the flanks of many of the high peaks in the Smokies including Clingmans Dome.

Andrews Bald is the highest and the most accessible grassy bald in the park. The 1.8-mile moderate but rocky Forney Ridge Trail begins from the parking lot at Clingmans Dome and climbs 1,200 ft. You will pass through lush spruce and fir forests on your way tothe bald where you will find excellent views of Fontana Lake.

Fraser magnolia trees, of which the largest one known in the park is 120 ft. tall and 9 ft. in circumference, can be found on the Albright Grove Loop. To reach this trail, drive 15.5 miles east from Gatlinburg on US 321. Turn right onto Baxter Rd. just past the Yogi Campground. The trail is accessed from the Maddron Bald trailhead. (Photo: Richtid†)

Stop 18. **Eastern Entrance to the Park and Cherokee, NC:** There is a small museum at the Oconaluftee Ranger Station about two miles north of the entrance to the Blue Ridge Parkway. The town of Cherokee, a few miles south of the Blue Ridge Parkway entrance, is situated in the Cherokee Indian Reservation. The Museum of the Cherokee Indian is located here. Special events staged for tourists and the American Indian festivals held here are major attractions.

Southwest of the Great Smoky Mountains

Much of the Blue Ridge Geological Province continues south of the Great Smoky Mtns., across North Carolina and into Georgia and Alabama. The northern part of this area includes a large part of the Blue Ridge Highlands, but the Blue Ridge continues to the south as a region of low topographic relief that resembles the eastern part of the Northern Blue Ridge. As in the Northern Blue Ridge, millions of years of erosion have reduced the region to one of low relief that resembles the inner part of the Piedmont.

A number of guidebooks, most of which are written for professional geologists are available on the geology of North Carolina, Georgia, and Alabama (Tull, 2007 and Kish, et al., 1975). Contact the Geological Surveys of these states for book lists.

The Blue Ridge ends south of Atlanta and east of Birmingham where gently dipping sedimentary rocks of Mesozoic-age overlap the Precambrian and Paleozoic rocks of the Piedmont, Blue Ridge, and Valley and Ridge. This overlap took place late in the Mesozoic Era as the Atlantic Ocean opened and invaded the continental margin. How far it reached into the area now making up the Appalachian Mtns. remains unclear, but it clearly covered large parts of the Piedmont all along the Southern Appalachians from Pennsylvania to Alabama where it continued into the Mississippi Embayment, invading all of Louisiana and most of Mississippi and southeastern Arkansas. It is not hard to believe that deposits from this marine invasion covered large areas in the southern most section of the Blue Ridge Geological Province. Evidence from wells and geophysical surveys indicates that the Appalachians continued under this embayment and emerged again in western Arkansas as the Ouachita Mtns. Details of this connection remain shrouded beneath the rocks of the coastal plain and are known mainly through deep wells and geophysical studies made by oil companies. ❖

Young black bear

Part 3:
Identification Guides

Chapter 9

Rocks and Minerals

Commonly Found in the Blue Ridge

All three of the major rock types, igneous, sedimentary, and metamorphic, occur in the Blue Ridge. Almost all of the rocks in the Valley and Ridge and those exposed along the western edge of the Blue Ridge are sedimentary. Igneous and metamorphosed rocks make up the core of the Blue Ridge, and metasedimentary rocks are present along the eastern margin of the mountains. Most of these rocks are composed of a few common minerals, which are solid, naturally occurring inorganic elements or compounds that have an orderly internal structure and a characteristic chemical composition and physical properties. Although large, single crystals are shown in some of the photographs in this chapter, most of the minerals you will find in the mountains are in rocks and so small that you will need to use a magnifying glass to see them. Individual mineral grains in some rocks are so small they cannot be identified visually. ❖

Common Rock-Forming **Minerals**

Quartz crystals like these are rarely found.

Minerals found in igneous, metamorphic, and sedimentary rocks:

Quartz: SiO_2

Potash Feldspar: $KAlSi_3O_2$

Plagioclase Feldspar: $NaAlSi_3O_8 - CaAl_2Si_2O_8$ the ratio of Ca/Na varies

Amphibole Group (Hornblende): Complex Iron-magnesium silicate

Pyroxene Group (Augite): Complex Ca/Na/Fe/Mg/Al/Si

Mica Group (Muscovite (light), Biotite (dark)): Complex silicate minerals

Minerals found mainly in metamorphic rocks:

Chlorite: Complex Fe-Mg-Al silicate

Garnet: Ca-Mg-Fe-Al silicate

Andalusite: Al_2SiO_5

Staurolite: Fe-Mg-Al silicate

Sillimanite: Al_2SiO_5

All of the minerals listed above may occur in sediments and sedimentary rocks.

Quartz

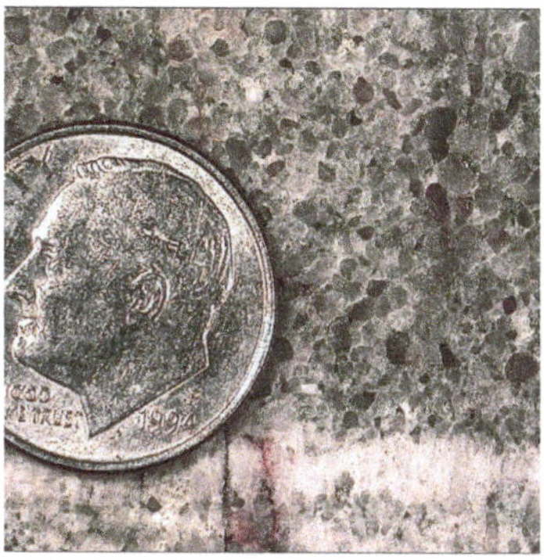

Quartz in quartzite rock
Quartzite is metamorphic. Above: quartz sandstone cemented by quartz cement.

Quartz vein
This milky white quartz is common in veins.

Quartz (silicon dioxide, SiO2) Quartz resembles glass and may be of any color but it is usually clear/translucent. It crystallizes from cooling molten rock (magma) and also forms in metamorphic rocks at high temp., but may also crystallize from silica-rich water circulating near ground surface. Hard and durable, quartz is nearly insoluble in water and highly resistant to chemical alteration. Thus quartz becomes a major component of sediments such as sand. It also occurs mixed with clay and other products of weathering. Some marine animals build their hard skeletal parts out of silica and occasionally silica is deposited from water, replacing the shells of invertebrate animals and wood in petrified wood. Silica also occurs as veins in all types of rocks and as cement in sediments composed of fragments. Rarely, quartz crystals line rock cracks where silica has precipitated from water moving through them. Quartz can scratch glass, distinguishing it from many other minerals.

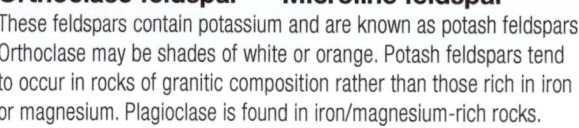

Orthoclase feldspar **Microline feldspar**
These feldspars contain potassium and are known as potash feldspars. Orthoclase may be shades of white or orange. Potash feldspars tend to occur in rocks of granitic composition rather than those rich in iron or magnesium. Plagioclase is found in iron/magnesium-rich rocks.

Plagioclase feldspar
These contain calcium and/or sodium. Note the striations found in plagioclase.

Feldspars are the most abundant minerals in the Earth's crust. Most feldspars originate by crystallization from magma or result from metamorphic processes. Unlike quartz, feldspars slowly decompose through chemical alteration to form clay minerals. For this reason they are not as common as quartz in sedimentary rocks formed in water. Some feldspars (orthoclase) contains potassium ($KAlSi_3O_8$) and is referred to as potash feldspar. It is usually pink in color. Plagioclase feldspars, which contain sodium and/or calcium are white or gray in color, and may exhibit striated surfaces.

Amphibole
Hornblende variety shown above. This black crystal is shiny because the mineral has near perfect mineral cleavage.

Pyroxene (Augite)
Augite, a common member of the pyroxene group, is usu. a dark greenish color. It is not a shiny black like hornblende and is not as prismatic in shape.

Muscovite mica (above) **and Biotite mica** (below)
These micas commonly occur as flakes in igneous and metamorphic rocks.

Pyroxene, amphibole, and olivine are silicate minerals that contain iron and magnesium. All of these minerals form in magmas or through metamorphism. Pyroxene (black to greenish black), and hornblende (a shiny black), are common in the Blue Ridge. Olivine (not shown), is a major constituent of Earth's mantle, but it is rare at the surface. Because olivine decomposes rapidly when exposed to water, it is not commonly found in sedimentary rocks. **Micas** are easily identified by their thin, sheet-like structure. Both black mica (biotite) and white mica (muscovite) form in igneous and metamorphic rocks.

Garnet crystals
Garnet crystals in a garnetiferous schist. Garnet often occurs as small crystals in metamorphic rocks and one of the most common varieties has a deep purple color.

Kyanite crystals
This mineral is found in high grade metamorphic rocks (Table 9-1). Note where these occur in the Blue Ridge Highlands (Fig. 7-13).

Staurolite crystals
Staurolite is found in high grade metamorphic rocks (Table 9-1). Twinned crystals such as the one illustrated here are rare. They can be found around the Philpott Reservoir (Ch. 8C).

Igneous Rocks

How they are formed: Igneous rocks form when molten rock, called magma cools and crystallizes. The rock bodies formed when this takes place inside the Earth are called **plutons**. When magma crystallizes at the surface the resulting rocks are called **volcanic rocks**. Cooling and crystallization that occurs beneath the ground surface takes place slowly, allowing time for crystals to form; but at the surface cooling is rapid so crystals have little time to form, resulting in rock that is very fine-grained. The texture of an igneous rock is a good indicator of the conditions under which molten material became a solid rock. Natural glass (obsidian), in which there are no crystals forms as a result of very rapid cooling. The size of grains generally increases with the length of the cooling process and ranges from grains so fine that they cannot be seen to very large crystals. When cooling takes place at two different rates the rock may become a porphyry. These rocks contain large crystals that formed slowly surrounded by a groundmass of material that cooled at a more rapid rate.

Rock types are defined on the basis of their mineral (or chemical) composition and texture. Table 9-1 shows minerals that are present in each rock type. For example, granite contains quartz, potash feldspar, and perhaps small amounts of biotite mica and hornblende. If the mineral grains are visible then the rock is called a granite. If the grains are too small to see then the rock would be called a rhyolite.

Lava eruption in Hawaii. The lavas of the Catoctin Fm. erupted in a similar way. (Photo: USGS Hawaiian Volcano Observatory)

Granite
Coarse-grained rock that cooled slowly, allowing feldspar and quartz to form large crystals. It is granitic in composition.

Granite Pegmatite
This pegmatite from Spruce Pine, N.C. contains quartz which has a gray glassy appearance and potash feldspar which is light in color.

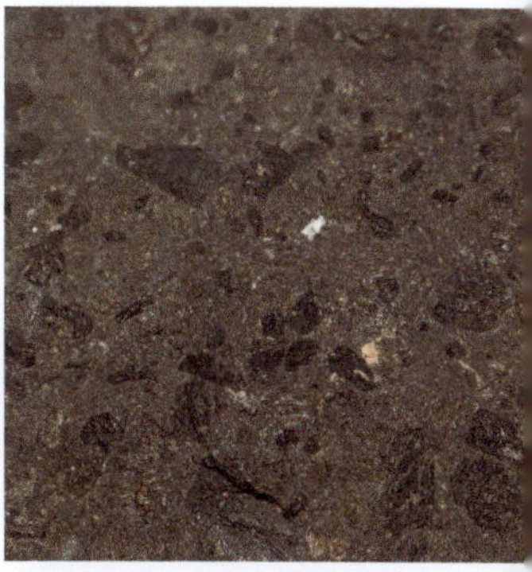

Andesite Porphyry
Large crystals (phenocrysts) are hornblende minerals formed while magma was still liquid. The groundmass is andesitic in composition.

Granite is a coarse- to medium-grained equigranular rock composed of potassium feldspars (usually orthoclase) and quartz, with small amounts of biotite mica and hornblende. Other minerals may be present but in small amounts. A granitic rock containing pyroxene formed under high temperature and pressure conditions, called charnockite, is present in many of the oldest rocks in the Blue Ridge. **Granite pegmatite** is a coarse-grained rock with granitic composition. A great variety of rare minerals may be present. Pegmatites are usually found near the borders of large plutons. The large crystals in pegmatites are thought to have formed from water-rich solutions present during the late stages of crystallination of plutons.

Diorite
The high percentage of dark minerals (hornblende) combined with quartz gives this rock a composition of diorite.

Gabbro
This gabbro is very coarse-grained which indicates very slow cooling. It probably formed deep in the continental crust or in the upper mantle.

Eclogite
Forms under high pressure below 19 miles and temp. above 300ºC, mainly in subduction zones when basaltic rocks are carried to great depths.

Diorite is a coarse- to medium-grained rock composed of plagioclase and biotite, hornblende, or pyroxene. Small quantities of quartz may be present. The proportion of dark minerals (11-36% of the total) distinguishes it from granite. **Gabbro** is an equigranular rock composed of coarse- to medium-grained pyroxene, hornblende, and biotite which give it a dark color. Olivine may be present; quartz is absent. Pegmatites of gabbroic composition, probably form from residual patches of water-rich magma.
Peridotite, a dark, coarse-grained, equigranular rock containing a lot of olivine and/or pyroxene or hornblende, but no quartz or feldspar, is an important part of Earth's mantle, but rare in the Blue Ridge.

Dunite
The vaguely greenish color of this rock is caused by olivine.

Basalt
Dark colored, very fine-grained rock. As flows cool the rock contracts and may form roughly shaped columns.

Rhyolite
Rhyolite exhibiting flow structure.

Felsite (rhyolite) is a general term applied to igneous rocks of fine-grained texture and light color indicating a granitic composition. If large crystals are present in a felsitic groundmass, it is called a felsite porphyry. Felsites range in color from white, pink, light green, and yellow to purple and brown.
Basalt is a dark-colored, fine-grained rock composed primarily of plagioclase and pyroxene. Basalts are dark because they contain large percentages of pyroxene, biotite, olivine, or other dark minerals. It is often an extrusive igneous rock and commonly contains cavities formed by bubbles of gas. Basalt dikes are present in the Blue Ridge, but most basalts have been metamorphosed to form greenstone.

Unakite

A rare rock containing quartz (gray), potash feldspar (pink) and epidote (green) caused by the hydrothermal alteration of granitic rocks.

Rhyolite Porphyry

The dark colored pieces contain large crystals, called phenocrysts. They are composed of feldspar.

Pegmatite

Coarse-grained with minerals that crystallized late in the cooling of igneous intrusions. This one from near the Philpott Reservoir, Va., contains mica w/ shiny black hornblende crystals.

Sedimentary Rocks

How they are formed: Most sedimentary rocks form in **layers** of sediment called **beds**. Sedimentary bedding is formed as loose fragments of materials that have been weathered out from preexisting rocks settle out from water or air and become cemented, forming a solid. Rocks that are formed by precipitation from surface or near-surface solutions, as well as from the bones and shells of animals, decomposing plant matter, or as a result of other types of organic activity are all considered sedimentary.

Fragments of older rock make up the bulk of sedimentary rocks. The general terms **clastic** (pyroclastic in the case of volcanic rocks) and **fragmental** apply to such materials. Clastic materials are subdivided on the basis of the size and shape of the fragments. The size categories and names associated with them are: **clay** (smaller than 1/256mm), **silt** (1/256-1/16 mm), **sand** (1/16-2 mm), **granules** (2-4 mm), **pebbles** (4-64 mm), **cobbles** (64-256 mm), and **boulders** (larger than 256 mm).

Sandstone showing cross-bedding which forms when the sand is deposited by a stream or wind current. Also shown are worm tracks often found in the Antietam (Erwin) Fm.

Ripples and worm tracks left in the once soft sediments of an ancient aquatic environment are evident in this sandstone.

Conglomerate
This conglomerate contains rounded fragments of Precambrian rock deposited by streams at the base of the Unicoi Fm.

Pebble Conglomerate
This pebble conglomerate contains some irregularly shaped shale fragments (black) that were ripped up by strong currents where the conglomerate was deposited.

Breccia
Breccia is a rock containing angular rock fragments generally of mixed sizes larger than 2mm.

Conglomerates consist of rounded rock fragments of usually mixed sizes in excess of 2 mm. Most conglomerates form where stream gravel or beach sands settle and are deposited—places where rock fragments have become rounded and eroded from constant movement. **Breccia** is a rock containing angular rock fragments generally of mixed sizes in excess of 2mm. Most breccias form along faults or where rock masses collapse as in caves or along cliff faces. **Sandstone** is a general name for clastic rocks composed of fragments that fall in the size range of 2 mm to 1/16 mm. In this range individual grains are clearly visible. Many types of sandstone consist of grains of pure quartz, but sandstone can be of any composition. Most sandstones in the Blue Ridge are quartz sandstones that originated as beach or shallow water marine deposits.

Siltstone

Shale
Siltstone differs from shale in the slightly larger size of the fragments of which it is composed. Shale is composed of exceedingly fine grains less than 1/256 mm in diameter. Siltstone feels like a very fine sandpaper while shale is smooth. Often fine-grained sediments contain a mixture of sizes.

Tuff (ash)
This water-laid ash was deposited in the area near Mt. Rogers, Va. about 700 million years ago.

Siltstone is a general name for clastic rocks composed of fragments occurring in a size range of 1/16 mm to 1/256 mm in diameter, the size of very fine sandpaper. Mixtures of silt, clay, and sand are common. **Shale** consists of particles less than 1/256 mm in diameter. Clay is the predominant constituent, but quartz and limestone may be present. Shale is formed as a result of the compaction and removal of water from clays. It is a sediment with the texture of peanut butter, deposited in marine environments often far from the original source of the clay. **Tuff** is solidified volcanic ash. The ash settles out of the air but may fall into water and then settle out from the water. Fine ash may travel great distances in the air before it falls.

Metamorphic Rocks

Metamorphism is the process by which solid rock undergoes changes in mineralogy and texture that occurs in response to changes in temperature, pressure, and the presence of fluids, notably water. If the temperature rises high enough then rocks melt and magma forms. But if the conditions are not that extreme then the rock gradually undergoes various changes, short of melting. These changes may occur close to igneous intrusions or on a regional scale as a result of crustal subsidence and burial especially during mountain building.

How they are formed: During metamorphism, the texture and mineral composition of rocks changes because some of the minerals in the rock are chemically or physically unstable under the new temperature and pressure conditions. New minerals may form by a recombination of elements or by re-crystallization of the original minerals. The fabric of the rock often changes as new crystals grow. Minerals in the original rock that are stable under the temp. and pressure that prevails during metamorphism may survive and even grow to form larger crystals. But minerals that are unstable under the new conditions undergo chemical reactions that establish a new assemblage of minerals that *is* stable. The composition of the minerals in rocks that undergo metamorphism determines what minerals can form. If the original is composed of quartz (e.g. quartz sandstone, quartz siltstone, vein quartz, or quartzite), the resulting metamorphic rock will also be composed of quartz but called quartzite. In a similar manner, metamorphism of limestones that are composed of calcite produces marble containing enlarged crystals of calcite. In contrast, rocks composed of minerals that contain many elements (e.g. clay) may undergo many changes in mineral composition during metamorphism. Which new minerals form during metamorphism depends on temp. and pressure conditions as well as on the chemical composition of the original rock. Certain key minerals are used as indicators of what is called the **grade of metamorphism**. These key minerals are chlorite, biotite, garnet, andalucite, and sillimanite (p. 314). Geologists also use a concept known as **metamorphic facies** to note the degree to which rocks have been metamorphosed. A **facies** is characterized by the assemblage of minerals formed when rocks containing a certain bulk chemical composition are metamorphosed under a particular set of temp., pressure, and water content conditions. From low to high the facies are termed greenschist, amphibolite, and granulite (p. 314). The granulite facies found in parts of the Blue Ridge basement forms at great depth in the crust at temps. in excess of 650° C.

New minerals that grow in metamorphic rocks often exhibit near perfect crystal form but if the rock is being deformed during metamorphism some of these new minerals such as chlorite, mica, or hornblende may grow in such a way that their flat or elongated shape is nearly perpendicular to the directed pressure. This produces an alignment of platy or flakey minerals like mica or elongate minerals such as hornblende. In some metamorphic rocks, these alignments, called foliation, are so well developed that the rock breaks along the planes in which minerals are aligned. Slate, phyllite, and schists all exhibit strong mineral alignment. Gneisses contain thin bands in which different minerals concentrate.

Original rock	After rock has been metamorphosed
Quartz sandstone	Quartzite
Limestone	Marble
Shale	**Slate** (metamorphosed under high directed pressure) **Phyllite** (metamorphosed at low temperature) **Schist** (metamorphosed at medium temperature) **Gneiss** (metamorphosed at high temperature)
Volcanic ash (tuff), Basalt or Gabbro	If the tuff is rich in magnesium and iron, it may become **Amphibolite Greenstone** (metamorphosed at low temp.) or **Amphibolite** (metamorphosed at med. to high temp.)

Prefix: "Meta-"
Igneous or sedimentary rocks that have been metamorphosed are commonly referred to with the prefix "meta-" (e.g. metabasalt, metasedimentary, metagraywacke.

Slate

Phyllite

Schist

Slate is a fine-grained rock characterized by a tendency to break along nearly perfectly parallel foliations, called slaty cleavage. Slate is derived from shale. During metamorphism the clay minerals are altered to small strongly aligned micaeous minerals. **Phyllite** is a fine-grained rock characterized by a lustrous, silky sheen caused by light reflected from the chlorite and muscovite micas of which it is composed. Quartz and plagioclase feldspar are often present. The grain size of phyllites is larger than that of slates, but finer than that of schists. Phyllites are usually greenish or red and may show the initial stages of segregation of some mineral constituents into layers. **Schist** is a foliated rock (with minerals that are aligned) of medium to coarse crystalline texture. Unlike phyllites, mineral constituents of most schists exhibit nearly parallel alignment of minerals, esp. of micas. Quartz, feldspars, and micas commonly occur in schists. If one of the constituents makes up 50% or more of the rock, its name is attached as a modifier (for example, mica schist, quartz schist, or hornblende schist). If no constituent comprises 50%, the names of the two most abundant constituents may be used (for example, garnetiferous-mica schist).

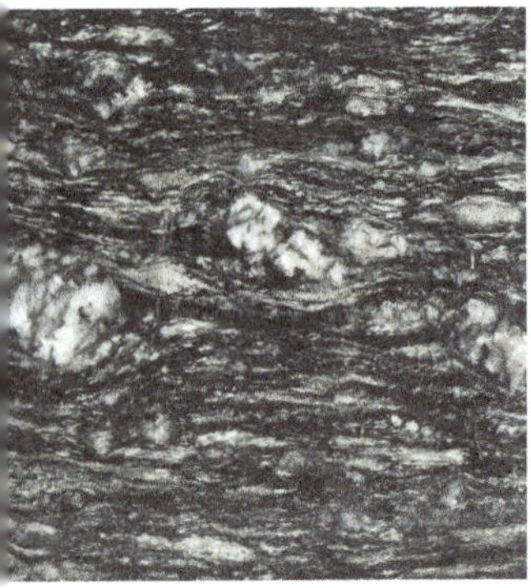

Augen Gneiss
Contains large feldspar crystals (white) that formed as the original rock was recrystallized. Often the feldspar crystals show signs of having been rotated.

Granitic Gneiss
Composed mainly of quartz, feldspar, biotite mica and hornblende. It was folded under high temperature and pressure conditions, but it was not as intensely deformed as the mylonites shown on pp. 153, 266, and 321.

Amphibolite Gneiss
Amphibolite with quartz veins.

Gneiss is a medium to coarse-grained metamorphic rock that exhibits compositional layering that produces a banded appearance. Most gneisses contain quartz and feldspar interlayered with thin layers rich in hornblende or mica. The quartz, feldspar, and other constituents usually have interlocking boundaries. The layering is thought to develop as a result of segregation of mineral constituents during metamorphism. Gneiss that contains large eye-shaped crystals (usually feldspar) is called augen gneiss.

Amphibolite

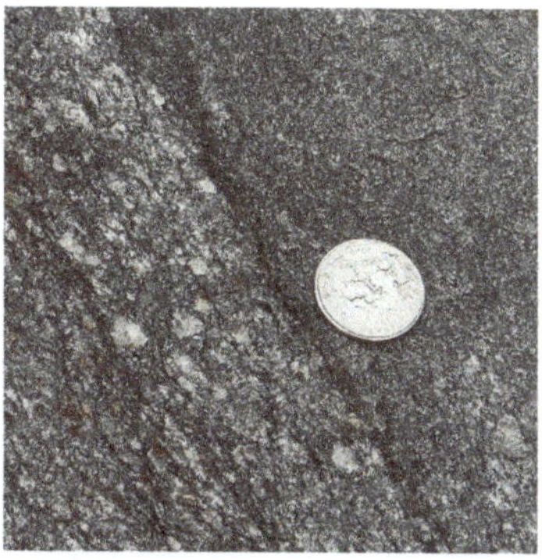

Quartzite
Coarse-grained (granule) quartzite.

Marble
Marble is a recrystallized limestone.

Amphibolite is composed of plagioclase, hornblende, and biotite. Prismatic hornblende crystals may be aligned. Quartz-free amphibolite is usually derived from basalt or iron and magnesium-rich volcanic tuffs. The presence of quartz suggests the amphibolite is derived from sedimentary rocks. **Marble** is composed of the mineral calcite. Marble forms as a result of the metamorphism of carbonate rocks such as limestone or dolostone. The grain size in these sedimentary rocks is often so small that one cannot see individual crystals. In contrast, the crystals in marble are usually visible to the unaided eye. Impurities may give marble a distinct foliation, but generally the texture consists of an interlocking mosaic of crystals. Marbles occur in a few places in the Blue Ridge, less often than schists and gneisses. **Quartzite** is a quartz-rich rock in which the grains of sand are either fused together at high temperatures (metaquartzite) or so tightly cemented by quartz that they resemble metaquartzites, in that both often break across sand grains.

Greenstone
Greenish colored, fine-grained igneous rocks that form as a result of low grade metamorphism of basaltic rocks. They are one of the main components of the Catoctin Fm. (see Ch. 6A).

Catoctin Greenstone
A variety of samples are shown above. Greenstone occurs in a variety of colors depending on the degree to which it has been altered by chemical weathering.

Mylonite
From Rockfish Valley ductile deformation zone. Texture forms when rocks with high percentage of quartz and feldspar are strongly deformed at high temp. and pressure. Minerals are drawn out and recrystallized.

Greenstone is a fine-grained greenish rock that is produced by the metamorphism of basalt. Most of the greenstones in the Blue Ridge were extruded in the later part of the Precambrian and were metamorphosed during early the part of the Paleozoic. Some of the greenstones contain rounded cavities that were originally bubbles in the lava. Most of them have been later filled with quartz or other minerals deposited from solutions circulating through the lava.

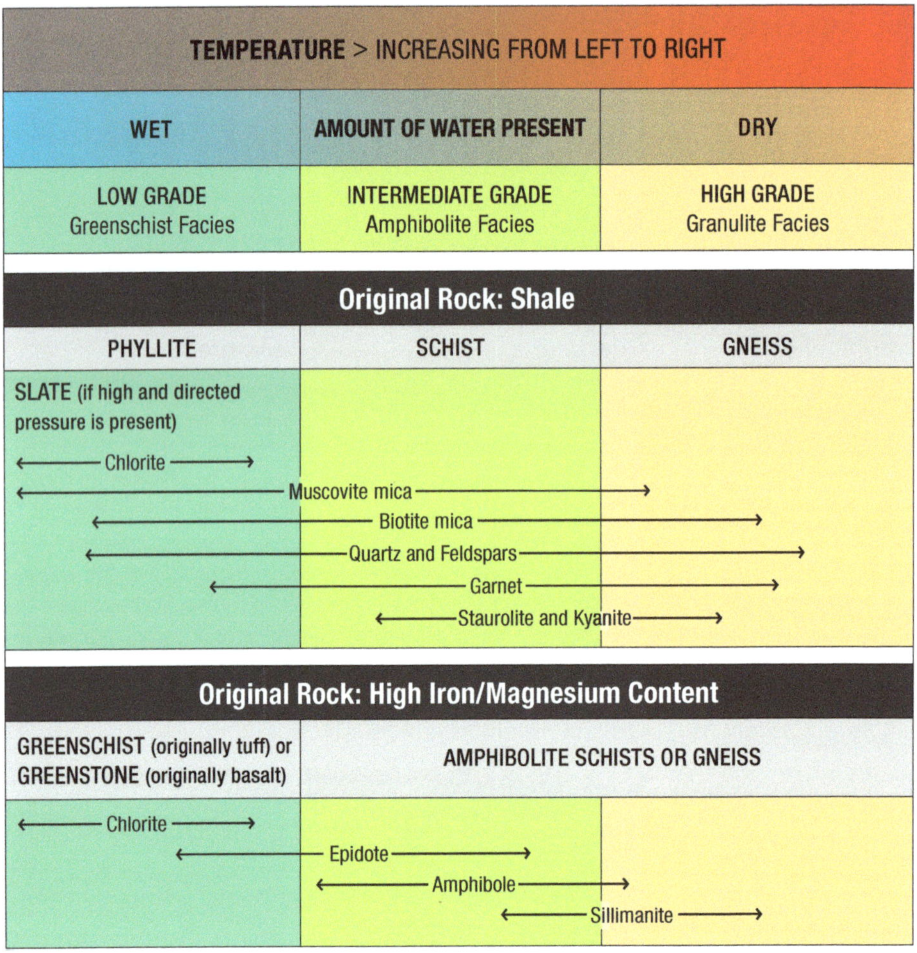

Table 9-1: The assemblage of minerals present in a metamorphic rock depends on the original composition and the conditions under which metamorphism takes place. The conditions (expressed in terms of temperature and amount of water present) are indicated at the top of this chart. Two types of original rock composition are shown: 1) shale and 2) rocks with high iron-magnesium content. The minerals formed under various conditions for each type are indicated. For example, if a shale is metamorphosed under low grade conditions (with high moisture and low temperatures) the resulting rock should be a phyllite composed of chlorite, mica, quartz and feldspar, and possibly some garnet. However, if the metamorphism moves into a higher grade condition (dry with high temperatures), then staurolite and kyanite will form, but not chlorite.

Chapter 10

Plants

Commonly Found in the Blue Ridge

This chapter on plant identification is divided into separate sections for deciduous trees, conifers, blossoms (by color), ferns, ground covers, vines, and mosses. Fungi (including mushrooms and lichen), although not true plants due to lack of photosynthesis, are also covered. Photos, latin names, seasonal growth periods and brief descriptions with key distinguishing features are provided for each entry.

For more on the environments of the Blue Ridge see Ch. 4 and the overviews of the ecological communities in the northern and southern sections of the Blue Ridge in Chs. 6 and 8.

The Blue Ridge contains one of the largest and most diverse plant populations in the world, matched only by parts of central China, Japan, and the Caucasus Mtns. This biodiversity results from its location in the mid-latitudes, its wide ranging elevations, diversity of rock and soil types, abundant rainfall, and an array of topographic configurations resulting in large variations in the amount of solar radiation reaching the ground. These diverse features have given rise to over 4,000 species of plants, including 2,000 fungi, over 1,400 flowering plants, 158 trees, and 500 mosses and lichens. Due to space considerations only those plants most common to the Blue Ridge Mtns. are included here. For those seeking a more comprehensive approach, a list of recommendations for guidebooks follows. For trees, the *National Geographic Pocket Guide to Trees & Shrubs of North America* by Crowder and *The Sibley Guide to Trees* are recommended. Wildflowers of the Blue Ridge are the focus of many excellent guidebooks by Gupton & Swope, Adkins & Cook, and Alderman. For more detailed information on mosses see *Common Mosses of the Northeast and Appalachians* by McKnight & Rohrer. Trusted publishers for plant identification guidebooks include National Geographic, Audubon, and Peterson.

Among the best websites for plant identification are the USDA Natural Resources Conservation Service's Plants Database at **http://plants.usda.gov/**, the Lady Bird Johnson Wildflower Center at University of Texas at **http://www.wildflower.org/plants/**, and the Southeastern Flora Plant Identification Resource at **http://www.southeasternflora.com**. All are excellent searchable plant databases.

Plants by Category: PAGE

Trees: Deciduous 327
Conifers 331
Tree Bark 333
Blossoms by color 334
Ferns 347
Ground covers 348
Vines 350
Mosses 351
Fungi: Mushrooms 352
Lichen 353

Plant Fundamentals

The plant kingdom is made up of organisms that use photosynthesis to convert solar energy, water, carbon dioxide, and nutrients into chemical energy (sugar) and oxygen. Plant cells, like human cells are eukaryotic; composed of a nucleus and organelles, each contained within membranes. The outermost plant cell membrane is enclosed by a wall composed of somewhat rigid cellulose. All plants fall into one of two basic structural categories:

Vascular plants, called the "higher" land plants, have special tissues called **xylem** and **phloem** that act like pipes to conduct water, minerals, and photosynthetic products through the plant. Xylem cell walls are partially composed of **lignin**, an organic polymer that lends extra stiffness (woodiness) to cellulose walls, giving them the strength and rigidity to grow to large sizes and the ability to avoid rot. Phloem transports photosynthetic products. Vascular plants include angiosperms (flowering plants), gymnosperms (naked seeded plants), and also ferns and clubmosses, which reproduce by minute spores.

Nonvascular plants, sometimes called the "lower" plants, lack xylem and phloem tissues characteristic of vascular plants and hence have no true roots, stems, or leaves. They are small and simple when terrestrial, but can be quite large when aquatic (e.g., marine kelps). Lacking xylem and phloem, nonvascular plant cells must be quite close to water to avoid desiccation. They form two main groups: **bryophytes**, including mosses, liverworts and hornworts; and diverse lineages of **algae**. Nonvascular plants, along with bacteria and microorganisms, are often among the first "pioneer species" to inhabit inhospitable environments.

Plant Reproduction

Vascular plants reproduce sexually in one of two ways. **Angiosperms**, the dominant plant form on land, reproduce sexually by flowers, usually relying on insects and birds for pollination. Once mature, the seeds of angiosperms are always enclosed in a fruit.

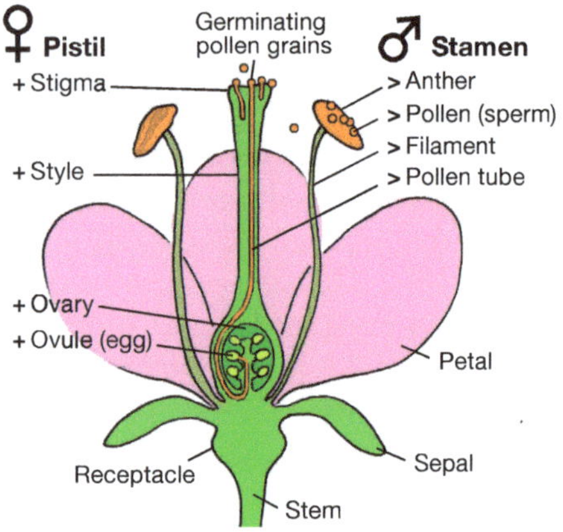

Sexual reproduction in angiosperms: Pollen (that produce sperm) from a blossom lands on the stigma and grows a pollen tube down to the ovule (egg). After the egg is fertilized it forms an embryo that matures into a seed encased inside a fruit. The fruit detaches and a new plant sprouts from it.

Gymnosperms arc non flowering vascular plants whose eggs are usually pollinated via the wind. Their seeds are "naked," often partially concealed in cones, but not completely enclosed in a fruit. Gymnosperms include conifers, ferns, cycads, and ginkgos. The most common asexual reproduction in vascular plants involves rhizomes and runners.

Most **nonvascular plants** reproduce sexually, though sometimes asexually (via cell division or plant fragments that fall to the ground and bud). Sexual reproduction here requires water that sperm must swim through to reach the eggs. ❖

Common Plant Definitions

Ferns are vascular plants, having water-conducting vessels such as stems, leaves, and roots. They reproduce via spores rather than seeds or flowers. Most have fronds that expand from a central head.

Forbs are herbaceous flowering plants that are neither grasses, sedges, nor rushes. Sunflowers, milkweed, red clover, and ragweed are common examples.

Fungi: (Plural of fungus.) A large group of organisms that do not photosynthesize and hence are not considered true plants. They are parasitic, obtaining nutrients from living plants or decaying organic matter. They decompose organic matter. Fungi include mushrooms and dry rot. Lichens contain a fungal component.

Grass: These herbaceous plants, often have hollow stems and narrow open leaf-sheaths that usually grow in two rows along the stem. Sedges and rushes are often mistaken for grasses but are not. True grasses prefer dry, open habitats.

Herbs are seed plants that have leaves and stems, but no supportive, woody tissue. They average only up to 3 ft. in height. Many of them are soft succulents that die down to the soil level at the end of the growing season. They may be annual, biennial, or perennial. Many are used for medicinal purposes or for their scent or flavor.

Lichens result from a symbiotic relationship between two organisms: green or blue-green algae and colorless fungal threads called hyphae. Lichens commonly appear on the trunks of dead trees but can also survive on solid rock surfaces.

Mosses are small, non-woody (nonvascular) plants that usually have stems but no roots. The stem has ribbed leaves that lack lobes or segments. They reproduce by spores rather than by flowers and seeds. Mosses usually grow in clumps or mats in damp or shady locations, and are anchored often to rocks by threadlike structures.

Mushrooms are fungi. They do not undergo photosynthesis and so are not true plants. They are the fleshy, spore-bearing, and fruiting bodies of a fungus. Most grow above ground on soil.

Nonvascular plants: See p. 324.

Sedges: Resembling grasses, most sedges have stems with triangular cross-sections, and typically closed leaf-sheaths arranged spirally in three ranks. They prefer colder and wetter habitats than grasses. 5,000+ species are in this plant family.

Shrubs or bushes are vascular plants that differ from herbs because of their woody stems and are distinguished from trees because of their shorter height (less than 18 ft.) and multiple woody stems (trees tend to have one main trunk). Some plants may qualify as both trees and shrubs. Smaller shrubs are called bushes and are less than 6 ft. high. Some very low shrubs, such as periwinkle and thyme, are called subshrubs.

Rhizome: A horizontal stem that grows underground, sending out shoots and roots at intervals. Runners are above-ground versions of rhizomes.

Rushes resemble grasses but are different. Their solid stems are round, rolling easily between the fingers. Their leaf sheaths grow from the base and are typically closed. They often thrive in cold and wet northern regions.

Trees are perennial vascular plants most of which have branches supported off the ground and growing outwards from a main woody stem. They grow to greater heights in competition for light in forests.

Vascular plants: See p. 324.

Vines Any plant with a growth habit of trailing runners that may climb. Some plants such as poison ivy can grow as low shrubs where support for a vine is not available.

Wildflower is not a precise term. It is often used as a synonym for "blossom," the reproductive structure of angiosperm (flowering) plants.

Tree Identification Tips

Leaf types: A leaf is a photosynthetic appendage found on a **stem node** (an imaginary ring around the stem). Leaves that have just one flat blade are called **simple leaves** while those with two or more discrete blades or leaflets are called **compound leaves**. At the base of every leaf, and in the upper angle between it and the stem, will be a **lateral bud** (leaflets do not have lateral buds).

Compound leaves that are organized like a feather, with one main axis and leaflets emerging from it, are **pinnately compound**. Those with leaflets radiating from a single point, like fingers radiating from the palm of a hand, are **palmately compound**.

Leaf arrangements: Whether a leaf is simple or compound, it will be found growing on a stem, usually in one of three arrangements: alternate, opposite, or whorled. If there is one leaf at a node then the arrangement is **alternate**. If there are two leaves at a node it is **opposite**. If there are three or more leaves at a node it is **whorled**.

Vein patterns: Parallel veined leaves have conspicuous veins that run roughly parallel (e.g., grasses). **Pinnately** veined leaves have one main central vein off of which many secondary veins emerge, in a pattern like that of a feather (e.g., birch). **Palmately** veined leaves have three or more main veins radiating from a single point at the base of the leaf blade (e.g., yellow poplar).

Leaf edges or margins vary greatly from completely smooth (entire) to slightly indented (dentate) to serrated (serrate) to extremely indented (lobed). (Compare hickory with redbud.)

Conifer leaves: Pine needles are simply the leaves of the pine tree. Other conifers, such as junipers have pointy, awl-shaped or scale-like leaves. Look for the number of needles in a bunch, the length of the needles, and the shape of the tips.

Cones, berries, and nuts: Seeds are another identifying feature of trees. Compare the small cones located at the end of hemlock leaves with the cones of Table Mtn. and white pines. The acorns of oak trees are quite distinct from the seed balls of the sweetgum.

Distinctive bark: Some trees have such distinctive bark that the tree can be identified by the bark alone. See the examples shown on p. 333.

Autumnal Colors: Deciduous trees (trees that lose their leaves during winter) change dramatically in October on the Blue Ridge when their color goes from shades of green to a mixture of reds, yellows, oranges, and purples. Dogwood, sourwood, and black gum turn a deep red. Poplar trees and hickories turn bright yellow. Red maples and sweet gums exhibit multiple colors, and oaks tend to turn russet and maroon. Chlorophyll, the agent of photosynthesis, is responsible for the green color we see during the spring and summer. In fall the amount of chlorophyll in plants decreases as the circulation of water between the leaves and the tree ceases. The intensity of the green fades and other colors caused by pigments produced in the leaves (carotenes, anthocyanin, and xanthophylls) begin to appear. The colors and their intensities are closely related to weather conditions. If severe frost occurs early in the fall the chlorophyll breaks down quickly and most of the leaves turn brown and fall off. If cold weather progresses slowly, then the colors emerge slowly, creating beautiful, autumnal panoramas.

Deciduous Trees

Elevation Key:
Low - 900-2,500 ft.
Mid - 2,500-4,500 ft.
High - 4,500-6,684* ft.
*Mt. Mitchell

Ash, Mountain *(Sorbus americana)* Pinnately compound leaves. White blossoms develop into distinctive red berry clusters in fall (shown). Grows to 35 ft. on peaks above 5,000'/elev. Fall foliage: yellow.

Ash, White *(Fraxinus americana)* Pinnately compound, fine-toothed leaves. Thin, purplish flower clusters; green seedpods. Grows to 130 ft. in moist soils, near water at high elev. Fall f.: yellow/purple.

Basswood *(Tilia americana)* Alternate, heart-shaped, simple leaves. Winged fruit. Grows to 100 ft. near streams at low elev. Fall f.: yellow.

Beech, American *(Fagus grandifolia)* Alternate, slightly serrated, simple leaves. Smooth, gray bark. Cigar-shaped buds. Grows to 70 ft. in shady, moist soils up to high elev. Lives to 300-400 yrs. Fall foliage: yellow/brown. (Photo: John Knox)

Birch, Yellow *(Betula alleghaniensis)* Alternate, simple leaves. Shiny, bronzed bark peels off in papery curls. Grows to 80 ft. in moist, cold ravines and north slopes at high elev. Fall f.: yellow.

Buckeye, Yellow *(Aesculus flava)* Palmately compound. Thick nut husks. Grows to 80 ft. in moist rich soils. Fall f.: orange.

Cherry, Black *(Prunus serotina)* Alternate, simple leaves. Small black cherries. Trees are smaller when near streams at low elev. Grows to 100 ft. at high elev. Fall foliage: red/yellow.

Cherry, Sweet *(Prunus avium)* Import. Alternate, ovoid, simple leaves. Grows to 100 ft. at low elev. Fall foliage: orange/red. (Photo: Des Colhoun†)

Chestnut, American *(Castanea dentata)* Alternate, toothed, simple leaves. Spiny, husked bur (shown). Grew to 100 ft. in moist uplands; few remain. (Photo: Am. Chestnt. Fdn.)

Dogwood, Flowering *(Cornus florida)* Opposite, simple leaves. White, 4-petaled blossoms; red berry clusters. Bark has small squarish cracks. Grows to 33 ft. below 3,000'/elev. in moist, rich soil. Fall foliage: deep red.

Elm, American *(Ulmus americana)* Alternate, toothed, simple leaves. Yellowish blossoms. Grows to 75 ft. in moist soil below 2,200'/elev. Threatened by Dutch elm disease. Fall foliage: purple/yellow.

Hawthorn *(genus Crataegus)* (*C. monogyna* shown) Alternate, deeply lobed leaves, long thorns. "Haws" edible but tart. Grows to 30 ft. in moist soil to 5,000'/elev. Fall f.: firey. (Photo: Stanzilla†)

Hickory, Mockernut *(Carya tomentosa)* Pinnately compound leaves, (hairy below). Stringy clusters of small flowers; thick-shelled nuts. Grows to 100 ft. in rich soil on slopes up to 3,000'/elev. Lives up to 500 years. Fall f.: yellow.

Hickory, Shagbark *(Carya ovata)* Pinnately compound leaves (5 leaflets). Bark partially-peels in long strips. Grows to 100 ft. on well-drained slopes. Lives over 350 yrs. Fall foliage: yellow.

Hop-Hornbeam *(Ostrya virginiana)* Alternate, double-serrated, simple leaves. Fruit clusters. Grows to 40 ft., often along streams. Fall f.: yellow.

Locust, Black *(Robinia pseudo-acacia)* Odd-pinnately compound leaves, large thorns. White flower clusters and curvy, flat seedpods. Grows to 100 ft. in dry limestone-soil. Fall f: yellow.

Magnolia, Umbrella *(Magnolia tripetala)* Huge, shiny, palmate, simple leaves. Large, white blossoms; red fruits. Grows to 50 ft. in moist, rich soil in woods and swamps. Fall f.: yellow.

Maple, Sugar *(Acer saccharum)* Opposite, 5-lobed, simple leaves. Grows to 120 ft. on cold, moist slopes up to high elev. Fall f.: yellow/red.

Maple, Red *(Acer rubrum)* Opposite, 5-lobed (v-shaped), toothed, simple leaves. Fluffy red flowers; bi-winged seedpods. Grows to 110 ft. in moist soil on slopes up to high elev. Fall foliage: various (red shown).

Maple, Striped *(Acer pensylvanicum)* Opposite, 3-lobed, toothed, simple leaves. Green flowers; bi-winged seedpods. Grows to 50 ft. on cold slopes, near streams above 3,000'/elev. Fall f.: yellow/red.

Oak, Black *(Quercus velutina)* Large, alternate, 5-lobed leaves (orangey underside). Grows to 100 ft. on drained slopes at low elev. Fall foliage: red.

Oak, Chestnut *(Quercus prinus)* Alternate, curvy-lobed, simple leaves. Deeply furrowed, dark gray-brown bark. Acorns favored by wildlife. Grows to 90 ft. on dry, rocky slopes up to 4,500'/elev. Lives up to 250 yrs. Fall foliage: yellow.

Oak, Northern Red *(Quercus rubra)* Alternate, 7-lobed, simple leaves. Smaller acorn caps. Tallest oak in B.R. Grows to 130 ft. in fertile coves above 1,500'/elev. Lives up to 500 yrs. Fall f.: red.

Oak, Scarlet *(Quercus coccinea)* Alternate, 7-lobed leaves. Grows to 100 ft. on dry soil below 4,000'/elev. (Photo: Jean-Pol Grandmont†)

Oak, White *(Quercus alba)* Alternate, 6-8-lobed, rounded, hairless, simple leaves. Longish acorns; whitish bark. Grows to 120 ft., often in rocky soils. Lives to 600 yrs. Fall f.: red/brown.

Oak, Willow *(Quercus phellos)* Alternate, smooth, narrow, simple leaves. Acorn cap covers nut. Grows to 100 ft. near water. Fall foliage: yellow/red.

Pawpaw *(Asimina triloba)* Large, alternate leaves (with tar-like odor when crushed) cluster at branch end. Large edible fruit. Grows to 30 ft. on floodplains at low elev. Fall foliage: yellow.

Poplar, Tulip (also Yellow Poplar) *(Liriodendron tulipifera)* Alternate, 4-lobed, simple leaves. Flower (inset) is followed by cone-shaped fruit. Grows to 190 ft. in moist, well-drained soils. Fall foliage: yellow.

Princess Tree *(Paulownia tomentosa)* Asian import. Huge, opposite, heart-shaped, simple leaves. Fragrant, lavender, tubular flowers. Grows to 80 ft. on rocky slopes, disturbed areas below 2,500'/elev. Fall foliage: yellow.

Redbud, Eastern *(Cercis canadensis)* Alternate, heart-shaped leaves. Pink blossoms; long peapod-like seedpods. Grows to 15 ft. along edges below 2,000'/elev. Fall f.: yellow.

Sassafras *(Sassafras albidum)* Alternate, 1 to 3-lobed, simple leaves, clustered at branch ends. Grows to 40 ft. in many soils and conditions (tolerates dry, sandy soils) up to 4,000'/elev. Fall foliage: yellow/orange/red (shown).

Serviceberry, Allegheny *(Amelanchier laevis)* Alternate, toothed, simple leaves. White blossoms in early spring; red berries. Grows to 25 ft. on moist, partially sunny slopes up to 6,000'/elev. Fall foliage: yellow/red.

Sweetgum *(Liquidambar styraciflua)* Alternate, pointed, 5-lobed leaves. Aromatic resin; spiky seed balls. Grows to 15 ft. at low elev. Fall f.: yellow/reds.

Sycamore, American *(Platanus occidentalis)* Alternate, 3-lobed, toothed, simple leaves. Ball-shaped flower and fruit. Grows to 110 ft. in moist soil, often near streams at low elev. Fall f.: gold.

Tree of Heaven *(Ailanthus altissima)* Invasive. Alternately pinnate compound leaves. Thick seed clusters (shown). Grows rapidly to 40 ft. under many conditions. Fall foliage: yellow.

Walnut, Black
(Juglans nigra) Pinnately compound, spicy-scented leaves. Green-husked nuts. Grows to 110 ft. in moist soil. Fall f.: yellow.

Witch-hazel, American
(Hamamelis virginiana) Alternate, toothed, simple leaves. Yellow ribbon-like flowers become woody capsules that eject black seeds. Grows to 15 ft. in moist, rich soil up to 4,000'/elev. Fall f.: yellow (shown).

Conifers

Elevation Key:
Low - 900-2,500 ft.
Mid - 2,500-4,500 ft.
High - 4,500-6,684* ft.
*Mt. Mitchell

Cedar, Red (also Juniper) *(Juniperus virginiana)* 2 leaf types: scaley, and awl-shaped spreaders. Aromatic. Has small, purplish-black, fleshy cones. Grows to 50 ft. in high, rocky places with shallow, limestone-derived soil.

Cedar, Northern White (also Arborvitae) *(Thuja occidentalis)* Soft, smooth, scale-like needles cover flat, fan-like branchlets. Grows to 70 ft. in cool, moist, rich soil near cool streams and bogs.

Fir, Fraser (also Balsam) *(Abies fraseri)* Flat, aromatic, round-tipped, 0.5"-1" needles. Grows to 60 ft. at high elev.

Hemlock, Eastern *(Tsuga canadensis)* Flat, rounded, 0.3" needles; tiny cones. Grows to 140 ft. on cold, moist, slopes above 1,000'/elev. Lives up to 500 yrs. Wooly adelgid has killed 80%.

Pine, Loblolly *(Pinus taeda)* 4"-8" needles grow in bunches of 2 to 5. Grows to 80 ft. in damp lowlands and drier uplands.

Pine, Pitch *(Pinus rigida)* Twisted 3"-6" needles grow in bundles of 3 at right angles to branch. Rounded cone. Grows to 90 ft. at lower elev.; taller if on dry, exposed slopes.

Pine, Shortleaf *(Pinus echinata)* Dark, blue-green, 3"-5" needles grow in bundles of 2 or 3. Narrow cone with prickly scales. Grows to 100 ft. at low elev. in dry, well-drained soil.

Pine, Table Mtn. *(Pinus pungens)* Stiff, pointed, 1"-3" needles grow in bundles of 2 or 3. Sharp spines on cone. Grows to 70 ft. on dry, rocky ridges from 1,000' to 5,800'/elev.

Pine, Virginia *(Pinus virginiana)* Stiff 1"-3" needles grow in bundles of 2. Abundance of 1"-3" rounded cones. Grows to 70 ft. in dry, sandy or rocky soil at low elevations.

Pine, White *(Pinus strobus)* 3"-7" needles usually grow in bundles of 5. Long (up to 8"), slender cone. Grows to 180 ft. in well-drained soil. Tolerates sandy, rocky, and wet soil from 1,700'-4,000'/elev. Lives up to 400 yrs.

Spruce, Red *(Picea rubens)* Short (0.5") needles grow singly with sharp tips. A hardy northerner, it survives severely cold conditions. Grows to 30 ft. only on the highest (cold and wet) peaks of the Blue Ridge.

Deciduous Tree Bark

Beech, American

Birch

Buckeye

Cherry

Dogwood

Hackberry

Hickory, Shagbark

Maple

Oak, Red

Oak, White

Poplar

Redbud, Eastern

Sweet Gum

Sycamore, Am.

Witch Hazel

Conifer Tree Bark

Cedar, Red

Hemlock

Pine, Loblolly

Pine, Table Mtn.

Pine, Virginia

Pine, White

Spruce, Red

Plants with **White Blossoms**

Note: Bloom times may vary with latitude and elevation.

Elevation Key:

Low - 900-2,500 ft.

Mid - 2,500-4,500 ft.

High - 4,500-6,684* ft.

*Mt. Mitchell

Anemone, Wood
(Anemone quinquefolia)
Spring - Early Summer. Has 4-9 (usu. 5) petal-like sepals. Grows to 10" in open woods and at wood edges. (Photo: NPS)

Aster, American (also Symphyotrichum) *(genus Symphyotrichum)* **Early Fall.** Grows to 5" in moderately dry fields and woods at all elevations. There are 90 species in the genus.

Autumn Olive
(Elaeagnus umbellata)
Spring. Invasive. Bush with sharp thorns, fragrant flowers, red berries. Grows to 20 ft. in fields and at wood edges.

Black Bugbane (also Black Cohosh or Black Snakeroot) *(Cimicifuga racemosa)* **Summer.** Grows to 6 ft. in woods with moist, rich soil and in coves at low to mid elev.

Blackberry, Common
(Rubus allegheniensis)
Early Summer. Shrub. Grows to 6 ft. in open areas at all elevations. Prospers after fires or tree blowdowns.

Bloodroot
(Sanguinaria canadensis)
Early - Late Spring. Grows to 10." Has red-orange sap. Prefers moist woods at low to mid elev.

Blueberry, Lowbush
(Vaccinium angustifolium)
Early Spring. Shrub. Grows to 2 ft. in well-drained, acidic soils in fields and open woods. (Photo: Albert Herring†)

Boneset
(Eupatorium perfoliatum)
Summer - Fall. Dense, hairy plant. Grows to 4.5 ft. in moist to wet open areas at low elev.

Bowman's Root
(Gillenia trifoliata)
Late Spring - Mid Summer. Grows to 3 ft. in dry to moist soils in fields and open woods at low to mid elev.

Buttonbush
(Cephalanthus occidentalis)
Summer. Shrub. Fragrant, fuzzy, creamy white blossoms. Grows to 12 ft. in wetlands and rivers at low elev. (Photo: Sten Porse†)

Chickweed, Star
(Stellaria pubera)
Early - Late Spring. Small flowers with 5 notched petals, red stamen. Grows to 16" on wooded and rocky slopes.

Clematis (also Virgin's Bower) *(Clematis virginiana)*
Summer - Early Fall. 4-petals. Climbing vine (p. 350). Grows to 20 ft. in fields and open woods. (Photo: SB Johnny†)

Cut-leaf Toothwort
(Cardamine concatenata)
Early - Late Spring. Grows to 16" in moist, rich soil on floodplains and wooded slopes at low to mid elev.

Deerberry
(Vaccinium stamineum)
Early Spring. Bell-shaped, white or greenish flowers. Grows to 6 ft. in dry, rocky habitats but also in moist, boggy areas.

Dutchman's Britches
(Dicentra cucullaria)
Early - Late Spring. Toxic. Fragrant. Grows to 1 ft. in moist soil on rocky slopes, nr. streams.

Elderberry, Common
(Sambucus canadensis)
Summer. Shrub. Black fruits used in wine, pies. Grows to 10 ft. in dry or moist soils in open areas up to high elev.

Fleabane
(Erigeron philadelphicus)
Spring - Summer. Can be pinkish. Grows to 2.5 ft. in fields and open woods at low elev.

Flowering Spurge
(Euphorbia corollata)
Summer - Early Fall. 5-petaled flowers. Grows to 3 ft. in dry fields and open woods. (Photo: Michele R. Fletcher, 2011)

Goat's Beard
(Aruncus dioicus)
Spring - Mid Summer. Grows to 6 ft. in dense colonies in moist, rich soils in fields, on wooded slopes, and along streams up to high elev.

Grass of Parnassus
(Parnassia asarifolia)
Late Summer - Fall. Flower stalks grow to 1.5 ft. in wet fields, bogs, moist woods up to high elev. (Also p. 211, 348) (Photo: Alan Cressler)

Honeysuckle
(Lonicera japonica) **Early Summer.** Invasive. Has black berries, unlike natives' red ones. Vines grow to 30 ft. in floodplains, woods, and edges at low elev.

Hydrangea, Wild
(Hydrangea arborescens)
Spring - Fall. Shrub. Large, opposite, toothed leaves. Flowers usu. tiny and clustered. Grows to 6 ft. on moist, rocky, shaded slopes up to highest elev.

Indian Pipe (also Ghost or Corpse Plant) *(Monotropa uniflora)*
Summer. Grows to 10." Rare parasite (esp. on beech trees) but not a fungus. Can grow in very dark areas. Seen in shady woods in the Smokies.

Mayapple *(Podophyllum peltatum)* **Early - Late Spring.** Fruits used in jelly; other parts toxic. Grows to 1.5 ft. in moist, rich soil in open woods, edges.

Mayflower
(Maianthemum canadense)
Early Summer. Grows to 8" in moist woodland soil at mid to high elev. (Photo: Albert Herring†)

Mountain Sandwort
(Arenaria montana)
Late Spring - Early Summer. Import. Grows to 8" on open rocky areas at high elev. (Photo: Mtiffany71†)

Ox-eye Daisy
(Leucanthemum vulgare)
Spring - Fall. Import. Grows to 3 ft., often in large colonies, in fields and open woods at low to mid elev. (Photo: H. Zell†)

Partridgeberry
(also Squaw vine) *(Mitchella repens)*
Late Spring - Summer. Hardy, low, evergreen vine (p. 349). 4-petals, blooms in pairs; red berries. Spreads 1 ft. in moist woods. (Photo: Fritz Flohr Reynolds†)

Pokeweed
(also Pokeberry) *(Phytolacca americana)* **Mid Summer - Fall.** Purple berries are poisonous to humans. Grows to 10 ft. in moist, open woods and edges at low elev.

Queen Anne's Lace
(Daucus carota)
Late Spring - Fall. Import. Grows to 3 ft. in sunny fields and along wood edges at low to mid elev.

Rattlesnake Plantain
(Goodyera pubescens)
Summer. (Closeup) Evergreen orchid with variegated leaves (p. 349) and up to 18" flowering stalk. Grows in dry to moist, acidic soil in upland woods.

Sand Myrtle, Allegheny
(Kalmia buxifolia) **Early Spring - Early Summer.** Low, evergreen shrub. Grows from 5" to 3 ft. on moist, rocky, heath balds up to highest elev. (See p. 267)

Solomon's Plume
(Maianthemum racemosum)
Spring - Summer. Grows to 3 ft. in moist, open woods.
(Photo: Walter Siegmund†)

Solomon's Seal
(Polygonatum odoratum)
Spring - Early Summer. Grows to 3 ft., on moist, wooded slopes and in coves at low to mid elev.

Spring Beauty
(Claytonia caroliniana)
Early - Late Spring. Grows to 8" in rich soil in open woods, thickets, and slopes up to high elev.

Stonecrop, Allegheny
(Hylotelephium telephioides)
Summer - Early Fall. Succulent leaves (p. 349). Grows to 1.5 ft. on rocky slopes at mid/high elev. (Photo: Fritz Flohr Reynolds†)

Trillium, White
(Trillium grandiflorum)
Spring - Early Summer. Largest of the trilliums. White to pink. Has 3 leaves and 3 large petals. Grows to 20." Likes rich soil in woods up to high elev.

Water-lily, Am. White
(Nymphaea odorata)
Summer. Fragrant. White flower grows to 3" height. Round leaf is reddish underneath. Grows in shallow lakes, ponds, and slow streams.

Wild Strawberry
(Fragaria virginiana)
Spring - Early Summer. Favorite of black bears. Grows to 8" in fields and open woods at low elev. (Inset: Walter Siegmund†)

Wineberry (also Wine Raspberry) *(Rubus phoenicolasius)* **Late Spring - Early Summer.** Asian import. Shrub. Prickly with reddish stems. Grows up to 9 ft. Spreads rapidly, forming thickets in fields and along wood edges.

Wintergreen, American
(also Teaberry) *(Gaultheria procumbens)* **Summer** Evergreen leaves. Berries are source of oil of wintergreen. Grows to 8," often under conifers in dry to moist soil. (Photo: Jomegat†, inset: John Delano†)

Yarrow
(Achillea millefolium) **Summer - Fall.** Flat-topped flower clusters; fern-like leaves. Grows to 3 ft. in dry, well-drained soil in fields and along wood edges.

Plants with **Pink Blossoms**

Note: Bloom times may vary with latitude and elevation.

Elevation Key:

Low - 900-2,500 ft.

Mid - 2,500-4,500 ft.

High - 4,500-6,684* ft.

*Mt. Mitchell

Bleeding Heart

(Dicentra eximia)

Spring - Summer. Heart-shaped blooms hang in rows. Grows to 1.5 ft. in moist soil in wooded or open, rocky slopes, cliffs, and streams at low to mid elev.

Crown Vetch

(Securigera varia)

Summer. Invasive. Legume vine. Grows to 2 ft. in fields and along wood edges at low elev.

Dame's Rocket

(Hesperis matronalis)

Spring - Early Summer. Pale pink to purple. Grows to 3 ft. in moist, well-drained soil in woods and along wood edges.

Joe Pye Weed

(also Trumpetweed) *(Eupatorium fistulosum)* **Late Summer.** Smells like vanilla when crushed. Grows to 7 ft. in moist fields, woods, and marshes.

Lousewort, Canadian

(Pedicularis canadensis)

Spring - Early Summer. Parasitic on roots. Grows to 1.5 ft. in fields and open woods at low to mid elev. (Photo: Craig Van Boskirk)

Milkweed

(Asclepias syriaca)

Summer. Pink to green. Monarch caterpillar's only food. Grows to 5 ft. in sunny fields.

Mountain Laurel

(Kalmia latifolia)

Spring - Early Summer. Evergreen shrub grows to 15 ft., often in thickets in moist, shaded, rocky woods up to high elev.

Pink Lady's Slipper

(Cypripedium acaule)

Spring - Early Summer. Grows to 15" in wet woods at low to mid elev. (Photo: Michele R. Fletcher, 2011)

Raspberry

(Purple Flowering) *(Rubus odoratus)* **Summer.** Thornless native shrub. (Cultivars produce tastier fruit.) Grows to 6 ft. in fields, edges at low to mid elev.

Rhododendron

(Mountain or Catawby) *(Rhododendron catawbiense)* **Late Spring-Early summer.** Evergreen native grows to 10 ft. in large thickets on moist slopes and near streams at mid to high elev.

Showy Orchis

(Galearis spectabilis) **Spring - Early Summer.** Pink to purple color. Grows to 10" in moist, rich soil in deciduous woods, often near water at low to mid elev.

Maiden's Tears

(also Bladder Campion) *(Silene vulgaris)* **Spring - Summer.** Import. Clusters of 5-20 blossoms. Grows to 2.5 ft. in fields and along wood edges.

Smartweed

(also Knotweed or Knotgrass) *(genus Polygonum)* **Fall.** Blossoms grow in dense clusters from many-jointed stems. 220 species grow up to 10 ft. in dry fields and open woods.

Spirea, Dwarf

(Spiraea corymbosa) **Summer.** Shrub. Pink to white blossoms on flat-topped spire. Grows to 3 ft. on rocky slopes and streambanks. (Photo: Michele R. Fletcher, 2011)

Spotted Knapweed

(Centaurea maculosa) **Summer - Early Fall.** Invasive. Thornless. Grows to 4 ft. in dry fields and open woods.

Trillium

(Trillium grandiflorium) **Spring - Early Summer.** Pink to white. 3 leaves; 3 large petals. Grows to 20." Likes rich soil in woods up to high elev.

Wild Geranium

(Geranium maculatum) **Spring - Early Summer.** Grows to 2 ft. in moist soil on floodplains and open woods up to mid elev. (Photo: NPS)

Plants with
Blue-to-Purple Blossoms

Note: Bloom times may vary with latitude and elevation.

Elevation Key:
Low - 900-2,500 ft.
Mid - 2,500-4,500 ft.
High - 4,500-6,684* ft.
*Mt. Mitchell

Aster, Bigleaf
(also Large-leaf Wood Aster) *(Eurybia macrophylla)* **Summer - Fall.** Has 4"-8" heart-shaped leaves. Grows to 4 ft. in dry to moist soil in woods and along wood edges. Favorite of butterflies.

Bluebells, Virginia
(Mertensia virginica)
Spring. 1"-long bluish trumpets. Grows to 2 ft., often in colonies along streams and in moist woods at low elev.

Blue Phlox, Wild
(also Wild Sweet William) *(Phlox divaricata)* **Spring.** Grows to 1 ft. in rich soil on rocky, wooded slopes and wood edges at low to mid elev.

Bluets
(Houstonia caerulea)
Spring - Mid Summer. Grows to 6" in moist fields and woods, along streams, and on wet outcrops up to the highest elevations.

Blue Violet, Common
(Viola sororia)
Early Spring. Stems grow directly from ground up to 8" in moist fields and open woods at low to mid elev. (Photo: Michele R. Fletcher, 2011)

Chicory
(Cichorium intybus)
Summer - Fall. Import. Grows to 4 ft. in fields and along wood edges at low elev.

Dayflower, Asiatic
(Commelina communis)
Summer - Fall. Import. Blossoms last only one day. Grows to 2" in moist woods and along wood edges.

Dwarf Crested Iris
(Iris cristata)
Spring. Flowers have 3 blue petals with a yellow crest. Grows to 8," often along streams at low elev.

Fringed Phacelia
(Phacelia fimbriata)
Spring - Early Summer. Petals turn purpler with age. Grows to 16," spreading in moist forests nr. streams at mid to high elev.

Gentian, Stiff
(Gentianella quinquefolia)
Late Summer - Late Fall. Grows to 2 ft. in fields, open wooded slopes, rock ledges, and along steams from low to high elev. (Photo: Irvine T. Wilson)

Great Blue Lobella
(Lobelia siphilitica)
Late Summer - Fall.
Grows to 3 ft. in moist soil in woods and fields at low elev.

Heal-All
(also Self-heal) *(Prunella vulgaris)* **Spring - Early Fall.** Grows to 1 ft. in sunny fields, and along wood edges.

Hepatica, Round-lobed
(Hepatica americana)
Early Spring. Rounded, 3-lobed leaves. Grows to 6" in moist limestone-derived soil in shady woods or on sunny hillsides at low to mid elev.

Horsenettle
(Solanum carolinense)
Summer. Prickly (but not a true nettle). Berries are highly toxic. Grows to 3 ft. in fields, esp. in sandy soils. (Inset: Susan Sweeney†)

Ironweed, Tall
(Vernonia gigantea)
Fall. Grows to 7 ft. in moist fields and woods, low elev.
(Photo: Michele R. Fletcher, 2011)

Jack in the Pulpit
(Arisaema triphyllum) **Early Spring - Early Summer.** Color varies from purple to green. Grows to 2 ft. in moist woods and bogs at low elev.

Morning Glory
(also Bindweed) *(Ipomoea purpurea)*
Summer. (Non-native of the Convolvulaceae family.) Climbing vine. Grows in fields and along wood edges.

Purple Dead Nettle *(Lamium purpureum)* **Early Summer.** Import. Non-stinging. Grows to 1 ft. in disturbed areas, fields, and wood edges.

Southern Harebell (also Appalachian Harebell, Bellflower) *(Campanula divaricata)* **Summer - Fall.** Grows to 2 ft. on dry slopes near rock outcrops and in dry woods at low to mid elev. (Photo: Gerald C. Williamson)

Spiderwort, Zigzag *(Tradescantia subaspera)* **Spring - Summer.** Bloom dies after mid-day; replaced at night. Grows to 2 ft. in fields and open woods at low to mid elev.

Teasel, Common (also Wild Teasel) *(Dipsacus sylvestris)* **Mid Summer - Fall.** Import. Pointy, tube-shaped bracts encircle the flower head. Grows to 8 ft. in fields and along wood edges.

Vernal Iris (also Dwarf Violet Iris) *(Iris verna)* **Spring.** Small iris, grows up to 6," often in colonies, in open, dry, conifer forests with low-nutrient soil at low elev. (Photo: NPS)

Wild Ginger *(Asarum canadense)* **Spring.** Aromatic like ginger but can cause kidney damage. Grows to 1 ft. in dense colonies in rich, moist woods at low to mid elev. (Inset: Fritz Flohr Reynolds†)

Plants with **Yellow Blossoms**

Note: Bloom times may vary with latitude and elevation.

Elevation Key:
Low - 900-2,500 ft.
Mid - 2,500-4,500 ft.
High - 4,500-6,684* ft.
*Mt. Mitchell

Black-eyed Susan
(Rudbeckia hirta)
Summer - Fall. Yellow petals with a brown center are distinctive. Grows to 2.5 ft. in dry to moist soil in fields and along wood edges.

Bluebead Lily
(Clintonia borealis)
Late Spring - Early Summer. 6-petaled flowers. Grows to 16" often in conifer forests at high elev. (Photo: Peter Coxhead†; inset: Circeus†)

Bulbous Buttercup
(Ranunculus bulbosus)
Spring - Early Summer. Import. Poisonous. Stem is bulblike at base. Grows to 2 ft. in dry, rich soil in fields and along edges.

Dandelion, Common
(Taraxacum officinale)
Early Spring - Fall. Import. This type, found in the eastern U.S., was imported as a food crop. Grows to 1 ft. in fields and along wood edges.

Evening Primrose, Northern *(Oenothera parviflora)*
Summer. Opens fully only after dark. 4 heart-shaped, yellow petals. Grows to 4 ft. in dry, poor soil in fields on slopes up to high elev.

Flannel Mullein
(Verbascum thapsus)
Summer. Import. Grows to 7 ft. in open, sunny fields.

Golden Alexanders
(Zizia aurea)
Spring. Blooms in umbrels. Has 3-lobed leaf-ends. Grows to 3 ft. in fields. (Photo: Albert Herring†)

Golden Ragwort
(Packera aurea)
Early Spring - Early Summer. Grows to 2.5 ft. in moist soil along streams at low elev.

Goldenrod
(genus Solidago)
Late Summer - Fall. (120 species) Grows to over 3 ft. in fields and open woods.

Green and Gold
(Chrysogonum virginianum)
Early Spring - Early Summer. Grows to 1 ft. in well-drained soils in open woods and along wood edges. (Photo: Derek Ramsey)

Heliopsis (also Ox-eye Sunflower) *(Heliopsis helianthoides)*
Summer. (Daisy family) Grows up to 6 ft. Tolerates dry soil in fields and open woods. (Photo: Michele R. Fletcher, 2011)

Sneezeweed
(Helenium autumnale)
Late Summer - Early Fall. Flower clusters. Grows to 4 ft. in moist soil in open areas. (Photo: Irvine Wilson)

Spicebush
(Lindera benzoin)
Spring. Shrub. Leaves are aromatic when crushed. Red berries. Grows to 12 ft. on floodplains and in open woods at low elev.

Squaw-root (also Bearcorn) *(Conopholis americana)* **Spring.** Non-photosynthetic, parasitic plant favored by deer and bears. Feeds on oak, beech tree roots. Grows to 8" in damp forests with rich soil.

Sunflower, Common
(Helianthus annuus)
Summer. Domestic cultivar from a wild native plant. Grows to 10 ft. in open areas.

Touch-me-not
(also Yellow Jewelweed) *(Impatiens pallida)* **Summer - Early Fall.** Grows to 6 ft. in moist to wet soil at low elev.

Yellow Stargrass
(Hypoxis hirsuta)
Spring. Grows to 1 ft. in moist to dry soil in fields and open woods at low elev.

Plants with **Red and Orange Blossoms**

Note: Bloom times may vary with latitude and elevation.

Elevation Key:

Low - 900-2,500 ft.

Mid - 2,500-4,500 ft.

High - 4,500-6,684* ft.

*Mt. Mitchell

Bee Balm, Scarlet

(Monarda didyma) **Mid - Late Summer.** Crushed leaves have a minty aroma. Grows to 5 ft. in moist, acidic soil up to high elev. (Photo: Gerald C. Williamson)

Butterfly Weed

(Asclepias tubesrosa) **Summer.** Poisonous. Intense orange-red color. Grows to 2 ft. in dry, sandy soil in fields and open woods at low elev.

Cardinal Flower

(Lobelia cardinalis) **Late Summer.** Favored by hummingbirds. Grows to 4 ft. along streams and other moist areas at low elev.

Columbine, Eastern Red

(also Wild Columbine) *(Aquilegia canadensis)* **Spring - Early Fall.** Grows to 3 ft. in thin soil on rocky slopes and along ledges and streams at low to mid elev.

Daylily, Common

(Hemerocallis fulva) **Late Spring - Mid Summer.** Import. Has 3 petals and 3 sepals. Grows to 4 ft. in clumps along roadsides, spreading from homes.

Flame Azalea

(Rhododendron calendulaceum) **Spring - Summer.** Native shrub grows to 10 ft. up to high elev. (Photo: Michele R. Fletcher, 2011)

Fire Pink

(Silene virginica) **Spring - Summer.** Grows to 2 ft. in dappled light on open, rocky, wooded slopes. (Photo: Michele R. Fletcher, 2011)

Indian Paintbrush

(Castilleja coccinea) **Spring - Summer.** Grows to 2 ft. in fields and moist, open woods. (Photo: NPS)

Poppy, Common
(also Corn Poppy) *(Papaver rhoeas)* **Late Spring - Early Summer.** Import. Has 4 petals. Grows to 2 ft. in open fields and disturbed areas.

Staghorn Sumac
(Rhus typhina)
Late Spring - Mid Summer. Shrub. Dull flowers mature into red, fuzz-covered berries (shown). Grows to 35 ft. in dry, infertile soils in open and rocky areas.

Turk's Cap Lily
(Lilium superbum)
Mid - Late Summer. Grows to 7 ft. in moist fields and along streams. (Photo: Michele R. Fletcher, 2011)

Ferns

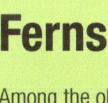

Among the oldest on Earth (360 million yrs.), these vascular plants have stems with pinnate leaves and usually reproduce via spores from sacs on leaf undersides, rather than by flowers and seeds. Most fern fronds begin as fiddleheads that slowly unfurl. Most like moist, shady woods, but many have adapted to various environments.

Cinnamon Fern
(Osmundastrum cinnamomeum)
Spore-bearing frond has a cinnamon color. Alternately pinnate, deeply-lobed frond grows up to 5 ft., usually in moist woods and swampy areas.

Common Bracken Fern
(Pteridium aquilinum) Large alternately bi-pinnate, triangle-shaped frond. Spores rim underside edge. Grows to 6 ft. in wide range of habitats. Prefers dry, acidic soil.

Ebony Spleenwort
(Asplenium platyneuron) Erect pinnate fronds grow to 20" on rocks or soil. Tolerates many habitats. (Photo: Jim Conrad)

Rattlesnake Fern (also Mtn. Lace Fern) *(Botrypus virginianus)*
Named after a rattlesnake's erect tail (spore-spike not shown). Frond (3-4 times pinnate) grows to 30" in rich soils.

Rock Cap Fern
(Polypodium virginianum)
Altrnt. pinnate frond (rounded leaf) grows to 1 ft. in colonies. Tolerates dry, rocky areas. (Photo: Jaknouse†)

Ground covers

This category includes plants from the low growing herbaceous layer in natural habitats. Ground covers reproduce or spread in various ways including: lateral growth, low branching, base growth, and rhizomes.

Note: Bloom times may vary with latitude and elevation.

Burdock, Common

(also Lesser Burdock) *(Arctium minus)*
Summer - Fall. Import. Large leaves, thistle-like flowers. Its small burrs stick to fur/clothing. Used medicinally. Grows to 6 ft. in fields and along wood edges.

Clover, Red

(Trifolium pratense)
Spring - Summer. Imported to increase soil fertility for crops. Grows to 2 ft. in dry to moist soils in fields, edges, open woods.

Coltsfoot

(Tussilago)
Early Spring. Import. Dandelion-like blossoms on 1 ft. stalks. Plant grows to 6" in moist, open areas, streambanks up to high

Galax

(also Wandplant) *(Galax urceolata)*
Late Spring - Summer Wide, leathery, evergreen leaves. Flower stalks grow to 2 ft. Grows in moist to dry open woods up to high elev. (Photo: Ted Bodner†)

Garlic Mustard

(Alliaria petiolata)
Spring - Summer. Invasive. Imported as a culinary herb. White blossoms. Grows to 3.5 ft., often in disturbed areas.

Grass of Parnassus

(Parnassia asarifolia) (p. 336)
Late Summer - Fall. Low plant in wet areas to high elev. (Photo: Margie Hunter†; inset: Alan Cressler)

Ground Ivy

(Glechoma hederacea)
Spring - Early Summer. Import. Aromatic evergreen creeper. Grows to 2 ft. in moist, shady woods, but also tolerates sun.

Liverwort, Crescent Cup

(Lunularia cruciata)
Import. Low, nonvascular plant. 0.5" leaves with crescent "cups" for buds. Grows in damp areas in deep shade.

Mayapple (also Appalachian Mandrake) *(Podophyllum peltatum)* **Early - Late Spring.** Grows to 1.5 ft. in moist, rich soil in open woods and along wood edges. (Blossom: p. 336)

Mountain Sandwort *(Arenaria montana)* **Late Spring - Early Summer.** Import. Grows to 8" on moist, well-drained, sandy to loamy soils in open rocky areas at high elev. (Photo: User: Mtiffany71†)

Partridgeberry (also Squaw vine) *(Mitchella repens)* **Late Spring - Summer.** Hardy, low, evergreen vine. Spreads up to 1 ft. in moist woods. (Blossom: p. 337)

Rattlesnake Plantain *(Goodyera pubescens)* **Summer.** Evergreen orchid (Blossom: p. 337). Leaves grow to 3" in dry/moist soil in upland woods. (Photo: Jason Hollinger†)

Running Cedar (also Crow's Foot, Fan Clubmoss) *(Diphasiastrum digitatum)* Glossy evergreen spreads underground sending up shoots that resemble tiny trees with spore-laden stalks. Grows to 4" in moist, shady, conifer forests.

Sand Myrtle, Allegheny *(Kalmia buxifolia)* **Early Spring - Early Summer.** Evergreen shrub. Grows from 5" to 3 ft. on moist, rocky, heath balds up to highest elev. (See p. 267)

Saxifrage (also Rockfoil, Stonebreaker) *(genus Saxifraga)* **Summer.** Small alpine. Toothed leaves, white flowers. Grows near wet rocks at high elev.

Skunk Cabbage *(Symplocarpus foetidus)* **Early Spring.** Foul-smelling, purple-brown, bulb-like flower. Grows to 3 ft. in wetlands, streams. (Inset: Sue Sweeney†)

Stonecrop, Allegheny *(Hylotelephium telephiodes)* **Summer - Early Fall.** Succulent. Grows to 1.5 ft. on rocky slopes at mid to high elev. (Blossom: p. 338)

Vines

Vines are defined by their climbing (or trailing) stems or runners that allow them to reach sunlight without having to spend energy on building a sturdy support structure. Some can grow as shrubs when there is no rock or other plant to climb; others always grow as vines.

Note: Bloom times may vary with latitude and elevation.

Clematis (also Virgin's Bower) *(Clematis virginiana)* **Summer - Early Fall.** Seeds shown. Climbs to 20 ft. in rich soil along wood edges, streambanks, and in thickets. (Blossom: p. 335)

Honeysuckle *(Lonicera japonica)* **Early Summer.** Invasive. Has black berries, unlike red of natives. Climbs to 30 ft. in trees and shrubs on moist floodplains, woods, and wood edges. (Also p. 336)

Kudzu *(Pueraria montana var. lobata)* **Summer.** Invasive. Purple blossoms. Climbs to 100 ft. (up to 1 ft./day). Smothering Blue Ridge forests, esp. along wood edges.

Morning Glory, Big-root (also Wild Potato-vine) *(Ipomoea pandurata)* **Summer.** Native climbing vine with heart-shaped leaves, white blossoms. Climbs to 20 ft. in fields and along open wood edges. (Photo: Cody Hough†)

Poison Ivy, Eastern *(Toxicodendron radicans)* **Spring.** 3 leaflets; small, green flowers. May have aerial roots in trees (climbing to 60 ft.) Found in fields, rocky slopes, woods, edges. (Also see p. 2.)

Trumpet Honeysuckle *(Lonicera sempervirens)* **Summer.** Native. Climbs to 20 ft. in rich, moist soils in many habitats. (Photo: Stan Shebs†)

Trumpet Vine (also Trumpet Creeper) *(Campsis radicans)* **Spring - Early Fall.** Native. Climbs to 30 ft. in open woods and thickets, along streambanks and esp. wood edges.

Virginia Creeper *(Parthenocissus quinquefolia)* **Summer.** Resembles poison ivy but has 5 leaves, tiny blue berries. Spreads 50 ft. in fields, wood edges.

Mosses

These non-vascular plants have simple, one-cell thick leaves that often form whorls around a short, thin stem. Lacking tissues to conduct water, they need moist environments like shady woods or streamsides. They grow year-round, photosynthesizing their food from sunlight and reproduce via spores. Approx. 350 species grow in the Blue Ridge.

Silk Moss, Waved
(also Snake Moss) *(Plagiothecium undulatum)* A "hanging moss" that forms large mats resembling shag rugs. Likes moist woods.

Carpet Moss
(genus Hypnum)
Forms dense, velvety, carpet-like mats on the ground, often along streambanks, and on the base of tree trunks.

Haircap Moss
(genus Polytrichum)
(70 species) Has water-conducting tissues, unusual for moss, thus tolerates drier conditions.

Leucodon Moss
(Leucodon brachypus)
Tends to grow in large mats on tree trunks at about 5 ft. above the ground.

Peat Moss
(genus Sphagnum)
Holds 16-26 times its weight in water. Thick mats grow in mtn. bogs, moist conifer forests. (Photo: User: External Affairs†)

Stair Step Moss
(also Feather Moss) *(Hylocomium splendens)* Grows one "step" per year up to high elev. (Photo: Walter Siegmund†)

White Cushion Moss
(Leucobryum glaucum)
Thick, ball-shaped cushions or mats grow on acidic soil, logs. Pale when drier. (Photo: Jerzy Opioła†)

White-tipped moss
(Hedwigia ciliata)
Sun-tolerant. Grows on soil and boulders in open woods. (Photo: Hermann Schachner)

Fungi

The Fungi kingdom includes an estimated 1-5 million species, from simple unicellular forms (yeasts and molds) to complex fruiting forms (mushrooms and lichens). Like animals, fungi absorb food molecules by secreting digestive enzymes–but fungi do this directly into their environment. Hence, they are known as the primary decomposers in ecosystems.

Mushrooms:

Mushrooms are the fruiting bodies (spore-producing organs) of fungi. They typically grow above ground and feed on soil, tree roots/trunks or rotting logs in moist, warm conditions. Usually they have a stem growing out of the ground with a cap on top and gills underneath that produce spores by which they reproduce and spread. However, there is a great deal of variation among the many species in the Blue Ridge. Some grow without a cap, others without a stem, etc.

Black Morel
(Morchella angusticeps)
Spring. A delicacy. Species vary from tan to gray to black. Grows after rains in loamy soil near trees.

Coker's Amanita (also Solitary Lepidella) *(Amanita cockeri)* **Spring.** Highly toxic. Dry but sticky white cap with warts, white gills, bulbous base. Likes oak and pine woods.

Death Cap Mushroom
(Amanita phalloides)
Fall. Fatally poisonous. Looks like honey mushroom. Greenish, sticky cap with white gills and a cup at the base. (Photo: Archenzo†)

Golden Chanterelles
(Cantharellus cibarius)
Summer. A delicacy. Orange-yellow, funnel-shaped, wavy edges and false gills underneath (appear melted, forking). (Photo: Strobilomyces†)

Hen of the Woods
(Grifola frondosa) **Summer - Fall.** Unlike similar "Chicken of the Woods," it commonly grows in clusters at the base of oak trees.

Honey Mushroom
(Armillaria mellea)
Fall. Multi-shaped, white to gold caps, white gills. Cluster on tree trunks. (Photo: Dan Molter†)

Jack-o'-lantern
(Omphalotus olearius)
Summer - Fall. Poisonous, resembles chanterelles but has true gills (sharp, thin blades). (Photo: A. Abbatiello)

Puffball, Common
(Lycoperdon perlatum)
Early Summer - Fall. Round with spines and warts. Color turns to brown as it matures.

Turkey Tail Fungus
(Trametes versicolor)
Year-round. Has concentric rings of differing colors. This mushroom grows on decaying logs and stumps.

Lichens:

Compound organisms, these fungi live symbiotically with an algae and/or cyanobacteria, often on rocks or tree bark year-round. **Three types are found in the Blue Ridge:** Most common are the **crustose** variety, named for their crust-like appearance. They are usually tightly attached to rock outcrops, trees or compacted soil and occur in various colors and distinctive shapes. **Foliose** lichens have two distinct sides (front/back), vary greatly in size and can be flat or leafy (lettuce-like), or convoluted with bumps and ridges. **Fruticose** lichens are often hair-like. They may be upright and shrubby or cup-like.

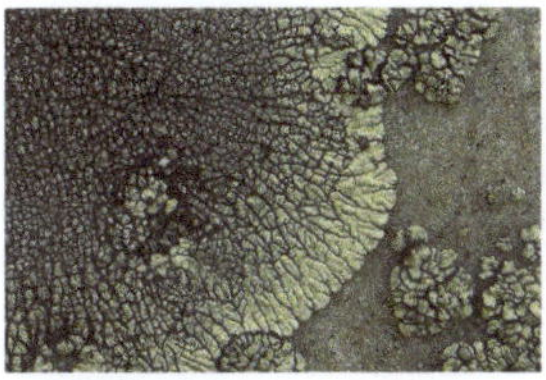

Crustose
has cracked crusts that grow closely attached to surfaces. (Golden Moonglow Lichen shown)

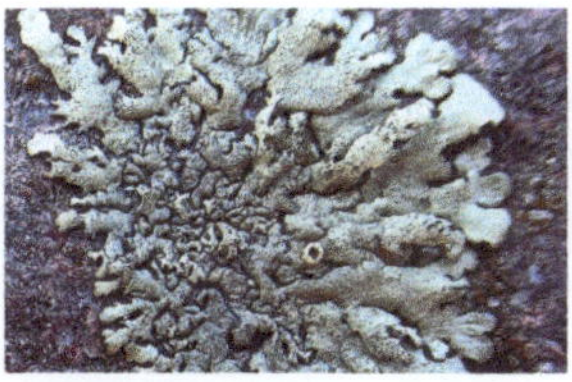

Foliose
has leaf-like lobes with flat sheets of tissue that extend from surfaces. (Greenshield Lichen shown)

Fruticose
often has shrubby, hair-like branching tubes that tend to hang down. (Old Man's Beard Lichen shown)

Greenshield Lichen
(Flavoparmelia baltimorensis)
Foliose. Greenish-gray. Grows on tree trunks and rocks. Likes shady, moist woods.

Reindeer Lichen
(Cladonia rangiferina)
Fruticose. Grows in well-drained, open areas. Cold-tolerant, often found on barrens at high elevations.

Rock Tripe
(genus Umbilicaria) (dark brown U. hyperborea shown) Crustose. Dark brown, tan, black, or white. Common on rocks and boulders.

Chapter 11

Birds

Commonly Found in the Blue Ridge

Bird Habitats

Over 400 bird species have been identified in the Blue Ridge. The great diversity of birds in these mountains is in part due to the varied ecosystems present here. The character of the forests and their inhabitants vary greatly from Southern Georgia and Alabama to the mid-latitudes in Maryland and Pennsylvania as well as from changes in elevation in any given area. Birds seen at lower elevations in the Blue Ridge also share the habitats of the Piedmont and Coastal Plain while those that are common in the more northern parts of North America also populate mountain peaks in the Blue Ridge Highlands.

Bird habitats in the Blue Ridge range widely from wetland bogs, lakes, swamps, rivers, and streams to fields, as well as coniferous, deciduous, and mixed forests. Many birds pass through almost all of these habitats, but they become very selective when it comes to choosing what to eat and where to build nests for offspring.

Nesting habits are included in the identification section. Their choice of food determines where you are most likely to find them. Many have evolved specialized beaks and other physical characteristics that are especially suitable for obtaining particular types of food. Anyone who has a bird feeder soon learns that some birds

Look for birds along trails in the early morning or late afternoon. Trail at Shenandoah Natl. Park.

such as cardinals and doves will feed on seeds put on the ground while finches, sparrows, tufted titmouse, and juncos prefer to eat above the ground. Hummingbirds feed from flowers, and have no interest in seeds. Nuthatches and woodpeckers look for food in tree bark, while the eastern blue bird passes up seeds altogether for berries or fruit.

Great horned owls

Some bird species are very selective and exhibit strong preferences for certain insects or seeds. Hawks feed on small mammals, toads, small birds, frogs, snakes, and lizards. Eagles feed on fish, small mammals, birds and carrion. Blackbirds, finches, grosbeaks, and sparrows feed on a wide variety of insects. Woodpeckers seek out beetles, weevils, grubs, ladybugs, lightning bugs, bees, wasps, ants, butterflies, moths, and caterpillars. Blue jays, chickadees, crows, and titmice prefer grains and fruits. Crows will also eat small fish, lizards, and snakes.

Where to Look for Birds

Popular places for birdwatching include: mountaintops devoid of trees and shrubs such as those at Big Meadows in Virginia at milepost 52 and Chimney Rock near Asheville, N.C.; trails in dense forests on mountain slopes, such as the South River Falls Trail at milepost 63; overlooks, including the Washington Monument near Boonsboro, Md., Rockfish Gap at milepost 0, Peaks of Otter at milepost 84-7, Harvey's Knob near Roanoke, Va. at milepost 95, Mahogany Rock at milepost 235, Alligator Back Overlook at milepost 242, Doughton Park at milepost 238-44; and lakes and reservoirs such as Cunningham Falls State Park in Maryland (p. 93); Claytor, Philpott and Smith Mountain lakes in Virginia; South Holston and Watauga Lakes near Bristol, Tenn. and Beaver Lake near Asheville, N.C.

Park rangers often lead field trips and give lectures about birds in many of the parks, and will direct you to popular viewing spots. Contact park headquarters for schedules. Additional information on where to look for birds can be found in the sectional field guides in Chapters 6 and 8. Also see "Birding Resources" on p. 359.

Birds by Category: PAGE

Birds of Prey 360
Perching Birds.................................... 362
Swallow-like Birds 370
Tree-clinging Birds............................ 371
Ground Birds 372

When to Look

Most birds, unless they are nocturnal are typically most active early in the morning and late in the afternoon. Walk along almost any trail in the Blue Ridge at these times and you will see a great variety of birds. Populations shift with the seasons. For any given locality some species remain throughout the year, while others leave for warmer climes, in search of food, or to return to their breeding grounds. Many species spend little time in the Blue Ridge and are most likely to be seen as they pass through during their spring or autumnal migrations. Others move to different elevations as the seasons and temperatures change. Still others seek refuge from storms or hurricanes that hit the southeastern coast.

The NPS provides guidance to the abundance, timing, and frequency of bird sightings in the Shenandoah and Great Smoky Mountains National parks on their website. The information at these sites is representative of which species are present in the northern and southern parts of the Blue Ridge.

Bird Migration

The seasonal migration of birds from one part of the world to another is a very complex phenomenon. It is almost certainly related to climate and to the climatic changes that have taken place during the last two million years. Over that time glaciers have come and gone across large parts of most continents. During each glacial advance ice sheets covered the northern parts of North America and Europe forcing most birds in those areas to search for food farther south. This was especially true of birds that depended on insects for their food supply. To this day birds still leave areas where insects are limited or lost during winter months in search of places where the food supply is more plentiful. Birds that feed on seeds, on the other hand, such as buntings and grosbeaks, may linger much longer into the fall before migrating and some may remain in place throughout the winter to search for seeds. These birds may appear at local bird feeders.

About half of all bird species migrate. In many species, part of the bird population migrates while the rest remain in place. Although migration routes are well established for many species, individual birds may migrate to one area one year and to some other far removed location the next. While most coastal birds migrate along the coasts many others, including birds of prey, swifts, swallows, martins, hummingbirds, woodpeckers, vireos, tanagers, warblers, finches, sparrows, and thrushes tend to migrate across or along the Blue Ridge as they move to warmer locations in the Southern U.S. or to Central or South America.

The distances traveled by some species during migration and the length of time they spend in the air can be truly impressive. Bobolinks that breed in Northern Canada migrate over 5,000 miles to central South America; turkey vultures, also strong fliers, move from the U.S. and Canada into South America; swifts may travel over 6,000 miles, and hummingbirds often journey several thousand miles. These trips are major undertakings and often extract a heavy a toll on bird populations, in extreme cases claiming the life of half the migrants. In preparation for such long trips, birds accumulate fat below the skin at the top of the breast.

Fig. 11-1: Flight styles

The excess fat may equal or exceed the weight of the bird. It is needed for the long journeys across the Gulf of Mexico and the Caribbean Sea. Some species stop along the way during their migrations; others, such as swifts and warblers, travel the entire distance without rest. Some warblers put on enough fat during the summer months to fly nearly 2,000 miles without stopping.

Most North American birds breed between April and October. Those that migrate do so in the spring and fall. The period of migration varies considerably, even within a particular species. Many species, such as blackbirds, gather and migrate as a flock. Some gather at sites in the South before leaving as a mixed group. Search local bird club websites or state game or fish and wildlife websites located in the area you plan to visit for more details about the timing of migrations and the species you may expect to find. Many localities have bird counts near the end of the year. These counts will give you a good idea of the resident species at those localities.

Flight Styles

The way a bird flies can help you make some identifications, especially for birds of prey that rarely come close to humans. Most birds fly in one of four styles: (Fig. 11-1) a) in a straight, horizontal line with continuous flapping of wings; b) in a vertical, sinuous path with flapping at the top of each wave followed by gliding with wings extended in between; c) in a vertical, sinuous path with flapping at the bottom of each wave, and wings folded in as it bounds or bursts upwards in between; and d) soaring, often combined with gliding. Some birds, notably turkey vultures and eagles, use updrafts and rising thermal cells to gain altitude, then glide on their wide wings in large circles around the thermals.

These flight styles are closely related to the shape and size of the bird, with larger wings providing more surface area to help it ride on top of the air. Hence, smaller birds tend to spend more time flapping to stay elevated. For example, the great blue heron, with its wide wingspan flaps very slowly in a straight path while the small pine siskin must rapidly flap and bound, flap and bound to stay afloat. Regardless of body size, however, most of the power for avian propulsion and lift generally comes from the downstroke of the wings. Tail feathers are used for additional lift and for steering and balance, like a rudder, as well as for braking.

Most birds do not flap their wings continuously while flying. Those that do tend to be heavier relative to their wing size, like ducks or geese. However, crows, sparrows, warblers and wrens also flap continuously (with occasional swooping).

Smaller birds tend to use some form of intermittent flapping along with gliding or bounding. The smallest tend to flap-bound but vary their style depending on speed, with more bounding at higher speeds and more gliding at lower speeds. Woodpeckers use a flap-glide style along with swooping. For finches, which are much smaller, this same style translates into an overall bouncing effect. Tree swallows are especially fun to watch as they soar, dive, bank, wheel, turn, pivot, and hover. Hummingbirds have a very unique kind of flight style—they can hover in one place for a long period of time with body upright and wings sweeping back and forth like a helicopter.

Bird Identification Tips

Aside from choosing the early morning or late afternoon to seek birds, it is also very helpful to have patience and to move slowly and quietly, blending in with your surroundings as much as possible. Standing or sitting in one place for awhile might be necessary to coax wary birds out of hiding. If you are with a group you should communicate with whispers. The edges of lakes are an especially good place to look for birds, as they are less obscured by trees and shrubs. Binoculars are essential for getting a good look. When you do locate a bird, before immediately turning to a field guide to identify it, instead fix your eyes upon it and study its features for it will likely fly away soon. Make note of its size and shape, wing and facial markings and beak shape, leg color and length, and tail shape. With these characteristics already in mind, it will be much easier to identify the bird from a guidebook while in the wild. Other things helpful to identification are flight style, bird calls or songs and exactly where and how it feeds.

Identifying Raptors on the Wing

You are most likely to see raptors in the air, and lucky to get a close look. Several can be distinguished by their flight style, the shape of their wings or tail, and their color. Their size and distinctive white head set the bald eagle apart, just as the soaring flight and bare reddish head make turkey vultures easy to spot. The red-tailed hawk has a broad fan shaped tail with white under their wings. The reddish color of the head and the outer edge of the wings help define the red-shouldered hawk. Red-tailed hawks fly with deep, regular wingbeats or soar in circles on their large wings. The broad-winged hawk has wide wings and a fan-shaped tail that is black with white strips. The coopers hawk is gray from above, with a reddish breast, and a distinctive long narrow tail with black bands. All these birds are carnivorous and hence have sharp curved claws used to seize prey and curved beaks used to tear flesh apart. Eagles, hawks, and vultures are evident during the daytime. Owls, though not categorized as raptors, are birds of prey that are more likely to be seen in the early morning or late in the evening.

Birders follow hawk migrations from mountain peaks and overlooks along the Blue Ridge Parkway and Skyline Drive. Eagles can also be seen from these places during their migrations, usually peaking in September.

Birding Resources

There are many excellent guidebooks to birding in the Blue Ridge and many online resources. The Virginia Dept. of Game and Inland Fisheries publishes the guidebook *Discover Our Wild Side: Virginia Birding and Wildlife Trail – Mountain Area* that contains many birding trail maps with detailed directions and descriptions. These are also available on their website at **http://www.dgif.virginia.gov/**. See the "Wildlife Watching" page for links. Also see Richard Rowe's "Birds of the Greater Rockbridge County, Va. Area" at **http://www.flickr.com/photos/vmibiology/sets/** for birds of central Virginia. For the Great Smoky Mtns. area, see the NPS website and other sites focusing on the Smokies. An excellent guidebook for that area is *Birds of the Smokies* by Fred Alsop.

For general birding information, The Cornell Lab of Ornithology (**http://www.allaboutbirds.org/**), The Audubon Society (**http://www.audubon.org/bird-guide**) and The American Birding Association (**http://www.aba.org/**) are excellent. In addition to having many useful links, the ABA publishes *A Birder's Guide to Virginia* by David Johnston. Another popular resource is the "Bird ID InfoCenter" at the USGS Patuxent Wildlife Research Center at **http://www.pwrc.usgs.gov/birds**. For information on bird habitats visit **http://www.birdnature.com/**. Additional sources can also be found in the Reference section.

Serious birders who would like to keep up with where specific birds are being sighted should seek information from local bird clubs or from national and regional birding websites. One of the most extensive, "eBird" (**http://www.ebird.org/**) is a database of observations about birds intended for use by professionals as well as amateurs. It was created in 2002 by the Cornell Lab of Ornithology in cooperation with the National Audubon Society. You will find there an extensive record of sightings from many parts of the world, and you can report your own sightings as well.

Bird Identification Guide

The following pages present photos with a brief description of some of the birds you are most likely to see in the Blue Ridge. They have been divided into the following categories: birds of prey, perching birds, swallow-like birds, tree-clinging birds, and ground birds. Many others, not shown here, cross the mountains during migrations. Some of these would be classified as water birds, likely to be seen along rivers and lakeshores. Although water birds are not covered in the identification guide several can be found on p. 221. ❖

Birds of Prey

(Organized loosely by color and/or family)

Note: The following icons indicate the seasons when each bird can be seen in the Blue Ridge Mtns.

YR = **Year Round**

S/S = **Spring/Summer to Early Fall**

W = **Winter only**

M = **Migrates through Blue Ridge**

Bald Eagle

(Haliacetus leucocephalus) W

Both sexes are large with a dark brown body and wings (wingspan usu. 6-7.5 ft.), a white head and tail, a yellow beak, and yellow eyes. Females are 25% larger. Numbers are slowly increasing since DDT was banned. Spends most of its time near lakes and rivers where it hunts for fish, but will also eat birds, small animals, and carrion. Builds a large nest from sticks high in tall trees. Its call is a squeaky cackle. (Photo: Hal Korber, PGC)

Peregrine Falcon

(Falco peregrinus) M Both sexes have a slate-colored back, a black and white barred underside, dark cap, a black-tipped yellow beak, and yellow eyerings. Females have cinnamon-colored tails with black tips (male shown). Rare, nearly extinct due to pesticides. Nests on cliff ledges and in hollow trees. Famous for its incredible speed when diving after prey (mostly birds). Its raspy call is heard only at its nest. (Photo: Joe Kosack, PGC)

Broad-winged Hawk

(Buteo platypterus) S/S

Both sexes have brown wings and cap, light barred breast, and a short, wide, barred tail. Inhabits deciduous and mixed forests near water. Feeds on small mammals, reptiles, amphibians, birds. Builds its nest high in trees. In winter large flocks go to tropical forests. Call is a long thin whistle. (Photo: Julie Waters)

Cooper's Hawk

(Accipiter cooperii) YR

Smaller than a crow, both sexes have a slate-gray back, light cinnamon and white barred underside and cheeks, and a long tail with bold black bars. Inhabits forests, open woods, and edges where it hunts for birds and small mammals. Builds its nest on tree branches 20-50 ft. above the ground. Call is a "cack-cack-cack." (Photo: H. Gilbert Miller†)

Red-shouldered Hawk

(Buteo lineatus) YR

Both sexes have black and white barred wings and back, a rusty patch on the shoulder, a white and rust striped head and underside, and a wide tail. Inhabits forests, esp. near water where it hunts small mammals, reptiles, amphibians, birds, and fish. Nests high in deciduous trees and has a shrill, scream-like call. (Photo: George Raiche, USGS)

Red-tailed Hawk
(Buteo jamaicensis) YR
Both sexes have a brown barred back and wings, a pale streaked underside, and a wide cinnamon-colored tail. Females are 25% larger than males. Widespread, it inhabits open woods where it hunts birds, small mammals, and reptiles, esp. snakes. Nests high in tall trees. Call is a high, rough scream. (Photo: Michele Fletcher)

Turkey Vulture
(also Buzzard) *(Cathartes aura)* YR
One of the largest scavengers. Both sexes have a dark brown body, 6 ft. wingspan (two-toned on the underside with distinctive wingtips), a small ugly red head, a white tipped beak, and a long, wide tail. Flies over fields looking for carrion. Builds no nest, lays its eggs in thickets, hollow logs, or cliff crevices. Often flies in wide circles on updrafts. Usually silent unless grunting and hissing while eating or nesting.

Barn Owl
(Tyto alba) YR Palest of the North American owls. Has light brown and gray barred wings and back, and is white underneath. Its white, heart-shaped face and square-shaped tail are distinctive. Female is bigger, darker, and more speckled. Inhabits open woods and fields where it hunts mostly small mammals. Nests in cavities. Call is a shrill hissing "ssssshpp." (Photo: Lori Richardson, PGC)

Barred Owl
(Strix varia) YR Both sexes are large, have gray-brown barred wings, a white streaked breast, and a distinctive face with large dark eyes. Inhabits forests near swamps where it hunts small mammals, birds, reptiles, and amphibians night and day. Nests in tree cavities. Call is a loud barking "hoo-hoo-hoo-aw." (Photo: Dick Rowe)

Eastern Screech-Owl
(Megascops asio) YR Small, rarely seen nocturnal owl. Coloring may be gray or light cinnamon with extensive barring. Both sexes have a lighter underside, small "horns," and huge yellow eyes. Inhabits open woods where it hunts at night for insects, small rodents, reptiles, fish, and amphibians. Nests in tree cavities. Given its name, its call is a surprisingly soft whinny or trill. (Photo: Christopher Bohinski, USGS)

Great Horned Owl
(Bubo virginianus) YR Large owl (up to 23" long, 52" wingspan) with feathery "horns." Inhabits forests and open woods near streams. Is a fierce nocturnal hunter of birds and mammals deep in the woods. Nests in abandoned stick nests, stumps, and cliff crevices. Call is a deep-toned "hoo-hoo-hoo". (Photo: Shannon Spencer)

Perching Birds

(Organized loosely by color and/or family)

Note: The following icons indicate the seasons when each bird can be seen in the Blue Ridge Mtns.

YR = **Year Round**

S/S = **Spring/Summer to Early Fall**

W = **Winter only**

M = **Migrates through Blue Ridge**

Cardinal, Northern

(Cardinalis cardinalis) YR Male cardinal (shown) is bright red with a black face and red beak; female is more grayish (inset). Both exhibit a prominent crest and a heavy bill. Inhabits thickets and wood edges where it feeds on insects, seeds, and berries and nests in shrubs or dense bushes. Has various songs, including "purty-purty-purty," and "sweet-sweet-sweet."

Scarlet Tanager

(Piranga olivacea) S/S Male (shown) has a brilliant red body with black wings and tail; female is greenish with black wings and a lighter underside. Inhabits deciduous or mixed forests and feeds mainly on insects, and some berries and fruits. Nest is loosely built on tree limbs. Song is a burry, repetitive warble. (Photo: Shawn Collins, 2011)

House Finch

(Haemorhous mexicanus) YR Male (shown) has a steaky, grayish-brown body, red head and throat; female is a softly streaked, grayish-brown all over. Inhabits trees near open country where it feeds on seeds and berries. Has proliferated in trees near human habitats. It has a chirpy call.

Purple Finch

(Haemorhous purpureus) W Male (shown) has a pinkish body with gray-brown barred wings and a raspberry red head and crest; female is brown with a white streak behind the eye. Both have white underparts with gray streaks. In open woods it feeds on seeds, berries, buds, insects. Prefers coniferous forests during breeding season. Songs include a rich, slurry warble. (Photo: Dick Rowe)

Rose-breasted Grosbeak

(Pheucticus ludovicianus) S/S Male (shown) has a black head and back, white underparts, white wing patches, and a red patch high on the breast; female is brown-black with light, streaked underparts and a white streak above the eye. Inhabits open forests where it feeds on insects, fruits, seeds, and buds. Nests on tree and shrub limbs. (Photo: Jacob Dingel, PGC)

Baltimore Oriole
(Icterus galbula) S/S Male (shown) has a bright orange breast, black head and wings; female is gray with a yellow breast. Inhabits open woods where it feeds on insects, spiders, and fruits. Builds a distinctive, purse-shaped nest that swings from tree limbs. Song is a series of piping, whistled notes. (Photo: Dave Neimeyer)

American Robin
(Turdus migratorius) YR Male has a dark-gray back and wings, a rust-colored breast, a black tail (white underneath), black head, yellow beak, and white eyerings; female is similar but with a dark gray head. Inhabits forests where it eats insects and berries. Lays its blue eggs in nests it builds on tree limbs from mud and grasses. Songs include a caroling, musical "cheer up, cheer up, cheerily, cheer up" and a chirpy whinny.

Eastern Towhee
(Pipilo crythrophthalmus) YR Male (shown) has a black head and back, an orange area on its side, and a white underbelly; female has a similar pattern but with a brown head and back. Inhabits the brush of open woods and edges. Feeds on insects, berries, and seeds found on or close to the ground. Nest is either on the ground or low in bushes. Song is highly varied.

Red-winged Blackbird
(Agelaius phoeniceus) YR Male (shown) is black with red-orange shoulders; female is brown with white streaks all over. Inhabits wet marshy areas and fields. Feeds on insects during summer; seeds during the rest of the year. Nests in grasses or bushes near water. Song is "o-ka-leee." (Photo: Peter S. Weber, USGS)

Cedar Waxwing
(Bombycilla cedrorum) YR Both sexes have a cinnamon crest, distinctive black mask, a yellowish underside, buff breast, gray wings with red spots on the tips, and bright yellow tail feather tips. Flocks fly in close formations. Inhabits open woods, feeding on insects in summer; berries in winter. Builds a cup-shaped nest (often many nests located close together) on tree branches. Call is a high thin "tseeee."

Ruby-throated Hummingbird *(Archilochus colubris)* S/S Male has a greenish head and back, light underparts, and red throat; female (shown) is similar, but without the red throat. Likes wood edges where it eats flower nectar and insects. Nest is often lichen-covered and saddle-shaped, hanging on a branch. Song is squeaky and chirping.

Yellow Warbler
(Setophaga petechia) S/S
Of 30 species of warblers in the Appalachians, this is the only one that is all yellow. Male (shown) is brighter in color, esp. in spring and summer. Inhabits thickets and woods near streams where it hunts insects and nests in small trees or shrubs. Song is a musical "sweet-sweet-sweet, sweeter-than-a-sweet."

Chestnut-sided Warbler
(Setophaga pensylvanica) S/S
Both sexes have a white breast and barred wings. The rust-colored streaks on its sides and yellow cap are distinctive. Male (shown) has a black streak through the eyes and down the cheek. Inhabits open woods and feeds on insects. Nests in dense scrubby areas a few feet off the ground. Winters in the tropics. Song sounds similar to "very very very pleased to meet ya." (Photo: Greg Schechter†)

Golden-crowned Kinglet
(Regulus satrapa) YR Both sexes are gray with a bright yellow and black striped cap and olive-gray back, lighter underparts, barred wings, a short tail, and a black stripe through the eye. Male has an orange streak down the center of the yellow cap (female shown). Inhabits conifer and mixed forests and feeds on small insects. Builds a cup-shaped nest under branches. Song is a "ti-ti-ti" that slowly rises, then falls. (Photo: Dick Rowe)

American Goldfinch
(Spinus tristis) YR Male is yellow (brighter in spring), has black and white barred wings and tail, and a black fore-cap; female is similar but more grayish. Inhabits fields and semi-open weedy areas where it eats seeds and some insects, and builds its compact nest near the ground. Call resembles "po-ta-to-chip."

Eastern Meadowlark
(Sturnella magna) YR Both sexes have a barred, dark brown and black back and wings, short, wide tail, a dark cap, and a long white streak above the eye. Notable also is its yellow breast marked with a large dark "v." Inhabits fields where it feeds on crickets, grasshoppers, larvae, seeds, and grains. Nests on the ground. Its song has a sweet, whistling quality. (Photo: A. Wilson, USGS)

Eastern Wood Pewee
(Contopus virens) S/S Medium-sized flycatcher. Both sexes are gray with a darker head, long dark tail, and two white bars on the wings. Inhabits deciduous and mixed forests where it hunts insects. Builds a shallow, cup-shaped nest on high tree limbs. Its song resembles "pee-ah-weeee-ah" with a low final note. (Photo: Kelly Azar)

Ovenbird
(Seiurus aurocapilla) S/S Large songbird of the warbler family. Both sexes have an olive-brown head and back, a white breast with streaky black spots, two dark streaks on its head, and white eyerings. Inhabits mixed forests and thickets where it builds dome-shaped nests. Male's song is a very loud "tea-Cher, tea-Cher, tea-Cher." (Photo: Mike's Birds†)

Wood Thrush
(Hylocichla mustelina) S/S Both sexes have a rusty brown cap, back, and wings, gray face, a white underside with streaky black spots, and white eyerings. Its dark eyes and rounded breast distinguish it from the brown thrasher. Inhabits damp deciduous or mixed forests where it forages on or near the ground for insects and berries. Builds its nest with mud, sticks, and leaves in trees. Its song is a melodious flute-like "ee-oh-lay." (Photo: Steve Maslowski, USF&WS)

Great Crested Flycatcher
(Myiarchus crinitus) S/S Both sexes are large with a short bushy brown crest, barred brown wings, gray face and throat, and a lemon yellow belly. Found in deciduous or mixed forests and at wood edges. Spends most of its time in tree canopies swooping after insects such as flies, bees, butterflies, and grasshoppers. Builds its nest in tree cavities. Song is a weepy whistle. (Photo: Meredith Lombard)

Eastern Phoebe
(Sayornis phoebe) S/S Both sexes are gray-brown with a lighter underside and barred wings. Inhabits open woods and edges, esp. near water. This flycatcher feeds mainly on flying insects and berries. Nests in rocky cliffs and man-made structures. Song is a soft "phee-bee." (Photo: Peter Wallack†)

House Wren
(Troglodytes aedon) S/S Both sexes are a plain brown-gray with delicate barring, and lighter underside. This small bird with a long curved beak moves very quickly often cocking up its short tail. Inhabits open woods and thickets. Nests in all manner of cavities, natural or man-made. Feeds on a wide variety of insects and spiders. Has a bubbly song that rises and falls. (Photo: Alan Vernon†)

Brown Thrasher
(Toxostoma rufum) S/S A large, cinnamon-brown bird with a curved bill, long tail, barred wings, white breast with black stripes, and yellow eyes. Inhabits thickets at wood edges where it feeds on insects, spiders, berries, and nuts. Builds its nest in thickets and bushes low to the ground. Mimics other birds using over 1,000 types of songs.

American Tree Sparrow
(Spizella arborea) W Both sexes have a buff-gray body with a dark spot on the breast, barred wings, a rusty cap, dark eye line, and yellow underbill. Inhabits shrubs in fields, wood edges, and marshes where it feeds on seeds and berries in winter. Flies north to breed in nests on the ground under shrubs. Winter song is "see-ler." (Photo: Steve and Wendy Richards)

Chipping Sparrow
(Spizella passerina) S/S Both sexes have a light gray body, brown barred wings, a wide rusty cap, and a long black streak through the eye. Inhabits open woods, edges, and shrubby fields where it forages mainly for insects and seeds near the ground. Often nests in conifers not far from the ground. Its loud monotone trilling songs and chip-like alarm call are common in spring. (Photo: Kelly Azar)

Field Sparrow
(Spizella pusilla) YR Both sexes are small and buff-colored with brown barred wings, cinnamon streaks on the head, white eyerings, and a pink beak. Inhabits shrubby fields and wood edges, esp. with tall grasses. Feeds on insects during warm seasons and seeds in winter. Builds its nest on or near the ground in shrubs. Songs are plaintive, ending with faster trills. (Photo: Kelly Azar)

House Sparrow
(Passer domesticus) YR Male has a rusty-brown head, back, and wings, a gray cap, a stout black bill, and white cheeks and underside. Male (shown) has a black bib; female has soft brown coloring overall. Found only in human habitats. Feeds on seeds, some insects. Nests in man-made cavities. Song is a sharp, monotone chirping. (Photo: Mdf†)

Song Sparrow
(Melospiza melodia) YR Both sexes have a gray and rust streaked back, a white breast with rusty streaks that converge to a central white bib, a rusty cap, and a brown streak behind the eye. Inhabits wood edges, thickets, brushy fields, and swamps. Eats insects, seeds and berries. Nests close to the ground. Song is a melodious series of tweets, followed by a trill with a flowery finish. (Photo: Kelly Azar)

White-throated Sparrow
(Zonotrichia albicollis) W Both sexes are buff-colored with brown barred wings, bold black and white streaks on the head, a white throat, and a bright yellow patch over the eye. Inhabits thickets and undergrowth in coniferous or mixed woods where it forages for seeds and insects on the ground. Migrates north to nests on the ground in thickets. Song is a long plaintive whistle.

Pine Siskin
(Spinus pinus) W Small finch. Both sexes have a streaky light brown head, a white underside with dark streaks, and dark brown and yellow barred wings. Flocks found in open conifer or mixed forests and edges. Feeds on seeds, esp. thistle. Builds its shallow, saucer-shaped nest on the branches of conifers. Breeds in the far north. Songs are rising and buzzy. (Photo: Dick Rowe)

Veery
(Catharus fuscescens) S/S Medium-sized thrush. Both sexes have a cinnamon cap, back, and wings (with gray barring), a white underside and a spotted throat. Inhabits moist, deciduous or mixed forests with a dense understory where it feeds on insects and berries. Breeds in the underbrush, often along streams, laying eggs in a cuplike nest on or near the ground. Winters in the tropics. Song is a descending "veer" with a delicate metallic whirling quality. (Photo: Kelly Azar)

Hermit Thrush
(Catharus guttatus) W Both sexes resemble the veery but are a duller brown. Has a rusty-colored tail, black spotted breast, and white eyerings. Tends to cock its tail up quickly then slowly lower it. Inhabits the underbrush of forests where it forages for insects and berries. Breeds mostly in the far north. Its song is a mournful see-sawing whistle. (Photo: Steve and Wendy Richards)

Carolina Wren
(Thryothorus ludovicianus) YR Both sexes have a brown head, back, and wings, a buff-colored underside, a short tail (cocked up), a long white streak over the eye, and a curved beak. Inhabits dense undergrowth in woods and edges. Eats insects and snails. Nests in cavities. Song is a repetitive "teakettle-teakettle, etc..." (Photo: Ken Thomas)

Mourning Dove
(Zenaida macroura) YR Wild pigeon. Both sexes are rounded, with a faintly iridescent gray-brown body, black spots on the wings and wingtips, a long tail, and blue-gray eye-ring. Inhabits fields and open woods where it feeds on seeds. Builds its nest in shrubs, trees, or on the ground. Both sexes feed the young "pigeon milk" from throat glands. The familiar male mating song is a soft, mournful "coo-OO-oo-oo-oo."

Rock Pidgeon
(also Rock Dove) *(Columba livia)* YR Both sexes have a gray body and wings (with two black bars), a green-purple iridescent neck, and orange eyes. Males puff up their necks and strut during courtship. Inhabits most human landscapes. Eats seeds, grains, berries. Roosts in groups on cliffs and in buildings. Coos are gentle and guttural. (Photo: Jörg Hempel†)

Red-eyed Vireo

(Vireo olivaceus) S/S Both sexes are small with an olive-green back and wings, white underside, gray cap, and a wide white streak with black borders above a dark red eye. Prefers open woods with undergrowth. Feeds mostly on insects and berries. Lays its eggs in a cup-shaped nest in trees. Male's song is a short, endlessly repeated up and down phrase. (Photo: Kelly Azar)

Gray Catbird

(Dumetella carolinensis) S/S Both sexes are light gray with a dark cap and rusty coloring under a long dark tail. Inhabits leafy thickets along wood edges or near streams where it feeds on insects and berries. Nests of grass and twigs are built in dense underbrush. Winters in warmer zones. Can sound like a cat mewing but has a wide range of vocal abilities similar to a mockingbird and can produce complex songs lasting many minutes. (Photo: Peter Massas†)

Dark-eyed Junco

(Junco hyemalis) YR Male (shown) is slate gray with a white belly and undertail and a pale pink beak; female is similar but dull brown. Inhabits open conifer or mixed forests, edges, and fields where it feeds on grass seeds, insects, and berries. Builds grassy nests in hidden spots on the ground. Male song is a loud trill with a rapid machine-gun quality. Its calls have a similar pulsing delivery.

Tufted Titmouse

(Bacolophus bicolor) YR Both sexes are small and gray with a tufted crest and a black spot over the beak, a white underside, and rusty tinge under the wings. Inhabits woods where it feeds on insects and seeds. Nests in natural cavities. Male feeds the female while she sits on the eggs. Song is a repetitive, whistled "peter, peter, peter."

Chickadee

(Poecile atricapillus, P. carolinensis) YR Carolina variant (shown): both sexes are light gray with a black cap and throat, bright white cheeks and a large head. The black-capped variant, more numerous in the N. Blue Ridge, is similar but with buff-colored sides. Inhabits deciduous and mixed forests and fields where it feeds on insects, seeds, and berries. Lays its eggs in tree cavities. Song is often a whistled "fee-bee"; call is a "chickadee-dee-dee."

Northern Mockingbird

(Mimus polyglottos) YR Both sexes have a gray back and head, white underside, black wings with white bars, long tail, and yellow eyes. Inhabits wood edges and fields where it eats insects and berries. Builds a thick nest of twigs with a soft lining in dense shrubs or trees. Both sexes have astounding vocal abilities and memory. Famously mimic the calls of many other birds.

Blue-gray Gnatcatcher
(Polioptila caerulea) S/S Both sexes are small and blue-gray with barred wings, a white underside, long tail and beak, and white eyerings. Males are darker, with a dark eye-streak (female shown). Inhabits open woods and wood edges. Hunts insects. Builds its nest in trees and covers it with spiderwebs. Call is a high, nasal "tzzzz." (Photo: Elaine R. Wilson†)

Brown-headed Cowbird
(Molothrus ater) YR Male (shown) is black (faintly iridescent) with a brown head and distinctive beak shape; female is light brown all over. Inhabits fields and edges where it feeds on seeds and insects. A brood parasite, it lays up to 40 eggs a year in the nests of over 200 other bird species. These often hatch and grow much faster than their nest mates. Male's spring song is an ineffable, liquid gurgle that drops, then ends in a thin, sliding whistle.

Common Grackle
(Quiscalus quiscula) YR Male (shown) has a glossy, dark, iridescent body, an iridescent blue head, long legs and tail, and a bright yellow eye; female is dull brown with darker wings, back, and cap. Inhabits open woods, edges, and fields where it feeds omnivorously on insects, amphibians, small animals, berries, seeds, and acorns. Its songs are squeaky and rough. (Photo: Meredith Lombard)

European Starling
(Sturnus vulgaris) YR Both sexes are iridescent black with light flecks (brighter in winter), purplish throat, and a long beak (yellow in summer). Generally inhabits human-occupied areas in flocks where it eats insects, berries, seeds. Nests in tree cavities. Song is highly varied, jumbled, often squeaky and raspy. (Photo: Daniel Plazanet†)

American Crow
(Corvus brachyrhynchos) YR Large, intelligent bird. Both sexes are all black, with long legs, and thick neck and beak. Known to use tools and to roost in large groups in winter. Inhabits woods and edges where it feeds omnivorously on insects, grain, seeds, amphibians, other birds, eggs, berries, and carrion. Builds large nests in trees. Call is a loud grating "caw-caw-caw" with many variations. (Photo: Jack Wolf†)

Common Raven
(Corvus corax) YR Enormous all-black bird (can be over 2 ft. long). Both sexes have pointed wingtips, a long, wedge-shaped tail, shaggy throat, and a thick curved beak. Uncommon in the Blue Ridge. It may inhabit forests and cliffs where it nests and hunts, feeding omnivorously. Call is a deep guttural croak. (Photo: Amy Langman)

Blue Jay

(Cyanocitta cristata) YR Both sexes are large and bright blue with a crested head, lighter face and underside, bold white wing bars, and a long, blue tail with thin black bars. Inhabits open mixed woods and edges where it feeds omnivorously, mostly on insects, seeds, berries (but also small animals, eggs). Builds its nest of twigs in tree crotches. Call is often loud and raucus.

Eastern Bluebird

(Sialia sialis) YR (Both male and female shown.) Male has a bright blue back and head with a cinnamon breast and white underside; female is dark gray where the male is blue. Inhabits open conifer forests and edges where it feeds on insects and berries. Roams in flocks and nests in tree cavities. Song has a low-pitched, warbling quality.

Indigo Bunting

(Passerina cyanea) S/S Male (shown) is bright iridescent blue all over with a slightly darker head, some black on wings; female is light brown. Both are small and stocky, with short tails. Inhabits open woods, edges, and brushy fields where it feeds on insects, seeds, and some berries. Builds a grassy nest in shrubs near the ground. Song has warbly paired tweets.

Swallow-like Birds

Small birds that tend to stay in the air most of the time hunting insects, have pointed wings and form large flocks.

- YR = **Year Round**
- S/S = **Spring/Summer to Early Fall**
- W = **Winter only**
- M = **Migrates through Blue Ridge**

Tree Swallow

(Tachycineta bicolor) S/S Male has a shiny blue-green back and cap, dark gray wings, a white throat and underside, and a black patch near the eye; female is gray where the male is blue. Both are small with pointed wings, a short, notched tail, and a tiny beak. Lives near lakes and marshes where it hunts flying insects. Builds its nest in tree cavities. Song is high and liquidly twittery. (Photo: Kelly Azar)

Chimney Swift

(Chaetura pelagica) S/S Both sexes are dark-gray with a thick pale-gray neck, flat head, and small beak. Unable to perch, it clings to vertical surfaces, often crossing its wings. Famous for its incredible sky maneuvers, it spends most of its life in the air hunting mainly flying insects. Nests in hollow trees and caves (also chimneys). Calls are a rapid-fire twittering. (Photo: NPS)

Tree-clinging Birds:

Note: The following icons indicate the seasons when each bird can be seen in the Blue Ridge Mtns.

YR = **Year Round**

S/S = **Spring/Summer to Early Fall**

W = **Winter only**

M = **Migrates through Blue Ridge**

Northern Flicker
(Colaptes auratus) YR Both sexes are a soft gray-brown with a lighter underside (with black spots), black, barred wings with yellow underneath (yellow-shafted var. shown), a black bib, flared tail, and a bright red crescent on the back of the neck. Inhabits open woods where it inserts its long beak into trees or anthills to withdraw insects with a sticky tongue. Also eats berries and seeds. Nests in tree cavities. Calls include a rattley "wik-wik-wik" (Photo: Jacob Dingel, PGC)

Downy Woodpecker
(Picoides pubescens) YR Smallest and most common woodpecker. Both sexes are covered with bold black and white streaks, and have white undersides. Males have a bright red patch on the back of the head (female shown). Inhabits forests and open woods where it works away at tree bark, hunting for grubs and insects. Nests in dead tree cavities. Calls include whinnying and a "pik, pik..."

Pileated Woodpecker
(Dryocopus pileatus) YR
Largest woodpecker. Both sexes are black with bold white streaks on the cheeks and neck, white under the wings, yellow eyes, and a bright red crest. Male (shown) also has a red "mustache." Inhabits mixed forests where it hunts insects, ants. Nests in tree cavities. (Photo: Jacob Dingel, PGC)

Red-bellied Woodpecker
(Melanerpes carolinus) YR Both sexes have a black and white barred back and wings, white face and underside, faintly reddish belly, and a red neck that in males (shown) extends into a full cap. Inhabits mostly deciduous or mixed forests where it often picks at bark for insects but is also omnivorous, eating nuts, berries, small animals. Nests in tree cavities. Calls include a rolling "querr, querr."

Red-headed Woodpecker
(Melanerpes erythrocephalus) S/S
Only woodpecker with an entirely red head and neck (in both sexes). Has a black body with white underside and partially white wings. Inhabits open pine woods and edges where it feeds omnivorously on insects, nuts, berries, etc. Nests in dead tree cavities. Call is a "churr, churr." (Photo: Francesco Veronesi†)

Tree-clinging Birds Continued

Nuthatch (White-breasted)

(Sitta Carolinensis) YR Both sexes have a blue-gray back and barred wings, a white face and breast, and a rusty area under its stubby tail. It has a wide neck, and a long slightly upturned bill with which it often lodges large seeds (such as acorns) into bark and then hammers away until the nut "hatches." Male (shown) has a black cap; female has a dark gray cap. Inhabits mixed coniferous forests where it feeds on insects and seeds and can be seen running head-first down tree trunks. Nests in high natural cavities. Songs include a nasal "wank, wank, wank."

Ground Birds

Note: The following icons indicate the seasons when each bird can be seen in the Blue Ridge Mtns.

- YR = **Year Round**
- S/S = **Spring/Summer to Early Fall**
- W = **Winter only**
- M = **Migrates through Blue Ridge**

Wild Turkey

(Meleagris gallopavo) YR Large dark bird with a brown-green iridescence, a bald blue-to-red head, and a red waddle. Male's coloring is stronger overall (shown, puffed up with fanned tail). Flocks inhabit deciduous woods. It often feeds on acorns, but is omnivorous. Builds its nest on the ground near trees, shrubs, or tall grasses. Male's mating call is the famous "gobble-gobble." (Photo: Joe Kosack, PGC)

Ruffed Grouse

(Bonasa umbellus) YR Both sexes are mottled brown with white and black barring and have a chicken-like shape. It inhabits deciduous or mixed forests and edges where it feeds on insects and seeds. Nests on the ground. Usually quiet, during mating season the male produces a drumming sound by beating its wings. (Photo: Jacob Dingel, PGC)

References

General Non-technical References

Nash, S., 1999, Blue Ridge 2020, An Owner's Manual: Chapel Hill, University of North Carolina Press.

Olson, T., 1998, Blue Ridge Folklife: Jackson, University Press of Mississippi.

Ruppert, J.F., 2010, Mountain Nature, A Seasonal Natural History of the Southern Appalachians: Chapel Hill, University of North Carolina Press.

Weidensaul, S., 1994, Mountains of the Heart: Golden, Colo., Fulcrum Publishing.

Zeitner, J.C., 1968, Appalachian Mineral and Gem Trails: San Diego, Lapidary Journal, Inc.

General Technical References

[Additional technical references are cited for Chapters 5 and 7.]

28th International Geological Congress Field-trip Guidebooks: Indexed by Trip Title, Volume, and Trip Number, 1989, International Geological Congress. [Includes a large number of guides to features of geological interest in the Appalachians. Many of these are best suited for use on geology field trips for students and professional geologists.]

Bailey, C.M., Sherwood, W.C., Eaton, L.S., and Powars, D.S., 2016: The Geology of Virginia, Martinsville, Va. Museum of Natural History, Special Publication 18.

Bailey, R.G., 1996, Ecosystem Geography: New York, Springer-Verlag.

Bally, A.W., and Palmer, A.R., 1989 Geology of North America - Vol. F-2, The Appalachian-Ouachita Orogen in the United States, Geological Society of America.

Edwards, L., Ambrose J., and Kirkman, L.K., 2013, The Natural Communities of Georgia: Athens, The University of Georgia Press.

Frankenberg, D., ed., 2000, Exploring N. Carolina's Natural Areas, Parks, Nature Preserves, and Hiking Trails: Chapel Hill, University of North Carolina Press.

Geological Excursions in Virginia and North Carolina, 1987, Guidebook – Field Trips No. 1-7: Geological Society of America Southeastern Section, Norfolk, Va., Old Dominion University.

Neathery, T.L., ed., 1986, Southeastern Section of the Geological Society of America: Boulder, Colorado, Geological Society of America Centennial Field Guide, v. 6. [This series of guides covers the southeastern section of the Appalachian Mountains, including the Blue Ridge in Georgia, Tennessee, North and South Carolina, Virginia, and West Virginia.]

Part 1

Chapter 1: The Appalachian Mtns.

Anonymous, 1981, Upland Archeology in the East - Symposium No. 1: USDA, Forest Service and Archaeological Society of America, Special Publication no. 38.

Bick, K.F., 1962, Geology of the Williamsville Quadrangle, Virginia: Virginia Division of Mineral Resources, Report of Investigations 2.

Clark, S.H.B., 2001, Birth of the Mountains: The geologic story of the southern Appalachian Mountains: U.S. Geological Survey, unnumbered report. [Available online at http://pubs.usgs.gov/gip/birth/]

Cook, F.A., and Vasudevan, K., 2006, Reprocessing and enhanced interpretation of the initial COCORP S. Appalachian traverse: Tectonophysics, v. 420.

Grow, J.A., 1980, Deep structure and evolution of the Baltimore Canyon trough in the vicinity of the COST No. B-3 well, in Scholle, P.A., ed., Geological Studies of the COST B-3 Well, U.S. Mid-Atlantic Continental Slope Area: U.S. Geological Survey Circular 833.

Hunt, C.B., 1974, Natural Regions of the United States and Canada: San Francisco, W.H. Freeman and Company.

Rankin, D.W., Espenshade, G.H., and Neuman, R.B. 1972, Geologic Map of the West half of the Winston-Salem Quadrangle, North Carolina, Virginia, and Tennessee: Miscellaneous Geologic Investigations Map I-709-A, scale 1:250,000, U.S. Geological Survey.

Rodgers, J., 1970, The Tectonics of the Appalachians: New York, Wiley-Interscience.

Chapter 2: Geological Evolution of the Blue Ridge

Bally, E.W., and Palmer, A.R., 1989, The Geology of North America: Geological Society of America.

Blakey, R., and Ranney, W., 2008, Ancient Landscapes of the Colorado Plateau: Grand Canyon Association.

Cook, F.A., and Vasudevan, K., 2006, Reprocessing and enhanced interpretation of the initial COCORP

S. Appalachian traverse: Tectonophysics, v. 420.

Damian N.R., and Linnemann, U., 2008, The Rheic Ocean: Origin, Evolution, and Significance: GSA Today, v. 18, no. 12.

Harris, L.D., 1979, Similarities between the thick-skinned Blue Ridge anticlinorium and the thin-skinned Powell Valley, Geological Society of America Bulletin, v. 90.

Harris, L.D., de Witt, W., Jr., and Bayer, K.C., 1982, Interpretative Seismic Profile Along I-64: USGS, Oil and Gas Investigations Chart OC-123.

Hatcher, R.D. Jr., and Butler, J.R., 1979, Guidebook for Southern Appalachian Field Trip in the Carolinas, Tennessee, and northeastern Georgia, The Caledonides in the USA: Raleigh, North Carolina Geological Survey.

Hatcher, R.D., Jr., 1989, Tectonic synthesis of the U.S. Appalachians: in Hatcher, R.D., Jr., Thomas, W.A., and Viele, G.W., eds., The Appalachian-Ouachita Orogen in the United States: Boulder, Colorado, Geological Society of America, The Geology of North America, v. F-2.

Hatcher, R.D., Jr., 2010, The Appalachian orogen: A brief summary: in Tollo, R.P., From Rodinia to Pangea: Geological Society of America Memoir 206, Geological Society of America: Falls Church, Virginia, Field Trip Guidebooks, American Geological Institute.

Hibbard, J., et al., 2005, Tectonic Lithofacies Map of the Appalachian Orogen: Geological Survey of Canada.

Merschat, A.J., and Hatcher, R.D., Jr., 2007, The Cat Square terrane: Possible Siluro-Devonian remnant ocean basin in the Inner Piedmont, southern Appalachians, USA: Geological Society of Am., Memoir 200.

Mitra, G., and Lukert, W.T., 1982, Geology of the Catoctin-Blue Ridge Anticlinorium in Northern Virginia: Central Appalachian Geology, Field Trip Guidebooks, Lyttle, P.T., ed., American Geological Institute.

Murphy, J.D., et al., 2006, Origin of the Rheic Ocean: Rifting along a Neoproterozoic suture?: Geology, v. 28

Nance, R.D., and Linnemann, U., 2009, The Rheic Ocean: Origin, Evolution, and Significance: GSA Today, v. 18, no. 12.

Prowell, D.C., 2006, Evidence for Late Cenozoic Uplift in the Southern Appalachian Mountains from Isolated Sediment Traps: abstract, Geological Society of America, Southeastern Section - 55th Annual Meeting, paper no. 27-11.

Schultz, A., and Compton-Gooding, E., 1991, Geologic Evolution of the Eastern United States, Field Trip Guidebook, Northeast-Southeast Geological Society, 1991: Martinsville, Virginia Museum of Natural History, Guidebook no. 2.

Schultz, A., and Henika, W.S., 1994, Field Guides to Southern Appalachian Structure, Stratigraphy, and Engineering Geology: Blacksburg, Virginia Polytechnic Institute and State University.

Stewart, K.G., and Roberson, M.R., 2007, Exploring the Geology of the Carolinas: Chapel Hill, University of North Carolina Press. [Includes field trips to Stone Mountain, Grandfather Mountain, Whiteside Mountain, Linville Falls, Mt. Mitchell, Chimney Rock, Table Rock.]

Thomas, W.A., 2006, Tectonic inheritance at a continental margin: Geology Today, v. 16, no. 2.

Tollo, R.P., Bartholomew, M.J., Hibbard, J.P., and Karabinos, P.M., 2010, From Rodinia to Pangea: The Lithotectonic Record of the Appalachian Region: Geological Society of America, Memoir 206.

Illustrations

Hatcher, R.D., Jr., Bream, B.R., and Merschat, A.J., 2007, Tectonic map of the southern and central Appalachians: A tale of three orogens and a complete Wilson cycle, in Hatcher, R.D., Jr., Carlson, M.P., McBridge, J.H., and Martínez Catalán, J.R., eds., 4-D Framework of Continental Crust: Geological Society of America Memoir 200, pp. 590-632.

Hatcher, R.D., Jr., 2002, Alleghanian orogeny, a product of zipper tectonics: Rotational transpressive continent-continent collision and closing of ancient oceans along irregular margins: in Martínez Catalán, J.R., Hatcher, R.D., Jr., Arenas, R., and Díaz García, M.J., eds., Variscan-Appalachian dynamics: the building of the late Paleozoic basement: Geological Society of America Special Paper 364.

Scotese, C.R., 2014, Paleomap Project, http://www.scotese.com. [A number of maps and animations of paleogeography are available.]

Spencer, E.W., 2003, Earth Science – Understanding Environmental Systems: New York, McGraw Hill Higher Education.

Chapter 3: Blue Ridge Landscape

Gallen, S.F., and Wegmann, K.W., 2015, Exploring the origins of modern topographic relief in the southern Appalachians: An excursion through the transient landscape of the Cullasaja River basin, North Carolina: The Geological Society of America Field Guide 39.

Gallen, S.F., Wegmann, K.W., and Bohnenstiehl, D.R.,

2013, Miocene rejuvenation of topographic relief in the S. Appalachians: Geology Today, v. 23.

Hack, J.T., 1982, Physiographic divisions and differential uplift in the Piedmont and Blue Ridge: U.S. Geological Survey Professional Paper 1265.

Pazzaglia, F.J., ed., 2006, Excursions in Geology and History: Field Trips in the Middle Atlantic States, Field Guide 8, Geological Society of America.

Spotila, J.A., et al., 2004, Origin of the Blue Ridge Escarpment along the passive margin of Eastern North America: Basin Research, v. 16.

Chapter 4: Natural Environments in the Blue Ridge

Bailey, R.G., 1996, Ecosystem Geography: from Ecoregions to Sites: New York, Springer-Verlag.

Drury, W.H., Jr., 1998, Chance and Change-Ecology for Conservationists: Berkeley, University of California Press.

Edwards, L., Amerose, J., and Kirkman, L.K., 2013, The Natural Communities of Georgia: Athens, The University of Georgia Press.

Gaddy, L.L., 2000, A Naturalist Guide to the Southern Blue Ridge Front: Columbia, University of South Carolina Press.

Heffernan, K.E., Coulling, P.P., Townsend, J.F., and Hutto, C.J., 2001, Ranking invasive exotic plant species in Virginia: Richmond, Virginia Department of Conservation and Recreation, Division of Natural Heritage, Natural Heritage Technical Report 01-12.

Huebner, C., Olson, C., and Smith, H., 2005, Invasive Plants Field and Reference Guide: An Ecological Perspective of Plant Invaders of Forests and Woodlands; USDA Forest Service, State and Private Forestry, Northeastern Area.

Linzy, D.W., 2008, A Natural History Guide to Great Smoky Mountains National Park: Knoxville, University of Tennessee Press.

Luoma, J.R., 1999, The Hidden Forest, The Biography of an Ecosystem: New York, Henry Holt and Company.

Ricketts, T.H., Dinerestein, E., Olson, D.M., and Loucks, C.J., 1999, Terrestrial ecoregions of North America: a conservation assessment: A World Wildlife Fund Ecoregion Assessment: Washington, D.C., Island Press.

Silver, T., 2003, Mt. Mitchell and the Black Mountains: An Environmental History of the Highest Peaks in Eastern North America: Chapel Hill, University of North Carolina Press.

Skeate, S., 2006, A Nature Guide to Northwest North Carolina: Winston-Salem, Parkway Publishers.

Spira, T.P. 2011, Wildflowers and Plant Communities of the Southern Appalachian Mountains & Piedmont, A Southern Gateways Guide: Chapel Hill, University of North Carolina Press.

Whittaker, R.H., 1956, Vegetation of the Great Smoky Mountains: Ecological Monographs v. 26, no. 1.

Young, J., Fleming, G., Townsend, P., and Foster, J., 2006, Vegetation of Shenandoah National Park in relation to Environmental Gradients: USGS-NPS Vegetation Mapping Program, Final Report, v. 1. 1.

Websites

Forests and Woodlands: U.S. Forest Service, Eastern Region. <http://www.fs.fed.us/r9/wildlife/nnis/invasive-species-field-guide.pdf>

Ecological Communities: The Natural Heritage Program. Search for <ecological communities> or <High Elevation Natural Communities>. The World Wildlife Fund also has good Websites about ecological communities.

National Biological Information Infrastructure. <http://www.nbii.gov/portal/community/Communities/NBII_Home/>

Part 2

Chapter 5: Northern Blue Ridge

Anonymous, 1993, Geologic Map of Virginia–Expanded Explanation: Division of Geology and Mineral Resources, Department of Mines, Minerals and Energy, Charlottesville, Virginia, scale 1:50,000.

Bailey, C.M., Southworth, S., and Tollo, R.P., 2006, Tectonic History of the Blue Ridge, north-central Virginia: in Pazzaglia, F. J., ed. Excursions in Geology and History: Field Trips in the Middle Atlantic States: Geological Society of America, Field Guides 8.

Bartholomew, M. J., Gathright, T.M. II, and Henika, W.S., 1981, A tectonic model for the Blue Ridge in central Virginia: American Journal of Science, v. 281.

Bartholomew, M.J., Schultz, A.P., Henika, W.S., and Gathright, T.M. II, 1983, Geology of the Blue Ridge and Valley and Ridge at the Junction of the Central and Southern Appalachians: Lyttle, P.T., ed., Central Appalachian Geology, Joint meeting of the Northeast-Southeast section, Geological Society of America, 1982, American Geological Institute.

Conley, J.F., 1989, Stratigraphy and Structure Across the Blue Ridge and Inner Piedmont in Central Virginia: in 28th International Geological, Field trip Guidebook T207, Washington D.C., American Geophysical Union.

Fleeger, G.M. and Whitmeyer, S.J., eds., 2010, The Mid-Atlantic Shore to the Appalachian Highlands: Field Trip Guidebook for the 2010 Joint Meeting of the Northeastern and Southeastern Geological Society of America Sections.

Lyttle, P.T., editor, 1982, Central Appalachian Geology, Northeast-Southeast Geological Society of America 1982, Field Trip Guidebook: Am. Geological Institute.

Rankin, W.D., 1993, Geologic Map of Precambrian (Proterozoic) rocks east of the Grenville Front and their geologic setting in the United States and adjacent Canada, Plate 5: v. 2 of the Geology of North America (GNA-C2), Geological Society of America.

Schultz, A., and Compton-Gooding, E., 1991, Geologic Evolution of the Eastern United States, Northeast-Southeast Geological Society of America 1991, Field Trip Guidebook: Martinsville, Virginia Museum of Natural History, Guidebook Number 2.

Schwab, F.L., 1986, Latest Precambrian-Earliest Paleozoic Sedimentation, Appalachian Blue Ridge and Adjacent Areas: Review and Speculation: in McDowell, R.C. and Glover, L. III, eds. The Lowry Volume: Studies in Appalachian Geology, Virginia Tech Department of Geological Sciences, Memoir 3, pp. 115-137.

Walker, D., and Simpson, E.L., 1991, Stratigraphy of Upper Proterozoic and Lower Cambrian Siliciclastic Rocks, Southwestern Virginia and Northeastern Tenn., in: Schultz, A. and E. Compton-Gooding, eds., Geologic Evolution of the Eastern United States, Field Trip Guidebook, NE-SE GSA, VMNH Guidebook 2.

Wehr, F., 1983, Geology of the Lynchburg group in the Culpeper and Rockfish River areas: Blacksburg, Virginia Polytechnic Institute and State University.

Websites

Fleming, G.P., 2003, Ecological Communities of the Northern Blue Ridge: Power Point Presentation, Division of Natural Heritage website.

Chapters 6A-6D: Catoctin Mtn. Park, Gambrill Park, Washington Monument, and Harpers Ferry

Anonymous, 2012, Geology of Gambrill State Park, Maryland: Md. Geological Survey, pamphlet 2012-01.

Brezinski, D.K., 2004, Geologic Map of the Frederick Quadrangle, Frederick, Maryland: Maryland Geologic Survey, scale 1:24,000.

Brezinski, D.K., 1992, Lithostratigraphy of Western Blue Ridge Cover Rocks in Maryland: Report of Investigations v. 55, Maryland Geological Survey.

Cloos, E., 1958, Structural Geology of South Mountain and Appalachians in Maryland: F. J. Pettijohn, ed., Johns Hopkins Univ., Studies in Geology, no. 17.

Fauth, J.L., 1977, Geologic Map of the Catoctin Furnace and Blue Ridge Summit Quadrangles, Maryland: Maryland Geological Survey, scale 1:24,000.

Means, J., 1995, Maryland's Catoctin Mountain Parks: Blacksburg, Va., McDonald & Woodward.

Schmidt, M.F., Jr., 1992, Maryland's Geology: Centreville, Maryland Tidewater Publishers.

Southworth, S., and Brezinski, D.K., 1996, Geology of the Harpers Ferry Quadrangle, Virginia, Maryland, and West Va.: U.S. Geological Survey Bulletin 2123.

Whitaker, J.C., 1955, Geology of Catoctin Mountain: Geological Society of America Bulletin, v. 66.

Websites

National Park Service website for Harpers Ferry Natl. Historical Park, <http://www.nps.gov>. [See links for History & Culture, incl. Stories, also Nature & Science.]

National Park Service website for Catoctin Mountain Park, <http://www.nps.gov>. [See "Bird Checklist," "History & Culture," and "Nature and Science".]

Chapter 6E: Shenandoah National Park

Badger, R.L., 1998, Geology along Skyline Drive, Shenandoah National Park, Virginia: Helena, Montana, Falcon Publishing, Inc.

Chirico, P., and Tanner, S., 2004 Shaded Relief Map of Shenandoah National Park: U.S. Geological Survey, Open File Report 04-13.

Espenshade, G.H., and Clarke, J.W., 1976, Geology of the Blue Ridge Anticlinorium in Northern Va.: Geological Society of America Field Trip Guidebook no. 5

Gathright III, T.M., 1976. Geology of the Shenandoah National Park, Virginia: Virginia Division of Geology and Mineral Resources, Bulletin 86.

Litwin F.R.J., Morgan, B.A., Eaton, L.S., and Wieczorek, G.F., 2004, Assessment of Late Pleistocene to Recent climate induced vegetation changes in and near the Shenandoah National Park: U.S. Geological Survey Open File Report 2004-1351.

Reed, J.C. Jr., 1969, Ancient Lavas in Shenandoah National Park, Virginia: USGS Bulletin 1265.

Simpson, A. and R., 2013, Nature Guide to Shenandoah National Park: Falcon Pocket Guide.

Southworth, S., Aleinikoff, J.N., Bailey, C.M., Burton, W.C., Cridesr, E.A., Hackley, P.C., Smoot, J.P., and Tollo, R.P., 2009, Geology of the Shenandoah National Park Region: Geologic Map of the Shenandoah National Park Regions: U.S. Geological Survey Open-File Report 2009-1153, scale 100,000.

Young, J., Fleming, G., Townsend, P., and Foster, J., 2009, Vegetation of Shenandoah National Park in Relation to Environmental Gradients: USGS.

Chapter 6F: Blue Ridge Parkway: Waynesboro to the James River

Carter, M.W., Southworth, S., and Aleinikoff, J.N., 2013 Proterozoic to Cenozoic geology above, within and beneath the Blue Ridge Composite Thrust Sheet as exposed along the Blue Ridge Parkway in the Peaks of Otter area, central Virginia: The 2013 Virginia Geological Field Conference.

Carter, M., 2012, Geologic Map of the Blue Ridge Parkway: U.S. Geological Survey, scale 1:48,000.

Chapter 6G: James River Gap

Brown, C.E., and Spencer, E.W., 1981, Geologic Map, James River Face, Wilderness, Virginia: U.S. Geological Survey Map MF-1337-A, scale 1:24,000.

Carter, M.W., Southworth, S., and Aleinikoff, J.N., 2013 Proterozoic to Cenozoic geology above, within and beneath the Blue Ridge Composite Thrust Sheet as exposed along the Blue Ridge Parkway in the Peaks of Otter area, central Virginia: The 2013 Virginia Geological Field Conference.

Freer, R.S., 1968, Plants of Central Virginia Blue Ridge: Castanea v. 33.

Ramsey, G.W., and others, 1993, Vascular Flora of the James River Gorge Watersheds in the Central Blue Ridge Mtns. of Virginia: Castanea, v. 58, no. 4.

Spencer, E.W., 1994, Structure of the Blue Ridge at the James River Gap: in Schultz, A., and Henika W.S., eds., Field guides to southern Appalachian structure, stratigraphy, and engineering geology, Va. Polytechnic Institute Dept. of Geological Sciences Guidebook 10.

Spencer, E.W., Bell, J.D., and Kozak, S.J., 1989, Valley and Ridge and Blue Ridge traverse, central Virginia: in 28th Int'l. Geological Congress, Field Trip Guidebook T157, Washington D.C., Am. Geophysical Union.

Chapter 6H: Blue Ridge Parkway – James River to Roanoke

Griffin, J.W. and Reeves, J.H., Jr., A Stratified Site at Peaks of Otter, Blue Ridge Parkway: Asheville, North Carolina, Southeast Archeological Center, National Park Service archives.

Henika, W.S., 1981, Geology of the Villamont and Montvale Quadrangles, Virginia: Virginia Division of Geology and Mineral Resources Publication 25.

Shallow, J.M., 2003, Geologic Map of the Blue Ridge Parkway: Boulder, Colorado, JMS Geologic, LLC.

Speer, J.H., The Peaks of Otter and the Johnson Farm on the Blue Ridge Parkway: Appalachian Cultural Resources Workshop Papers, NPS.

Spencer, E.W., Bell, J.D., and Kozak, S.J., 1989, Valley and Ridge and Blue Ridge Traverse, Central Virginia: Washington D.C., in International Geological Congress, Field Trip Guidebook T 157, American Geophysical Union.

Spencer, E.W., 1968, Geology of the Natural Bridge, Sugarloaf Mountain, Buchanan, and Arnold Valley Quadrangles, Virginia: Virginia Division of Geology and Mineral Resources Report of Investigations 13.

Hamilton, C.L., 1964, Geology of the Peaks of Otter Area, Bedford and Botetourt Counties, Virginia: Blacksburg, Virginia Polytechnic Institute PhD thesis.

Websites

Peaks of Otter: <http://www.nps.gov/history/history/online_books/sero/appalachian/sec3.htm>

Chapter 7: Southern Blue Ridge

Bartholomew, M.J., Schultz, A., Henika, W.S., and Gathright, T.M., II, 1982, Geology of the Blue Ridge and Valley and Ridge at the Junction of the Central and Southern Appalachians, in Lyttle, P.T., ed., Central Appalachian Geology, Northeast-Southeast Geological Society of America 1982.

Boyer, S.E., and Elliott, D., 1982, Thrust systems: American Association of Petroleum Geologists Bulletin, v. 66.

Conley, J.F., and Henika, W.S., 1973, Field trip across the Blue Ridge Anticlinorium, Smith River Allochthon, and Sauratown Mountains Anticlinorium near Martinsville, Va.: Virginia Minerals, v. 1., no. 4.

Carpenter, P.A. III, 1989, A Geologic Guide to North Carolina's State Parks: Raleigh, North Carolina Geological Survey.

Carter, M.W., Merschat, C.E., and Wilson, W.F., 1999,

A Geologic Adventure Along the Blue Ridge Parkway in North Carolina: North Carolina Geological Survey Section Bulletin 98.

Catlin, D.T., 1984, A Naturalist's Blue Ridge Parkway: Knoxville, University of Tennessee Press.

Carter, B., Hibbard, J., Tubrett, M., and Sylvester, P., 2006, Detrital zircon geochronology of the "Smith River allochthon and Lynchburg Group, southern Appalachians: Implications for Neoproterozoic-Early Cambrian paleogeography: Precambrian Research, v. 147.

Craig, J.R., Sears, C.E., Gilbert, M.C. and Hewitt, D.A., 1971: Virginia Polytechnic Institute Geological Sciences Guidebook no 5.

Cyphers, S.R., and Jubb, M.G.V., 2012, Tectonics of the Central and Eastern Blue Ridge: Geotraverse from Hayesville Fault to the Brevard Fault Zone: Geological Society of America, Southeastern Section Meeting Field Trip Guidebook.

Frankenberg, D., ed., 2000, Exploring N. Carolina's Natural Areas, Parks, Natural Preserves, and Hiking Trails: Chapel Hill, University of North Carolina Press.

Gilliam, D.R., and Henika, W.S., 1999, Engineering Challenges of Southwest Virginia: Field Trip Guidebook for the 50th Annual Highway Geology Symposium: sponsored by Virginia Department of Transportation, Radford University Institute for Engineering Geosciences and Virginia Division of Geology and Mineral Resources, Transactions.

Hibbard, J., Stuart, K. and Henika, W.S., 2001, Framing the Piedmont Zone in North Carolina and Southern Virginia: Raleigh, North Carolina, 50th annual Southeastern Section, Geological Society of America Field Trip Guidebook.

Henika, W.S., Beard, J., Tracy, R., and Wilson, J.R., 2000, Structure and Tectonics Field Trip to the Eastern Blue Ridge and Western Piedmont Near Martinsville, Virginia: Virginia Minerals, v. 46, no. 3.

Henika, W.S., 1994, The Eastern Blue Ridge Sequence: an Iapetan Rift Sequence, in Schultz, A. and Henika, B., eds. Field Guides to Southern Appalachian Structure, Stratigraphy and Engineering Geology: Virginia Tech Department of Geological Sciences Guidebook no. 10.

Henika, W.S., Beard, J., Tracy, R., and Wilson J.R., 2000, Structure and Tectonics Field Trip to the Eastern Blue Ridge and the Western Piedmont Near Martinsville, Virginia: Virginia Minerals v. 46, no. 3.

Henika, W.S. 1997, Economic and Environmental Geology across the Boundary Between the Blue Ridge and Valley and Ridge, Near Roanoke, Virginia: Virginia Geological Field Conference Oct. 11, 1997.

Hibbard, J., Henika, W.S., Beard, J., and Horton, J.W., 2012, The Western Piedmont: in Bailey, C. and Berquist, R., eds., The Geology of Virginia: Virginia Division of Geology and Mineral Resources.

Labotka, T.C., and Hatcher, R.D. Jr., 2006, Field Trip Guidebook: Geological Society of America, 2006 Southeastern Section Meeting, Knoxville, Tennessee.

Lemiszki, P.J., 2003, Geology of the Roan Mountain State Park, Tennessee: Division of Geology, State Park Series 3.

Merschat, C.E., and others, 2012, Tectonics of the Central and Eastern Blue Ridge: Geotraverse from the Hayesville Fault to the Brevard Fault Zone: Pre-meeting Field Trip Guidebook, Geological Society of America, 61st Southeastern Section Meeting.

Merschat, C.E., and Wiener, L.S., 1990, Geology of Grenville-Age Basement and Younger Cover Rocks in the West Central Blue Ridge, North Carolina: Carolina Geological Survey Field Trip Guidebook, September 29-30, 1990, North Carolina Geological Survey.

Merschat, A.J., Hatcher, R.D. Jr., Peterson, Virginia, and Stahr, III, S.R., 1999, An Inventory of the Significant Natural Areas of Ashe County, North Carolina: North Carolina Natural Heritage Program.

Misra, K.C., and Congte, J.A., 1991, Amphibolite of the Ashe and Alligator Back Formations, North Carolina: Geological Society of America Bulletin, v. 103.

Rankin, D.W., 1971, Guide to the Geology of the Blue Ridge in Southwestern Virginia and adjacent N. Carolina: VPI Guidebook no 5.

Rankin, D.W., 1971, Geology of the Blue Ridge in Southwestern Virginia and adjacent North Carolina: in Guidebook to Appalachian Tectonics and Sulfide Mineralization of southwestern Virginia: Guidebook no. 5, Virginia Polytechnic Institute & State University, Department of Geological Sciences.

Stewart, K.G., Adams, M.G., and Trupe, C.H., 1997, Paleozoic Structure, Metamorphism, and Tectonics of the Blue Ridge of Western N. Carolina: Carolina Geological Society 1997 Field Trip and Annual Meeting.

Stewart, K.G. and Roberson, M.R., 2007, Exploring the Geology of the Carolinas: Chapel Hill, University of North Carolina Press.

Trupe, C.H., and others, 2010, The Burnsville fault:

Evidence for the timing and kinematics of southern Appalachian Acadian dextral transform tectonics: Geological Society of America Bulletin, v. 115.

Weinburg, E.L., 1971, Sulfide mineralization in southwestern Virginia, Austinville-Ivanhoe mine: in Guidebook to Appalachian Tectonics and Sulfide Mineralization: Geological Society of America.

Weinberg, E.L., Misra, K.C., and Congte, J.A., 1991, Amphibolite of the Ashe and Alligator Back Formations, North Carolina: Geological Society of America Bulletin, v. 103.

Whisonant R.C. and Tso, J., 1982, Stratigraphy and Structure of the Lower Ashe Formation (upper Precambrian) along the Fries Fault in southwestern Virginia: Raleigh, Geological Society of America Southeastern Section Guidebook.

Websites

Raymond, Loren, 2004, <www.emporia.edu/earthsci/student/pfaff1/Blue_Ridge2.htm >
[This site provides a brief description of geologic features exposed along the Blue Ridge Parkway in the Blue Ridge Highland region. Sites are keyed to milepost markers.]

Chapter 8A: Upland Plateau

Prince, P.S., Spotila, J.A., and Henika, W.S., 2010, New physical evidence of the role of stream capture in active retreat of the Blue Ridge Escarpment, southern Appalachians: Geomorphology, v. 123.

Spotila, J.A., Bank, G.C., Reiners, P.W., Naeser, C.W., Naeser, N.D., and Henika, W.S., 2004, Origin of the Blue Ridge Escarpment along the passive margin of Eastern North America: Basin Research, v. 16.

Ver Steeg, K., 1946, The Teays River: Ohio Journal of Science, v. XLVI, no. 6.

Chapter 8B: Blue Ridge Parkway – Roanoke to Doughton Park

(This includes Roanoke Overlook, Buffalo Mountain, Rocky Knob, Meadows of Dan, Mabry Mill, Stewart's Creek, Galax, and Doughton Recreation Area. See Chapter 7 for other regional references.]

Carter, M., 2012 and 2014, Geologic Maps of the Blue Ridge Parkway. These maps show recent revisions in the geology along the Parkway: U.S. Geological Survey, scale: 1:48,000.

Chapter 8C: Philpott Reservoir and Fairy Stone State Park

Bartholomew, M.J., Henika, W.S., and Lewis, S.E., 1994, Geologic and Structural Transect of the New River Valley: Valley and Ridge and Blue Ridge Provinces, Southwestern Virginia: Virginia Polytechnic Institute & State University, Department of Geological Sciences, Guidebook 10.

Conley, J.F., and Henika, W.S., 1970, Geology of the Philpott Reservoir Quadrangle, Virginia: Virginia Division of Geology and Mineral Resources, Report of Investigations 22.

Conley, J.F., 1989, Stratigraphy and Structure Across the Blue Ridge and Inner Piedmont in Central Virginia: 28th International Geological Congress, Field Trip Guidebook T207: Washington D.C., American Geophysical Union.

Conley, J.F. and Henika, W.S., 1970, Geology of the Philpott Reservoir Quadrangle: Virginia: Report of Investigations 22, Virginia Division of Mineral Resources. [A field trip road guide for this quadrangle is available through the Virginia Div. Geology and Mineral Resources, Charlottesville, Va.]

Henika, W.S., 1997, Twenty-seventh Annual Virginia Geologic Field Conference, October 11, 1997: Virginia Division of Mineral Resources, Department of Geological Sciences, Virginia Tech, Blacksburg, Va.

Wilson, J., 2000, Structure and Tectonics Field Trip to the Eastern Blue Ridge and Western Piedmont Near Martinsville, Va.: Virginia Minerals, v. 46, no. 3.

Chapter 8D: Stone Mountain

Daubenmire, R., 1968, Plant communities; a textbook of plant synecology: New York, Harper & Row. [See discussion of Oosting and Anderson, 1939.]

White, W.A., 1945, Origin of Granite Domes in the Southeastern Piedmont: The Journal of Geology, v. 53, no. 4.

Chapter 8F: Mount Rogers Area

Diegel, F.A., 1986, Topological constraints on imbricate thrust networks, examples from the Mountain City window, Tennessee, U.S.A.: Journal of Structural Geology, v. 8.

Merschat, A.J., Southworth, S., McClellan, E, Tollo, R.P., Rankin, D.W., Hooper, S., and Bauer, S., 2014, Key structural and stratigraphic relationships from the northeast end of the Mountain City window and the Mount Rogers area, Virginia – N. Carolina – Tenn.: Geological Society of America Field Guide 35.

Rankin, D.W., 1967, Guide to the Geology of the Mount Rogers Area, Virginia, North Carolina and Tennessee: Durham, Carolina Geological Society, Field Trip Guidebook.

Rankin, D.W., Miller, J.M.G., and Simpson, E.L., 1994, Geology of the Mt. Rogers Area, Southwestern Virginia Blue Ridge and Unaka Belt: Virginia Polytechnic Institute, Dept. of Geological Sciences Guidebook 10.

Rankin, D.W., 1993, The Volcanogenic Mount Rogers Fm. and the Overlying Glaciogenic Konnarock Fm., Two Late Proterozoic Units in Southwestern Virginia: U.S. Geological Survey Bulletin 2029.

Woodward, N.B., 1989, Pulaski-Holston Mtn. thrust zone in northeastern Tennessee: in International Geological Congress, Field Trip Guidebook F167, Washington D.C., American Geophysical Union.

Woodward, N.B., 1989, Geometry and Deformation Fabrics in the Central and Southern Appalachian Valley and Ridge and Blue Ridge, in International Geological Congress, Field Trip Guidebook T357, Washington D.C., American Geophysical Union.

Woodward, N.B., 1985, Valley and Ridge thrust belt: balanced structural sections, Pennsylvania to Alabama: Appalachian Basin Industrial Associates, University of Tennessee Department of Geological Sciences Studies in Geology no. 12.

Chapter 8G. Grandfather Mountain Region

[This region includes Mt. Mitchell, Lineville Falls, Craggy Gardens, Mars Hill, and Spruce Pine. [See Chapter 7 for other regional references.]

Bryant, B. and Reed, J.C. Jr., 1970, Geology of the Grandfather Mountain Window and Vicinity, North Carolina and Tennessee: U.S. Geological Survey Professional Paper 615.

Berquist, P.J., and others, 2003, The Mars Hill Terrane: Extent, Age, and Origin of the Oldest Rocks in the Southeastern U.S.A.: abs. South-Central Section Geological Society of American Joint Annual Meeting.

Dennison, J.M. and Stewart, K.G., eds., 1992, Geologic field guides to North Carolina and vicinity: Geological Society of America.

Labotka, T.C., and Hatcher, R.D., Jr., eds., Field Trip Guidebook, 2006 Southeastern Section Meeting: Geological Society of America.

Lemiszki, P.J. Jr, 2003, Geology of the Roan Mountain State Park: Tenn. Division of Geology, State Park Series 3.

Ownby, S.E., and others, 2004, U-Pb geochronology and geochemistry of a portion of the Mars Hill terrane, North Caroline-Tennessee: Constraints on origin, history, and tectonic assembly: Geological Society of America, Memoir 197, North Carolina Geological Survey Field Trip Guidebooks.

Hatcher, R.D. Jr., Merschat, A.J. and Raymond, L.A., 2006, Geotraverse: Geology of northeastern Tennessee and the Grandfather Mountain region: Knoxville, Tennessee, Field Trip Guidebook, Geological Society of America 2006 Southeastern Section Meeting, Labotka, T.C. and Hatcher R.D. Jr., eds.

Hatcher, R.D. Jr. and Thomas, W.S., 1989, Southern Appalachian Windows: Comparison of Styles, Scales, Geometry and Detachment Levels of Thrust Faults in the Foreland and Internides of a Thrust-Dominated Orogen, in International Geological Congress, Field Trip Guidebook T167, American Geophysical Union.

Raymond, L.A., Neton, M.J., and Cook, T., 1992, The Grandfather Mountain window: Laurentian basement and Neoproterozoic, rift-related cover locks: in Dennison, J. M., and Steward, K.G., eds., Geologic field guides to North Carolina and Vicinity, Geological Society of America Southeastern Section Guidebook.

Schwab, F.L., 1977, Grandfather Mountain Formation: Depositional Environment, Provenance, and Tectonic Setting of Late Precambrian Alluvium in the Blue Ridge of North Carolina: Journal of Sedimentary Petrology, v. 47, no. 2.

Tager, M., 1999, Grandfather Mountain: A Profile, Winston-Salem, Parkway Publisher.

Trupe, C.H., Stewart, K.G, Adams, M.G., Waters, C.L., Miler, B. V., and Hewitt, L.K., 2003, The Burnsville fault: Evidence for the timing and kinematics f southern Appalachian Acadian dextral transform tectonics: Geological Society of America Bulletin, v. 115.

Weiner, L.S. and Merschat, C.E. 1990. Guidebook to the central Blue Ridge of North Carolina and the Spruce Pine mining district: North Carolina Geological Survey.

Chapter 8L: Blue Ridge Parkway – Asheville to Oconaluftee

Merschat, A.J., Hatcher, R.D. Jr., Peterson, Virginia, Stahr, D.W. III, Cyphers, S.R. and Jubb, M.G.V., 2012, Tectonics of the Central and Eastern Blue Ridge: Geotraverse from the Hayesville Fault to the Brevard Fault Zone, Field Trip Guidebook: Geological Society of America, southeastern section meeting, Asheville,

North Carolina. [The references cited in this guidebook include many of the best references to the geology of this region.]

Chapter 8M: Great Smoky Mountains National Park

Hatcher, R.D. Jr., and others, 2004, Field Trip Guide - Southern Appalachian Foreland Fold-Thrust Belt: Knoxville, University of Tennessee.

Hatcher, R.D. Jr., A.J. Merschat, and J.B. Whisner, 2004, Field Trip Guide Southern Appalachian Foreland Fold-Thrust Belt: Knoxville, University of Tennessee.

Kemp, S., 2006, Trees & Familiar Shrubs of the Smokies: Gatlinburg, Great Smoky Mtns. Association.

King, P.B., Neuman, R.B., and Hadley, J.B., 1968, Geology of the Great Smoky Mountains National Park, Tennessee and North Carolina: U.S. Geological Survey Professional Paper 587.

Linzey, D.W., 2008, A Natural History Guide to Great Smoky Mountains Nat'l. Park: Knoxville, Univ. of Tenn.

Moore, H.L., 1994, A Geologic Trip across Tennessee by Interstate 40, Outdoors Tennessee Series: Knoxville, University of Tennessee Press.

Moore, H.L. and Brown, F., 2001, Discovering October Roads, Fall Colors and Geology in Rural East Tennessee: Knoxville, University of Tennessee Press.

Moore, H.L., 1988, A Roadside Guide to the Geology of the Great Smoky Mountains National Park: Knoxville, University of Tennessee Press.

Schulz, A., Southworth, S., Fingeret, C., and Weik, T., 2000, Geology of the Mount Le Conte 7.5 minute quadrangle, Great Smoky Mountains National Park, Tennessee and North Carolina: U.S. Geological Survey Open File Report #00-261.

Sigalas, M. and Bradley, J., 2002, Smoky Mountains, Moon Handbooks: Emeryville, Ca., Avalon Travel Publishing.

Southworth, S., Chirico, P.G., and Putbrese, T., 2000, Digital Geologic Map of Parts of the Cades Cover and Calderwood Quadrangles, Tennessee and North Carolina, Great Smoky Mountains National Park: USGS Open File Report 99-175, scale 1:24,000.

Southworth, S., Schultz, A., and Denny, D., 2006, The Geology and natural history of Mount Le Conte, Great Smoky Mountains National Park, Tennessee: Southeastern Section meeting of the Geological Society of America Field Trip Guidebook 3.

Whittaker, R.H., 1956, Vegetation of the Great Smoky Mountains: Ecological Monographs, v. 26, no.1.

Website

History of the Grassy Balds in Great Smoky Mountains National Park: [search by subject]

[Maps: National Geographic has published a Trails Illustrated topographic map of the Great Smoky Mountains that is durable and frequently revised. This map is an excellent source for trail information.

A geologic map of the park area appears in U.S. Geological Survey Professional Paper 587, scale 1:125,000.

South of the Great Smoky Mountains

Kish, S.A., Merschat, C.E., Mohr, D.W., and Wiener, L.S., 1975, Guide to the Geology of the Blue Ridge South of the Great Smoky Mountains, North Carolina: Raleigh, Carolina Geological Survey.

Merschat, A.J., 2012, Tectonics of the Central and Eastern Blue Ridge: Geotraverse from the Hayesville Fault to the Brevard Fault Zone: Field Trip Guidebook, Geological Soc. of Am., Southeastern Section Meeting.

Tull, J.F. editor, 2007, Tectonics of the Georgia Blue Ridge: Georgia Geological Society Guidebooks, v. 27.

Part 3

General References

Simpson, A. and R., 2013, Nature Guide to Shenandoah National Park: Falcon Guide. [This guide includes most plants and animals found in the park.]

Rappole, J.H., 2007, Wildlife of the Mid-Atlantic: Philadelphia, University of Pennsylvania Press.

Websites

Pennsylvania Game Commission <pgc.state.pa.us/pgc/cwp/browse.asp?a=521&bc=0&c=70161>

Tennessee Wildlife Resources Agency www.tennessee.gov/twra/

Chapter 9: Rocks and Minerals

[Most introductory geology and earth science textbooks have descriptions of common rocks and minerals.]

Pellant, Chris, 2002, Smithsonian Handbooks: Rocks and Minerals, DK Publishing.

Schumann, Walter, 1993, Handbook of Rocks, Minerals, and Gemstones, Houghton Mifflin Harcourt.

Spencer, E.W., 2003, Earth Science – Understanding Environmental Systems: New York, McGraw-Hill.

Chapter 10: Plants

Adkins, L.M., Cook, J., and Cook, M., 1999, Wildflowers of the Appalachian Trail, Appalachian Trail Conservancy: Birmingham, Ala., Menasha Ridge Press.

Brodo, I.M., Sharnoff, S.D., and Sharnoff, S., 2001, Lichens of North America: New Haven, Yale Univ. Press.

Cobb, B., 2005, Peterson Field Guide to Ferns: Northeastern and Central North America, 2nd edition: New York, Houghton Mifflin.

Gupton, O.W. and Swope, F.C., 1979, Wildflowers of the Shenandoah Valley and Blue Ridge: Charlottesville, University of Virginia Press.

Hutson, R.W., Hutson, W.F., and Sharp, A.J., 1995, Great Smoky Mountains Wildflowers, 5th edition: Northbrook, Ill., Windy Pines Publishing.

Laessoe, T., Del Conte, A., Lincoff, G., 1996, The Mushroom Book: New York, D.K. Publishing Book.

Levine, C., 1995, A Guide to Wildflowers in Winter: New Haven, Yale University Press.

Medina, B. and Medina, V., 2002, Southern Appalachian Wildflowers: Guilford, Colo., A Falcon Guide, The Globe Pequot Press.

Miller, J., 2015, A Management Guide for Invasive Plants in Southern Forests, U.S. Dept. of Agriculture. [This book is available in PDF format on the U.S. Forest Service website <www.srs.fs.usda.gov/pubs/36915>]

Miller, J.H.; Chambliss, E.B.; Loewenstein, N.J., 2010, A Field Guide for the Identification of Invasive Plants in Southern Forest: U.S. Forest Service.

Medina, B. and Medina, V., 2002, Southern Appalachian Wildflowers: Guildford, Conn., A Falcon Guide.

Roody, W.C., 2003, Mushrooms of West Virginia and the Central Appalachians: Lexington, University Press of Kentucky.

Stupka, A.J. and Robinson, D., 1965, Wildflowers in Color: New York, Harper Perennial.

Weakley, A.S., Ludwig, J.C., and Townsend, J.F., 2012, Flora of Virginia: Richmond, Foundation of the Flora of Virginia Project Inc. and Fort Worth, Botanical Research Institute of Texas Press.

Weber, N.S., and Smith, A.H., 1985, A Field Guide to Southern Mushrooms: Ann Arbor, University of Michigan Press.

Trees

Little, E.L., 1980, National Audubon Society Field Guide to N. American Trees: New York, Alfred A. Knopf.

Rushforth, K. and Hollis, C., 2006, National Geographic Field Guide to the Trees of North America: Washington D.C., National Geographic.

Petrides, G.A., 1993, Peterson First Guides Trees: Boston and New York, Houghton Mifflin Company.

Sibley, D.A., 2009, The Sibley Guide to Trees: New York, Alfred A. Knopf.

Williams, M.D. and Clatterbuck, W.K., 2005, The All Season Pocket Guide to Identifying Common Tennessee Trees: Tennessee Dept. of Agriculture, Forestry.

Wojtech, M., 2011, Bark: A Field Guide to Trees of the Northeast: Lebanon, N.H., Univ. Press of New England.

Chapter 11: Birds

[Many excellent guides to birding and wildlife are available as well as a number of fine websites produced by state governments of the Appalachian region. The Virginia Department of Game and Inland Fisheries publish one, entitled "Birding and Wildlife Trail." It contains maps showing the locations of trails throughout the Virginia. The trail maps indicate sites of special interest with detailed directions and descriptions. More information is available on the web, especially at The Audubon Society's online field guide; <www.audubon.org/bird-guide>; Cornell University's "The Cornell Lab" <www.allaboutbirds.org>, and Birds of the Greater Rockbridge County Area, Virginia by Richard Rowe: http://www.flickr.com/photos/vmibiology/sets/.]

Anonymous, Discover Our Wild Side, Virginia Birding and Wildlife Trail, Mountain Area: Virginia Department of Game and Inland Fisheries: Richmond, Virginia. [To find changes in trail information go to www.dgif.state.va.us]

Elphick, J., ed., 2007, Atlas of Bird Migrations: Buffalo, N.Y., Firefly Books.

Peterson, T.P., 2010, Peterson Field Guide to Birds of Eastern and Central North America: 6th edition: New York, Houghton Mifflin Harcourt Publishing Co.

Poole, A.F., and Gill, F.B., 1996, The Birds of North America: The Academy of Natural Sciences, Philadelphia, Pa., and The American Ornithologists' Union, Washington, D.C.

Shomon, J.J., 1951, Birdlife of Virginia: Commonwealth of Virginia Commission of Game and Inland

Fisheries, Birds of the Blue Ridge Mountains: A Guide for the Blue Ridge Parkway, Great Smoky Mountains, Shenandoah National Park, and Neighboring Areas: The Blue Ridge Parkway Foundation.

Simpson, M.B., Jr., 1992, Birds of the Blue Ridge Mountains: A Guide for the Blue Ridge Parkway, Great Smoky Mountains, Shenandoah National Park, and Neighboring Areas: Raleigh, University of North Carolina Press.

Tekiela, S., 2002, Birds of Virginia Field Guide: Cambridge, Minn., Adventure Publications. [This book has superb photographs.]

Whitehurst, D.K., Wajda, R.K., and Trollinger, J.B., 2002, Discover Our Wild Side: Virginia Birding and Wildlife Trail - Mountain Area: Richmond, Va., Virginia Dept. of Game and Inland Fisheries. [For changes in trail information check <http://www.dgif.state.va.us>]

Websites

Blue Ridge Parkway Guide: Virtual Blue Ridge: <http://www.virtualblueridge.com/parkway/general/birds.asp

Illustrations

Figs. 2-2; 2-3; 2-4; 2-5. Spencer, E.W., 2003, Earth Science – Understanding Environmental Systems: New York, McGraw-Hill Higher Education.

Fig. 2-12. Hatcher, R.D., Jr, Bream, B.R., and Merschat, A.J. 2007, Tectonic map of the southern and central Appalachians: A tale of three orogens and a complete Wilson cycle, in Hatcher, R.D., Jr, Carlson, M.P., McBridge, J.H., and Martinez Satalan, J.R., eds, 4-D Framework of Continental Crust: Geological Society of America Memoir 200, p. 590-632.

Fig. 2-14. Hatcher, R.D., Jr., 2002, Alleghanian orogeny, a product of zipper tectonics: Rotational transpressive continent-continent collision and closing of ancient oceans along irregular margins: in Martínez Catalán, J.R., Hatcher, R.D. Jr, Arenas, R., and Díaz-Garcia, F., eds, Variscan-Appalachian dynamics: the building of the late Paleozoic basement: Geological Society of America Special Paper 364.

Fig. 2-14. Scotese, C.R., 2014, Paleomap Project. [A number of maps and animations of paleogeography are available. See website for the most recent work.]

Fig. 5-9. Rankin, W.D., 1993, Geologic Map of Precambrian (Proterozoic) rocks east of the Grenville Front and their geologic setting in the United States and adjacent Canada, Plate 5: v. 2 of the Geology of North America (GNA-C2), The Geological Society of America.

Fig. 6A-7 & 8. Fauth, J.L., 1977, Geologic Map of the Catoctin Furnace and Blue Ridge Summit Quadrangles, Maryland: Maryland Geological Survey, scale 1:24,000.

Fig. 6B-1. Brezinski, D.K., 2004, Geologic Map of the Frederick Quadrangle, Frederick, Maryland: Maryland Geologic Survey, scale 1:24,000.

Fig. 6D-4 & 5. Southworth, S., and Brezinski, D.K., 1996, Geology of the Harpers Ferry Quadrangle, Virginia, Maryland, and West Virginia: U.S. Geological Survey Bulletin 2123.

Whitaker, J.C., 1955, Geology of Catoctin Mountain: Geological Society of America, Bulletin, v. 66.

Fig. 6D-9. Cloos, E., 1958, Structural Geology of South Mountain and Appalachians in Maryland: Pettijohn, F.J., ed. Johns Hopkins University Studies in Geology, no. 17.

Fig. 6E-4. Gathright III, T.M., 1976, Geology of the Shenandoah National Park, Virginia: Virginia Division of Geology and Mineral Resources, Bulletin 86.

Fig. 6E-5. Southworth, S., Aleinikoff, J.N., Bailey, C.M., Burton, W.C., Cridesr, E. A., Hackley, P. C., Smoot, J.P., and Tollo, R.P., 2009, Geology of the Shenandoah National Park Region: Geologic Map of the Shenandoah National Park Regions, Open-File Report 2009-1153: U.S. Geological Survey, scale 1:100,000.

Fig. 6E-3. Young, J., Fleming, G., Townsend, P., and Foster, J., 2009, Vegetation of Shenandoah National Park in Relation to Environmental Gradients: U.S. Geological Survey.

Fig. 6H-7. Griffin, J.W. and Reeves, J.H., Jr., 1967, A Stratified Site at Peaks of Otter, Blue Ridge Parkway: Asheville, North Carolina, Southeast Archeological Center, National Park Service archives.

Fig. 6H-13 & 14. Henika, W.S., 1981, Geology of the Villamont and Montvale Quadrangles, Va.: Virginia Division of Geology & Mineral Resources, Publication 25.

Fig. 7-9. Boyer, S.E., and David, E., 1982, Thrust systems: American Association of Petroleum Geologists, Bulletin, v. 66.

Fig. 8A-4. Spotila, J.A., Bank, G.C., Reiners, P.W., Naeser, C.W., Naeser, N.D., and Henika, W., 2004, Origin of the Blue Ridge Escarpment along the passive margin of Eastern North Am.: Basin Research, v. 16.

Fig. 8F-9. Diegel, F.A., 1986, Topological constraints on imbricate thrust networks, examples from the Mountain City window, Tennessee, U.S.A.: Journal of Structural Geology, v. 8.

Fig. 8F-11. Woodward, N.B., 1985, Valley and Ridge thrust belt: balanced structural sections, Pennsylvania to Alabama: Appalachian Basin Industrial Associates, University of Tennessee Department of Geological Sciences Studies in Geology no. 12.

Fig. 8G-5. Hatcher, R.D. Jr., Merschat, A.J. and Raymond, L.A., 2006, Geotraverse: Geology of northeastern Tennessee and the Grandfather Mountain region: Field Trip Guidebook, Geological Society of America 2006 Southeastern Section Meeting, Labotka, T.C. and Hatcher R.D. Jr., eds, Knoxville, Tennessee.

Fig. 8G-5. Raymond, L.A., Neton, M.J., and Cook, T., 1992, The Grandfather Mountain window: Laurentian basement and Neoproterozoic, rift-related cover locks: in Dennison, J.M., and Steward, K.G., eds., Geologic field guides to North Carolina and Vicinity, Geological Society of America Southeastern Section Guidebook.

† Additional Photographic Attributions

P. 125 – Striped Skunk by Tom Friedel/http://www.birdphotos.com/CC-BY-3.0

P. 133 – Maleberry, Lyonia ligustrina © Fritz Flohr Reyonlds/Wikimedia Commons/CC-BY-SA-3.0

P. 133 – Deerberry, Vaccineum stamineum by Ted Bodner/CC-BY-3.0

P. 133 – Red Elderberries © M.J. Richardson/Wikimedia Commons/CC-BY-SA-2.0

P. 134 – Yellow Warbler, Setophaga aestiva © Geoff Clarke/Wikimedia Commons/CC-BY-SA-3.0

P. 197 – Eastern Coyote, Coyote-face-snow, Virginia © User: ForestWander/http://www.ForestWander.com/Wikimedia Commons/CC-BY-SA-3.0

P. 207 – Bog Turtle by USF&WS HQ/CC-BY-2.0

P. 211 – Mealy Bug © User: Mokkie/Wikimedia Commons/CC-BY-SA-3.0

P. 221 – Killdeer © Alan D. Wilson/Wikimedia Commons/CC-BY-SA-3.0

P. 221 – Wood Duck (Aix sponsa) by Frank Vassen/CC-BY-2.0

P. 221 – Bufflehead, Bucephala albeola © User: Mdf/Wikimedia Commons/CC-BY-SA-3.0

P. 268 – Roan Mtn. State Park by User: TranceMist/Wikimedia Commons/CC-BY-2.0

P. 293 – Appalachian Cove Forest on Baxter Creek Trail in Great Smoky Mountains National Park © Miguel Vieira/Wikimedia Commons/CC-BY-SA-3.0

P. 293 – Fire Pink, Silene virginica by Jason Hollinger/CC-BY-2.0

P. 294 – Black-chinned Red Salamander, Pseudotriton ruber © User: Athene cunicularia/Wikimedia Commons/CC-BY-SA-3.0

P. 303 – Cades Cove Panorama © Anthony Chavez/Wikimedia Commons/CC-BY-SA-3.0

P. 304 – Cable Mill by Chris Totsky/CC-BY-3.0

P. 305 – Hairy Beard Tongue, Penstemon hirsutus © Fritz Flohr Reynolds/Wikimedia Commons/CC-BY-SA-3.0

P. 309 – Fraser Magnolia, Magnolia fraseri © User: Richtid/Wikimedia Commons/CC-BY-SA-3.0

P. 327 – Wild Cherry in July © Des Colhoun/Wikimedia Commons/CC-BY-SA-2.0

P. 328 – Hawthorn, Crataegus monogyna Eure © User: Stanzilla/Wikimedia Commons/CC-BY-SA-3.0

P. 329 – Scarlet Oak, Quercus coccinea © Jean-Pol Grandmont/Wikimedia Commons/CC-BY-SA-3.0

P. 331 – Witch-hazel, Hamamelis virginiana © User: BotBln/Wikimedia Commons/CC-BY-SA-3.0

P. 334 – Blueberry, Lowbush, Vaccinium angustifolium by Albert Herring - Superior National Forest/CC-BY-2.0

P. 335 – Clematis, Clematis virginiana © User: SB Johnny/Wikimedia Commons/CC-BY-SA-3.0

P. 335 – Buttonbush, Cephalanthus occidentalis © User: Sten Porse/Wikimedia Commons/CC-BY-SA-3.0

P. 336 – Lily of the Valley, False, Maianthemum canadense by Albert Herring - Superior National Forest/CC-BY-2.0

P. 336 – Mountain Sandwort, Arenaria montana by User: Mtiffany71/Wikimedia Commons/CC-BY-3.0

P. 337 – Ox-eye Daisy, Leucanthemum vulgare © H. Zell/Wikimedia Commons/CC-BY-SA-3.0

P. 337 – Partridge Berry, Mitchella repens © Fritz Flohr Reynolds/Wikimedia Commons/CC-BY-SA-3.0

P. 337 – Solomon's Seal, False, Maianthemum racemosum © Walter Siegmund/Wikimedia Commons/CC-BY-SA-3.0

P. 338 – Allegheny Stonecrop, Hylotelephium telephioides © Fritz Flohr Reynolds/Wikimedia Commons/CC-BY-SA-3.0

P. 338 – Wild Strawberry, Fragaria virginiana (blossom inset) © Walter Siegmund/Wikimedia Commons/CC-BY-SA-3.0

P. 338 – Wintergreen, American, Gaultheria procumbens (blossom photo) © User: Jomegat/Wikimedia Commons/CC-BY-SA-3.0

P. 338 – Wintergreen, American, Gaultheria procumbens (berry inset) © John Delano/Wikimedia Commons/GNU GPL

P. 342 – Horsenettle berries (inset) © Susan Sweeney/Wikimedia Commons/CC-BY-SA-3.0

P. 343 – Wild Ginger, Asarum canadense (blossom inset) © Fritz Flohr Reynolds/Wikimedia Commons/CC-BY-SA-3.0

P. 344 – Bluebead Lily, Clintonia borealis © Peter Coxhead/Wikimedia Commons/CC-BY-SA-4.0

P. 344 – Bluebead Lily berries, Clintonia borealis © User: Circeus/Wikimedia Commons/CC-BY-SA-2.0

P. 344 – Golden Alexander, Zizia aurea © Albert Herring - Superior National Forest/Wikimedia Commons/CC-BY-2.0

P. 347 – Rock Cap Fern, Polypodium virginianum © User: Jaknouse/Wikimedia Commons/CC-BY-SA-3.0

P. 348 – Galax urceolata by Ted Bodner/CC-BY-3.0

P. 348 – Grass of Parnassus, Kidney-leaf foliage © Margie Hunter, https://hikinginthesmokies.wordpress.com

P. 349 – Mountain Sandwort, Arenaria montana by User: Mtiffany71/Wikimedia Commons/CC-BY-3.0

P. 349 – Rattlesnake Plantain, Goodyera repens by Jason Hollinger/CC-BY-2.0

P. 349 – Skunk Cabbage, Moss (blossom inset) © Sue Sweeney/Wikimedia Commons/CC-BY-SA-3.0

P. 350 – Wild Potato-vine, Ipomoea pandurata © Cody Hough/Wikimedia Commons/CC-BY-SA-3.0

P. 350 – Trumpet Honeysuckle, Lonicera sempervirens © Stan Shebs/Wikimedia Commons/CC-BY-SA-3.0

P. 351 – Peat Moss, Sphagnum Moss in May 2010 by User: External Affairs/Wikimedia Commons/CC-BY-2.0

P. 351 – Stair Step Moss, Hylocomium splendens © Walter Siegmund/Wikimedia Commons/CC-BY-SA-3.0

P. 351 – White Cushion Moss, Leucobryum glaucum © Jerzy Opiola/Wikimedia Commons/CC-BY-SA-3.0

P. 352 – Death Cap Mushroom, Amanita phalloides © User: Archenzo/Wikimedia Commons/CC-BY-SA-3.0

P. 352 – Golden Chanterelles, Chanterelle Cantharellus cibarius © User: Strobilomyces/Wikimedia Commons/CC-BY-SA-3.0

P. 352 – Honey Mushroom, Armillaria mellea © Dan Molter /Wikimedia Commons/CC-BY-SA-3.0

P. 360 – Cooper's Hawk, Accipiter cooperii © H. Gilbert Miller/Wikimedia Commons/CC-BY-SA-2.5

P. 364 – Chestnut-sided Warbler by Greg Schechter/CC-BY-2.0

P. 365 – Ovenbird in the Garden © User: Mike's Birds/Wikimedia Commons/CC-BY-SA-2.0

P. 365 – Eastern Phoebe at Cape May in the Fall © Peter Wallack/Wikimedia Commons/CC-BY-SA-3

P. 365 – House Wren in Full Song by Alan Vernon (http://flickr.com/people/32541690@N02)/CC-BY-2.0

P. 366 – House Sparrow, Passer domesticus © User: Mdf/Wikimedia Commons/CC-BY-SA-3.0

P. 367 – Rock Pigeon, Columba livia © Jörg Hempel/Wikimedia Commons/CC-BY-SA-3.0

P. 368 – Gray Catbird, Dumetella carolinensis © Peter Massas/Wikimedia Commons/CC-BY-SA-2.0

P. 369 – Blue-gray Gnatcatcher, Polioptila caerulea © Elaine R. Wilson/Wikimedia Commons/CC-BY-SA-2.5

P. 369 – European Starlings (Etourneau), Sturnus vulgaris © Daniel Plazanet/Wikimedia Commons/CC-BY-SA-2.5

P. 369 – American Crow, Corvus brachyrhynchos by Jack Wolf/CC-BY-2.0

P. 371 – Red-headed Woodpecker © Francesco Veronesi/Wikimedia Commons/CC-BY-SA-2.0

Index of Technical Terms

Acadian Orogeny 27
acid rain 59
Alleghanian Orogeny 30
angular unconformity 53
Antietam Fm. 80
anticline 15, 36
Appalachian Mtns. defined 7
Appalachian Plateau 11
asthenosphere 18
Avalonia 33
axis (of folds) 36, 288

balds 69
beds, bedding (sedimentary) 317
Blue Ridge, defined 11, 13
Blue Ridge as a geological province 73
Blue Ridge as a topographic feature 73
Blue Ridge basement complex 81

Candler Fm. 82
Carolinia 33
Catoctin Fm. 78
Catoctin volcanism 33
Central Appalachian Mtns. 11
Cenozoic Era 22, 23
chemical weathering 38
Chilhowee Group 78
clastic 317
cleavage, rock 53, 111
climate 56
Coastal Plain, Atlantic 8
colluvium 42
continental margin cross-section 8
core (of the Earth) 17, 18
core (of the Northern Blue Ridge) 75
cove, forests 68
columnar joints 124
crust, Earth's 17, 18

dew 57
dike 37, 38
disconformity 53

escarpment 15

facies 15, 319
fall line 15
fault 15, 36
fault line 36
flatirons 151
fogs 57
folds 36
footwall 36
formation (rock unit) 76
frost 57
fall line 15
floodplain 45

Ga (billions of yrs. ago) 22, 23
geologic time 22
Gondwana 16, 32
Great Smoky Group 296
Grenville Orogeny 25, 33
grade (in metamorphism) 319
graded bedding (of sediments) 306
groundwater 41
group (rock unit) 76

hanging wall 36
Harpers Fm. 80
headward erosion 48
Highlands (Blue Ridge) 188, 198

Iapetus Ocean 26, 32
igneous, rocks 15, 315
internal parts of a mountain belt 9
interior of the Earth 17

klippe 53

Laurasia 33
Laurentia 26, 32
large patch ecological communities 88
lithosphere 18
Lowlands (Blue Ridge) 187, 198
Lynchburg Group 85

Ma (millions of yrs. ago) 22, 23
mantle 17, 18
mass wasting (downslope movement) 42
matrix forests 87
meander 45
member (rock unit) 76
megaripples 150
Mesozoic Era 22, 23
metamorphism 9, 319
migmatite 196
mountain building (orogeny) 15, 21

natural environment, classifying 65
Natural Heritage Program Classif. 70
knickpoint 44
nonconformity 53
northern hardwoods 67
normal fault 36

oak-chestnut forest 66
oak-pine forest 67
Ocoee Supergroup 296
orogen (mountain belt) 7
orogeny 15, 21

Pangaea 16, 33
Paleozoic Era 22, 23
paleogeographic map 16
patchwork land covers 117
physical weathering 39
Piedmont Province 14
Piedmontia 25, 32
plate tectonic theory 18
plunging (inclined) fold axis 36
pluton 37, 38
pothole 43
Precambrian basement complex 118, 131

radiometric dating 24
reentrant 122
relief 15
reverse fault 36
Rheic Ocean 32
rock record, interpreting 76
rock units 76
Rodinia 25, 32

scree 15
sedimentary rocks 15, 317
shearing 15
shifting drainage divides 49, 50
slicken-fibers 136
small patch ecological communities 89
soil 62
soil acidity 63
spruce-fir forest 68
stream piracy, beheading 48
strike-slip fault 36
structure, of rock 35
subhorizontal plane 15
Swift Run Fm. 78
syncline 36

Taconica 25, 32
talus 15
temperatures in the Blue Ridge 56
temperature inversion 58
terrane 29
thrust fault 36
time (geologic) 22
trellis drainage pattern 46
type locality 15

Upland Plateau (Blue Ridge) 186, 197
unconformity 53

Valley and Ridge Province 13
volcanic island arc (subduction) 19

waterfall 44
weathering 38
Weverton Fm. 80
Wilhite Fm. 302
window, geologic 53

Index

Places of Special Interest in the Blue Ridge

Arnold Valley, Va. 168
Big Meadows, Va. 117
Brevard, N.C. 282
Buena Vista, Va. 150
Cades Cove, Tenn. 303
Catoctin Mtn. National Park, Md. 95
Chilhowee Mtn., Tenn. 301
Clingmans Dome, N.C. 308
Crabtree Falls, Va. 147
Craggy Gardens, N.C. 273
Cunningham Falls Park, Md. 96
Devil's Marble Yard, Va. 174
Doughton Park (Recreation Area), N.C. 217
Fairy Stone State Park, Va. 219
Fries, Va. 215
Gambrill State Park, Md. 99
Goose Creek window, Va. 180
Grandfather Mtn., N.C. 258
Great Smoky Mtns. National Park, Tenn./N.C. 258
Harpers Ferry, W. Va. 104
I-26 basement overlook, N.C. 278
James River Face Wilderness, Va. 168
James River Gap, Va. 154
Linville Falls, N.C. 264
Looking Glass Rock, N.C. 283
Mabry Mill, Va. 213
Mars Hill terrane, N.C. 268
Massanutten Mtn., Va. 120
Mt. Jefferson, N.C. 236
Mt. Mitchell State Park, N.C. 273
Mt. Rogers National Recreation Area, Va. 238
New River Basin, Va./N.C. 202
New River State Park, N.C. 207
Newfound Gap, Tenn./N.C. 307
Old Rag Mtn., Va. 129
Peaks of Otter, Va. 177
Philpott Reservoir, Va. 217
Roan Mtn. State Park, Tenn./N.C. 268
Rocky Knob Recreation Area, Va. 212
Roanoke Valley, Va. 209
Shenandoah National Park, Va. 113
Spruce Pine Mining District, N.C. 272
St. Mary's Wilderness, Va. 148
Stone Mtn. State Park, N.C. 230
Stewarts Creek Wildlife Management Area, Va. 214
Virginia Creeper Trail, Va. 252
Washington Monument State Park, Md. 102
Wintergreen Resort, Va. 146